Perspectives in Primate Biology

Photo by Michael Lyster, Zoological Society of London

This Symposium was held to mark the 75th birthday of
Professor Lord Zuckerman, OM, KCB, DSc, FRS

SYMPOSIA OF THE ZOOLOGICAL SOCIETY OF LONDON
NUMBER 46

T.M

Perspectives in Primate Biology

*(The Proceedings of a Symposium held at
The Zoological Society of London
on 31 May and 1 June 1979)*

Edited by

E. H. ASHTON

*Department of Anatomy, The Medical School,
Birmingham, England*

and

R. L. HOLMES

*Department of Anatomy, University of Leeds,
Leeds, England*

Published for

THE ZOOLOGICAL SOCIETY OF LONDON

BY

ACADEMIC PRESS

1981

ACADEMIC PRESS INC. (LONDON) LTD
24/28 Oval Road, London NW1 7DX

United States Edition published by
ACADEMIC PRESS INC.
111 Fifth Avenue, New York, New York, 10003

British Library Cataloguing in Publication Data

Perspectives in primate biology. — (Symposia of
the Zoological Society of London, ISSN 0084-5612;
no. 46)
1. Primates — Congresses
I. Ashton, E. H.   II. Holmes, R. L.
III. Series
599.8   QL737.P9

ISBN 0-12-613346-8

Printed in Great Britain at the Alden Press
Oxford London and Northampton

Contributors

ASHTON, E. H., *Department of Anatomy, The Medical School, Birmingham B15 2TJ, UK*

CROSS, B. A., FRS, *ARC Institute of Animal Physiology, Babraham, Cambridge, UK*

FINLAYSON, L. H., *Department of Zoology, University of Birmingham, Birmingham B15 2TJ, UK*

HARRISON, R. J., FRS, *Department of Anatomy, University of Cambridge, Downing Street, Cambridge CB2 3DY, UK*

HEALY, M. J. R., *London School of Hygiene and Tropical Medicine, Keppel Street (Gower Street), London WC1E 7HT, UK*

HERBERT, J., *Department of Anatomy, University of Cambridge, Downing Street, Cambridge CB2 3DY, UK*

HOLMES, R. L., *Department of Anatomy, Medical and Dental Building, University of Leeds, Leeds LS2 9JT, UK*

JOYSEY, K. A., *University Museum of Zoology, University of Cambridge, Downing Street, Cambridge CB2 3EJ, UK*

LEWIS, O. J., *Department of Anatomy, St Bartholomew's Hospital Medical College, Charterhouse Square, London EC1M 6BQ, UK*

MARTIN, R. F., *Department of Anthropology, University College, Gower Street, London WC1, UK*

MOORE, W. J., *Department of Anatomy, Medical and Dental Building, University of Leeds, Leeds LS2 9JT, UK*

OXNARD, C. E., *Departments of Anatomy and Biological Sciences, The Graduate School, University of Southern California, Los Angeles, California 91106, USA*

PARKES, A. S., CBE, FRS, *1 The Bramleys, Shepreth, Royston, Herts SG8 6PY, UK*

PASSINGHAM, R. E., *Department of Experimental Psychology, University of Oxford, South Parks Road, Oxford OX1 3UD, UK*

TANNER, J. M., *Department of Growth and Development, Institute of Child Health, 30 Guilford Street, London WC1N 1EH, UK*

WIDDOWSON, E. M., CBE, FRS, *Department of Medicine, Level 5, Addenbrooke's Hospital, Hills Road, Cambridge CB2 2QQ, UK*

YANG, H. C. L., *Department of Anatomy, Division of Biological Sciences, Pritzkes School of Medicine, University of Chicago, Illinois, USA*

YOUNG, J. Z., FRS, *The Wellcome Institute for the History of Medicine, 183 Euston Road, London NW1 2BP, UK*

ZUCKERMAN, S., OM, KCB, FRS, *The Zoological Society of London, Regent's Park, London NW1 4RY, UK*

Organizers and Chairmen of Sessions

ORGANIZERS

E. H. ASHTON and R. L. HOLMES, on behalf of the Zoological Society of London

CHAIRMEN OF SESSIONS

E. J. W. BARRINGTON, FRS, *2 St Margaret's Drive, Alderton, Tewkesbury, Glos GL20 8NY, UK*

A. J. E. CAVE, *The Zoological Society of London, Regent's Park, London NW1 4RY, UK*

J. HERBERT, *Department of Anatomy, University of Cambridge, Downing Street, Cambridge CB2 3DY, UK*

W. A. MARSHALL, *Department of Human Sciences, University of Technology, Loughborough, Leics., UK*

The following four plates portray contributors to the Symposium and chairmen of its several sessions, together with other close academic associates of Lord Zuckerman. (Top) Professor Lord Zuckerman receiving from Professor R. J. Harrison a commemorative medal subscribed by the participants in the symposium. (Bottom, left to right) Professor E. J. W. Barrington, F.R.S., Professor J. Z. Young, F.R.S., Professor A. J. E. Cave. The editors acknowledge most warmly the assiduous work of Dr D. Darlington in preparing these compilations.

(Top, left to right) Professor Sir Alan Parkes, F.R.S., Professor E. H. Ashton, Professor R. L. Holmes. (Centre, left to right) Dr J. Herbert, Professor W. J. Moore, Dr Elsie M. Widdowson, F.R.S. (Bottom, left to right) Professor W. A. Marshall, Dr. B. A. Cross, F.R.S., Professor R. J. Harrison, F.R.S.

(Top, left to right) Professor O. J. Lewis, Dr F. Yates, F.R.S., Professor C. E. Oxnard. (Centre, left to right) Professor L. H. Finlayson, Dr K. A. Joysey, Professor M. J. R. Healy. (Bottom, left to right) Dr R. E. Passingham, Dr R. D. Martin, Professor J. M. Tanner.

(Top, left to right) Professor P. L. Krohn, F.R.S., Dr Anita Mandl, the late Professor Sir Francis Knowles, F.R.S. (Centre, left to right) Dr. D. Darlington, Professor J. T. Eayrs, Mr T. F. Spence. (Bottom, left to right) Dr S. H. Green, Professor F. P. Lisowski, Dr W. P. Dallas Ross.

Foreword

I first met Solly Zuckerman in, I think, 1933; perhaps it was earlier than that. Anyway it was a very long time ago, when we were both young. One of the first things that came to me when I considered what to say was that Solly always seems young, always the same young enthusiast. I remember so well how he first taught me what little I know about primate anatomy as we were working on the dorsal roots of monkeys. I learned through the actual fact of dissecting tissues, the fact of contact with living tissues and further knowledge of what might be done to them. This was Solly in his prime at the time of the first work on the social and sexual life of monkeys and apes. I think that perhaps nowadays we sometimes forget what a tremendous discovery for mankind that was. It was the start of an entirely new era of primate research. We had had the Gestalt work of Köhler on apes, we had had the experimental work of various people on various other mammals, but before that I think I can say nobody had really begun to study animals in their social environments as themselves. Still less had anybody studied the hormonal background which regulates their social life. Far more than that, this was, I think I am right in saying, the beginnings of our understanding of human reproduction. Is it too much to say that until Solly showed us how to interpret the sexual cycle we had no knowledge of when ovulation occurred? It may be that later on we shall hear from one of the speakers that I have over-simplified the matter, but I think not. This was how I at least first realized the whole significance of the sexual cycle, human as well as primate. And what has flowed from that knowledge? The whole control of the human race really depends upon it. I hope I am not exaggerating; it is important that we see these things, if we can, in perspective. It is essential to try to understand the whole problem of human population and how to control it. I emphasize this because it tends to get forgotton nowadays among the many, many other things that Solly has promoted since. We are going to hear this morning about primate growth and evolution which was the central part of his work from then onwards. After that we shall hear about the hormonal aspects, which come from the same beginnings but spread out into many other channels. We shall, I hope, hear finally from Lord Zuckerman himself something of what he feels about the present state of all the areas of research that have developed from his own work. Indeed I hope that he can contribute throughout and tell us where we are going right and where we are going wrong. We are looking forward to a really exciting couple of days in which we can all meet each other and discuss these many matters. But we are here primarily to show Lord Zuckerman how very much we appreciate the tremendous stimulus he has given to so many fields of work: to thank him, and thank him again.

J. Z. YOUNG, *F.R.S.*

Preface

Lord Zuckerman's 70th birthday fell in 1974 and was marked by a dinner given, in Downing College, Cambridge, by a group of his former pupils and associates. Soon afterwards, thoughts turned towards a celebration of his 75th birthday, and it was felt that this might be suitably marked by an academic occasion involving more of those who have been associated with him in the biological fields towards which his interest has been directed.

As Secretary of the Zoological Society of London for many years, Lord Zuckerman played a major role in transforming that Society into a modern organization devoted to zoological research and education. One component of this development was the founding in 1960 of a series of Symposia held in the Society's meeting rooms at approximately six-monthly intervals. It was suggested and agreed that one of these should be timed to coincide with Lord Zuckerman's 75th birthday and should cover each of the biological fields to which he had contributed.

From the early stages of his scientific career, Lord Zuckerman's basic biological interest was in the biology of several species of primates, but notably the Chacma baboon which was indigenous to the areas of South Africa where he spent his earlier years. One of his first research interests was in fact the postnatal growth changes in the skull of the South African baboon, and was the start of a series of studies of growth changes in sub-human primates and man, that were developed throughout his scientific career. It was thus fitting that the first session of the present Symposium, held in the meeting rooms of the Zoological Society of London on 31 May and 1 June 1979, should centre upon comparative studies of growth.

It was during Solly Zuckerman's early years in South Africa that the first fossil (*Australopithecus africanus* — the Taungs Ape) of the group subsequently styled the Australopithecinae was discovered and described by Professor R. A. Dart. Zuckerman's doubts about the extravagant claims put forward of supposedly human-like features in this basically ape-like creature led to what was to prove to be a life-long interest in general studies of human and primate palaeontology and evolution, and more specifically in the application of quantitative methods to the study of primate fossils. Hence, a second session of the present symposium was appropriately devoted to primate and human evolution.

Again, as a young worker in South Africa, Solly Zuckerman initiated field studies of the behaviour and sociology of the South African Chacma baboon, which were to prove to be the foundation of many similar studies of sub-human primates. The social behaviour of monkeys and apes, previously enshrouded in anthropomorphic romanticizing, was shown to be conditioned by a scale of dominance among males coupled with the constantly recurring reproductive cycles of the females. On coming to London in the mid 1920s, he was able, by observation of the free-ranging baboon colony, then in the gardens of the Zoological Society, to continue and develop such studies of sub-human primate behaviour and to start what were to prove to be epoch-making analyses of the biology of the reproductive cycles in primates. Such studies led later to enquiries into endocrine control of the reproductive processes and this, in turn, to studies of the interaction between the nervous and endocrine systems. Further sessions of the Symposium were therefore devoted to reproductive biology, neuroendocrinology, primate behaviour and sociology.

In selecting the title "Perspectives in Primate Biology" the editors were conscious that a symposium of this type, commemorating the biological work of one of such great originality and intellectual energy, should not only review past work, but should also present new factual material and indicate the more fruitful pathways along which our thinking might in future be directed. Notwithstanding his dominant interest in primate biology, Lord Zuckerman has never hesitated to draw information from other animal groups to elucidate the problems posed by the study of prosimians, monkeys, apes and man. It is thus quite fitting that contributions to the present Symposium relate to widely diverse groups of animals.

The contributors to this Symposium have been drawn equally widely, some being former pupils or associates of Lord Zuckerman, others workers of distinction in the several fields who may not have had the opportunity of direct personal association. But it gave the greatest pleasure to the organizers that certain senior scientists (Professors J. Z. Young, A. J. E. Cave, E. J. W. Barrington and R. J. Harrison) who have for long been among Solly's personal friends, accepted invitations to introduce the Symposium, to chair two of its sessions, and to sum up the proceedings. The Chairmen of the other two sessions were former pupils and colleagues of Lord Zuckerman, and the editors are honoured to have had a similar association.

The academic editors have been greatly helped by the assiduous work of Unity McDonnell (Zoological Society of London). To her, their most grateful thanks are tendered.

November 1980   E. H. ASHTON
                                          R. L. HOLMES

Contents

PRIMATE GROWTH

Chairman's Remarks

W. A. MARSHALL

Growth of Creatures Great and Small

E. M. WIDDOWSON

Size and Shape in Relation to Growth and Form

M. J. R. HEALY and J. M. TANNER

Facial Growth in Primates with Special Reference to the Hominoidea

W. J. MOORE

PRIMATE EVOLUTION, BIOMETRICAL MORPHOLOGY, PALAEONTOLOGY

Chairman's Remarks

A. J. E. CAVE

The Australopithecinae: Their Biometrical Study

E. H. ASHTON

Beyond Biometrics: Studies of Complex Biological Patterns

C. E. OXNARD and H. C. L. YANG

Functional Morphology of the Joints of the Evolving Foot

O. J. LEWIS

Molecular Evolution and Vertebrate Phylogeny in Perspective

K. A. JOYSEY

REPRODUCTIVE BIOLOGY AND NEUROENDOCRINOLOGY

Chairman's remarks

E. J. W. BARRINGTON

The Pars Intermedia of the Mammalian Pituitary Gland: Problems and Some Answers

R. L. HOLMES

Form, Activity and Function in Insect Neurosecretion

L. H. FINLAYSON

Reproduction in the Green Sea Turtle, *Chelonia mydas*

ALAN S. PARKES

Sexual Physiology of the Brain

B. A. CROSS

PRIMATE SOCIOLOGY AND BEHAVIOUR

Chairman's Remarks

J. HERBERT

Field Studies of Primate Behaviour

R. D. MARTIN

Hormones and the Sexual Strategies of Primates

J. HERBERT

Primate Specialization in Brain and Intelligence

R. E. PASSINGHAM

Primate Growth

Chairman's Remarks

It is a little more than 50 years since a young man called Zuckerman presented to this Society a paper entitled "Growth-changes in the skull of the baboon, *Papio porcarius*". Since that time there have been many advances in the study of growth but relatively few of them have been of fundamental importance. A lot of progress has been made in the development of the techniques. These have been of great help in our work, but unless we use our new methods and instruments to answer worthwhile questions we may greatly increase the sophistication of our gadgetry whilst making very little progress with our science. We must recognise that electronic gadgets do not necessarily advance science any more than electronic organs improve music. Clear thought is the only sound basis for scientific progress. We must get things into perspective and identify the key issues in the field. This, of course, is what Lord Zuckerman taught us to do and this is what we shall try to do this morning. We are going to think about perspectives in primate growth and, of course, we cannot do this by considering primates alone. We shall therefore begin by asking Dr Widdowson to tell us about growth in a large number of creatures, and provide a background for our discussions of man and other primates.

W. A. MARSHALL

Symp. zool. Soc. Lond. (1981) No. 46, 5—17

Growth of Creatures Great and Small

E. M. WIDDOWSON

Department of Medicine, Addenbrooke's Hospital, Cambridge, UK

SYNOPSIS

Growth in the early stages of development is brought about entirely by cell division; this goes on more rapidly in some species than others, and sets the pace for the whole of growth before birth. The primates and the elephant start slowly, the small rodents grow very rapidly soon after conception and the growth of the carnivores and ungulates is intermediate.

Although the total weight of the offspring of small species is generally greater in relation to the weight of the mother than in large species, this does not necessarily apply to the weight of individual young. One newborn rat for ex- ample is the same proportion of the mother's weight (2.5%) as a newborn blue whale. The number of times the body substance multiples itself after birth is very variable and is not necessarily related to the size of the species.

The foetuses of man and the macaque monkey gain weight at the same rate until term in the macaque at 167 days from conception. Then the rate of growth of the human foetus increases so that at term it weighs four times as much as the macaque at 280 days from conception. The macaque foetus is longer than the human foetus. From a gestational age of 167 days, when the macaque is born, the human foetus accumulates a considerable amount of fat and this continues after birth; this does not occur in the macaque. The body of the macaque foetus contains less water per 100 g than that of the human foetus at similar ages from conception and this is evident also in its skeletal muscle. Its body composition is therefore more mature than that of the human foetus at each gestational age.

The weights of the brains of the macaque and human foetuses are similar until the macaque is born; thereafter the human brain grows more rapidly, and becomes larger in relation to the body weight than the macaque brain. The human brain has more DNA and therefore more cells than the brain of the macaque at the same gestational age.

INTRODUCTION

The phenomenon of growth is remarkable enough itself, but perhaps equally remarkable are the different rates at which, from similar beginnings, the various mammalian species increase in size. Some go on growing for much longer than others and this enables slow-grow- ing animals to become larger in the end than fast-growing ones. This is no new idea, for in 1891 Minot pointed out that from conception to maturity guinea-pigs grew at an average rate of 1.82 g a day, rabbits at a rate of 6.3 g a day and man at a rate of 6.69 g a day. Men therefore grew larger than rabbits because they grew for a longer time, but rabbits grew larger than guinea-pigs because they grew faster.

All growth depends on a supply of food, and from the nutritional point of view growth can be considered in two stages, the first before birth, when feeding is by infusion of nutrients from the mother's circulation through the placenta directly into the blood-stream of the foetus, and the second when feeding is by mouth, and nutrients have to be absorbed through the intestinal mucosa before they can be utilized.

This chapter is divided into three parts. The first will consider briefly growth before birth in a variety of species and show how primates fit into the picture. The second will discuss equally briefly growth after birth, and the third will take two primates, man and the macaque or rhesus monkey, about which we have a considerable amount of detailed information, and compare growth not only in terms of weight, but of length, and weights of brain and other organs. The chemical development of the tissues of these two species at the same ages from conception will also be compared.

GROWTH IN WEIGHT BEFORE BIRTH

Growth in the early stages of development after conception is brought about entirely by cell division, with little or no increase in cell size, so at first the embryo can only grow as fast as its cells divide. Cell division goes on more rapidly in some species than in others. Table I shows the ages from conception at which six species reach a weight

TABLE I

Time from conception at which six species reach a weight of 1 g

Species	Days
Hamster	14.5
Rabbit	17.0
Rat	17.5
Pig	26
Cat	36
Man	55

of 1 g (Adolph, 1972). The hamster, rat and rabbit grow to a weight of 1 g in about two weeks, the pig and cat in four to five weeks while the human foetus takes eight weeks to reach this weight. The rate of growth soon after conception sets the pace for the whole of growth before birth. Figure 1 shows the weights at birth of 20 species plotted

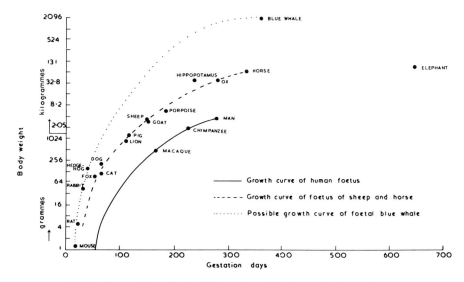

FIG. 1. The weight at birth of 20 species and their lengths of gestation.

against the length of gestation. The scale used for weight is a continuously doubling one since this seemed the most appropriate way of expressing growth in the early stages of development. Three curves are also shown. The lowest one depicts the growth of the human foetus, and the weights of the newborn chimpanzee and macaque fall on this curve. Other primates, not shown here, the baboon, the gorilla and orang-utan, are not far off it. The lower part of the middle curve represents the growth of the foetal lamb, which grows more rapidly than the human foetus in the early stages of development and maintains this difference throughout its growth *in utero*. It is born at about the same weight as the human infant after a gestation period only half as long. The second part of the middle curve, and probably the first part also, represents the growth of the foetal foal. It has a gestation period longer than that of the human foetus, and it weighs considerably more than the human infant at 280 days. The birth weights of the fox, cat, dog, lion, pig, goat, porpoise, hippo-potamus and ox appear to lie along the growth curves of the foetal sheep and horse. These species grow more rapidly than the primates in the early stages and are therefore heavier at all times from con-ception until they are born. The third curve is a hypothetical one. It takes in the mouse, rat, rabbit and hedgehog, and illustrates their very rapid gain in weight at the beginning; it is also a possible repre-sentation of the growth of the foetal blue whale which achieves the immense weight of 2000 kg in 1 year. The elephant has a longer gestation period than other mammals, and its birth weight is only a little more than that of the foal after a gestation lasting nearly twice as long. It looks as though the elephant, like the primate foetus, makes rather a slow start.

The weight of the young at birth must clearly bear some relation to the weight of the mother, because the mother has to deliver her young while they are still smaller than herself. Table II shows the

TABLE II

Relationship between the weight of the mother and that of her offspring at term

Species	Weight of mother (kg)	Weight as percentage mother's weight	
		Whole litter	One young
Lesser horseshoe bat	0.006	34	34
House mouse	0.025	40	5
Rat	0.200	25	2.5
Guinea-pig	0.560	68	17
Rabbit	1.175	19	4
Cat	2.75	16	4
Red fox	4.2	16	2.4
Macaque monkey	5.3	9	9
Dog: Cocker spaniel	8.4	4	0.8
Sheep	37	10	10
Chimpanzee	46	4	4
Woman	56	6	6
Gorilla	60	3	3
Lion	114	4	0.8
Black bear	100	0.3	0.3
Polar bear	258	0.2	0.2
Pig	250	5	0.5
Cow	600	7	7
Horse: Shetland	191	10	10
Horse: Shire	800	11	11
Elephant	2 540	4	4
Blue whale	79 000	2.5	2.5

weights of the adult females of 22 species and the weights of their newborn young expressed as a percentage of the mother's weight (Leitch, Hytten & Billewicz, 1959). The smallest mother in this series is the lesser horseshoe bat, which weighed 6 g; the largest is the blue whale which weighed 79 000 kg. The latter, therefore, weighs over 10 million times as much as the former. Both produce one young; the bat's young weighs 2 g, over 30% of the mother's weight, the whale's 2000 kg, which is 2.5% of the weight of the mother. This illustrates the general principle that although the weight of the off-spring increases as the size of the mother increases, it does not increase in proportion, and, generally speaking, the total weight of the offspring at birth forms a larger percentage of the mother's weight in small animals than in large ones. Most small mammals have large litters and the total weight of offspring is divided among a number of young. The mouse's litter, for example, which is 40% of the mother's weight, is divided among eight individuals, and the guinea-pig, which produces a greater weight of young in relation to maternal weight than any other species I know, 50–60% or even more, has an average litter of four. The single young of the horse and sheep are comparatively large, weighing about 10% as much as the mother, and this is true of both large and small breeds.

GROWTH IN WEIGHT AFTER BIRTH

The weight of a single young at birth as a percentage of the weight of the mother gives an indication of the amount of growth that goes on in the animal after birth. The lesser horseshoe bat has only to treble its weight to reach adult size, and the guinea-pig has to increase by a little over five times. At the other extreme are the bears, which increase their birth weight by 300–500 times, and the pig, 200 times. Man and other primates come in between with increases of 10–30 times. The newborn rat and blue whale are both 2.5% of the mother's weight and therefore have a forty-fold increase in weight ahead of them.

The length of the growth period after birth varies considerably between species. Generally speaking, small species have shorter growth periods and life-spans than large ones, and the rat achieves its forty-fold increase in birth weight in three or four months, where-as man takes 18 years to multiply his birth weight 20 times. The pig is a particularly rapid grower. At the end of its first year it is still gaining weight fast, but even at that time it may weigh 180 times as much as it did at birth.

Man is often regarded as an exception in not reaching puberty until he is almost fully grown. The rat and pig both reach puberty during their period of most rapid growth and this is true of many other species, but not of all. Beagle dogs are said to be fully grown before they reach puberty, the blue whale is over 90% of its adult length and the elephant is about 80% of its mature shoulder height at this time (McCance & Widdowson, 1978). What is peculiar to man and other primates is the rapid gain in weight that precedes and accompanies puberty, and this is associated with changes in the proportions of the body— in the human species an increase in muscle mass in boys and in fat in girls.

A COMPARISON BETWEEN THE GROWTH AND DEVELOPMENT OF THE HUMAN FOETUS AND THAT OF THE MACAQUE MONKEY

The macaque monkey is born weighing about 470 g after a gestation period of 167 days. The human baby weighs about 3400 g after a gestation lasting 280 days. We have already seen that the newborn macaque weighs the same as the human foetus at the same time from conception and this is shown in more detail in Fig. 2. The weights of the macaque foetus come from the work of Cheek (1975); those for the human foetus are from our own studies (Southgate & Hey, 1976). These small series confirm the observations of others (Dawes, 1968) that the foetal growth curves of the two primates are similar until the macaque is born. Figure 3 shows that after birth the growth of the macaque continues at about the same rate as before birth, while that of the human, still in the uterus, increases, so that by the time the human baby is born at 280 days it weighs nearly four times as much as the macaque at the same age from conception. The human infant, too, continues to gain weight after birth at approximately the same rate as it was doing before birth, so that the two curves fall farther and farther apart. By 16 weeks from birth the macaque has doubled its birth weight, and the human infant doubles its much larger birth weight in approximately the same period of time. The gain in weight of the human individual before and after birth is associated with, but not entirely due to, the deposition of fat in the body. This increases from about 1% of the body weight at 167 days to 15% at term (280 days) (Southgate & Hey, 1976) and continues to increase to about 26% at four months or so after birth (Fomon, 1974). In the macaque the percentage of fat in the carcass at term (167 days) is about 2% and this increases after birth to 6—9% at a conceptual age of 280 days (Cheek, 1975). The

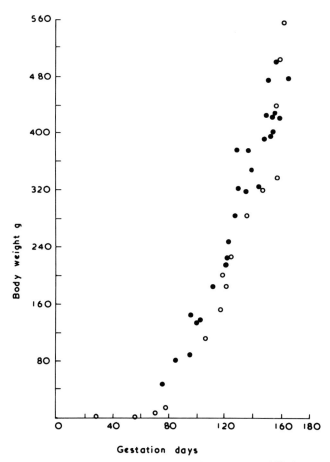

FIG. 2. Growth in weight of human and macaque foetuses up to 167 days gestation (term for the macaque).

human infant at birth is therefore considerably fatter than the macaque monkey at the same age from conception, and it differs in other ways too. Figure 4 shows that even over the period when the weights of the foetuses of the two species were increasing at similar rates the lengths of the body measured from crown to heel were not the same. At all ages from 70 to 167 days gestation the macaque foetus is longer than the human foetus, and this is true also of the crown–rump length. After the macaque is born, however, it gains weight more slowly than the human foetus, grows more slowly in length, and the curves cross over at about 180 days.

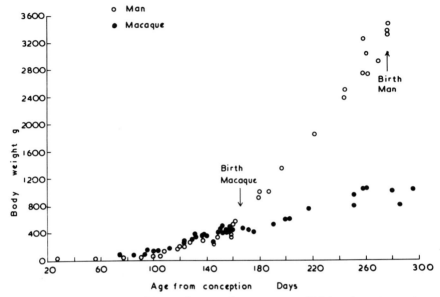

FIG. 3. Growth in weight of human foetus and macaque up to 280 days (term for man).

Growth implies change. In all species the different parts of the body increase in weight at different rates, but they increase in a consistent manner. The brain forms a larger proportion of the body weight early in development than later, whereas the contribution of skeletal muscle to the body weight increases as the organism grows.

Figure 5 shows the weights of the human and macaque brain related to age from conception. The increased rate of gain in body weight of the human foetus after 167 days is accompanied by an acceleration in brain growth. In the macaque the growth of the brain slows down soon after birth and by 280 days, when the human infant is born, the macaque brain has reached its adult weight although the body does not attain its mature weight until the age of five or six years. Similarly in man the growth rate of the brain slows down after birth, and mature weight is reached by four or five years of age whereas the body goes on growing until 18 years.

In Fig. 6 the weight of the brain is related to body weight in the two species, and this brings out a further point. At all weights above 470 g and ages from conception of 167 days, the brain is larger in relation to the body in man than in the macaque.

Much has been said and written about the number of cells in the brain as measured by the amount of DNA in it. If the results of Dobbing & Sands (1973) for the human brain are compared with

BODY (crown-heel) LENGTH

o   Man

•   Macaque

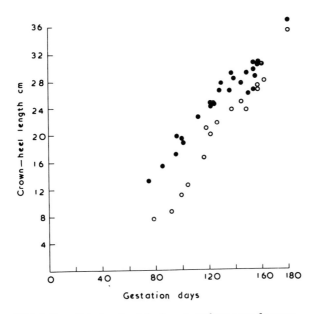

FIG. 4. Growth in length of the human and macaque foetuses.

those of Cheek (1975) for the macaque, it appears that even though the weights of the brain are similar in the two species up to 167 days from conception, the amounts of DNA in them are not. The human brain has more DNA in it, and therefore more cells than the brain of the macaque at all corresponding ages, body weights and brain weights.

The chemical composition of all parts of the body and of the body as a whole changes with development and, like the proportions of the body, it changes in a consistent way. The percentage of water falls, and that of protein increases. The fall in water is due largely to a decrease in the proportion of the body occupied by extracellular fluid and an increase in cell mass. The smallest human foetus that has been analysed weighed less than a gramme and contained 93% water (Widdowson & Dickerson, 1964). A macaque foetus of 1 g contained 91% water (Behrman, Seeds, Battaglia, Hellegers & Bruns, 1964). By 167 days gestation and a weight of 470 g the amount of

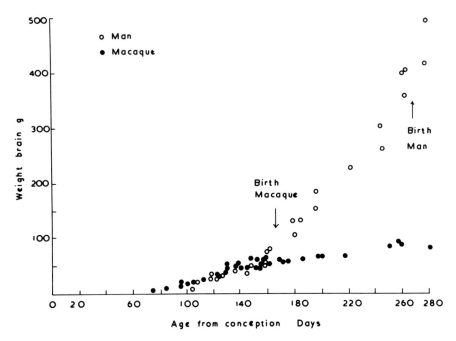

FIG. 5. Weight of brain of human foetus and of macaque.

water in the human foetus has fallen to 89%, but in the macaque it is
75%, so the body of the macaque is more mature chemically than the
human body, and even by term at 280 days the fat-free body tissue
of the human foetus is still 82% water. The percentage of water in
the fat-free body tissue continues to fall after birth to the mature
value of 73%, which is reached by about two years. At the same time
the percentage of protein in the body increases. The protein is largely
confined to the cells and, as we have seen, growth in the early stages
of development is brought about by an increase in the number of
cells without any appreciable increase in average cell size. As the cells
increase in number they occupy a larger proportion of the tissues at
the expense of the extracellular material surrounding them. Then at
an age from conception which is characteristic for each organ and
each species the individual cells begin to increase in size as well as to
increase in number. In the liver this stage is reached at 120 days in
the macaque (Cheek, 1975) and 210 days in the human foetus
(Widdowson, Crabb & Milner, 1972), that is about three-quarters of
the way through gestation in both species.

Skeletal muscle shows greater changes with development in
cellular—extracellular proportions than the internal organs. Table III

BRAIN WEIGHT RELATED TO BODY WEIGHT

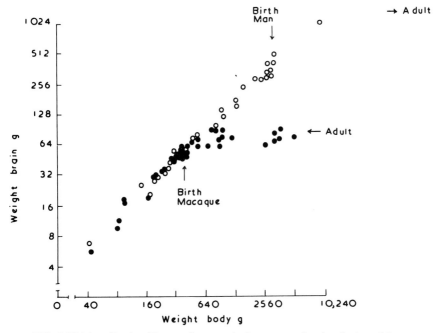

FIG. 6. Weight of brain of human foetus and of macaque related to body weight.

TABLE III

Composition of skeletal muscle

	Man		Macaque	
Age from conception (days)	160	280	160	280
Water (mg g^{-1})	887	804	794	763
Protein (mg g^{-1})	93	169	171	214

shows the concentrations of water and protein in the muscle of man
(Dickerson & Widdowson, 1960) and the macaque (Cheek, 1975) at
two ages — 167 days when the macaque is born and 280 days, which
is full term in the human infant. At both ages the muscle of the
macaque has less water and more protein than human muscle and,
like the body as a whole, it is therefore chemically more mature. In
man the full number of muscle fibres is present at birth or soon
afterwards (McCallum, 1898; Montgomery, 1962) and it seems likely
that the same is true of the macaque (Cheek, 1975). Growth of the

muscle after birth in both species is brought about almost entirely by an increase in the size of the existing fibres, and this is considerable, for not only does the muscle grow along with the rest of the body, but in both species it increases from about 25% of the body weight at birth to 40—45% in the adult.

There are many other ways in which the developing bodies of the two species might be compared. Although the macaque does not accumulate a great deal of fat in its adipose tissue before birth, linoleic and palmitic acids appear to cross the placenta with ease (Portman, Behrman & Soltys, 1969). Fatty acids cross the human placenta more readily than was at one time supposed, but even so glucose provides the substrate for synthesis by the human foetus of most of the fat that accumulates in its adipose tissue before birth. While the energy requirements for maintenance and growth of the two species are probably similar for the first 160—170 days of gestation, the period from then until term in the human foetus is characterized by greatly increased demands for energy as glucose for synthesis of fat.

In all species nutrition after birth is a less efficient process than nutrition before birth for two reasons. In the first place there is inevitably more wastage when the nutrients have to be absorbed through the intestine than when they are delivered direct into the blood-stream through the placenta. More important, however, is the fact that a great deal of the energy taken in in the milk must be used for maintaining the body temperature in an environment that is almost certainly cooler than the uterus. Heat is lost from the surface of the body to the environment; a small body has a greater surface area in relation to its weight than a large body and it therefore has a greater energy requirement per kilogramme for maintenance than a large one. In line with this the daily energy requirement of the macaque during the first weeks after birth $(238\,\text{kcal}\,\text{kg}^{-1})$ is approximately twice that of the human infant $(120\,\text{kcal}\,\text{kg}^{-1})$ (Jacobson & Windle, 1960; Fomon, 1974). In both species there is a loss of body weight for a few days after birth, but the birth weight is regained before the end of the first week, and so long as the energy requirements are met the gain in weight of both species proceeds along the same curve as it was doing before birth, and this is also true of non-primate species. Birth is the greatest physiological upheaval that takes place in the lifetime of an individual of any species, but so far as gain in weight, and indeed chemical development of the body are concerned, birth makes surprisingly little impact.

REFERENCES

Adolph, E. F. (1972). Development of physiological functions. In *Nutrition and development*: 1—25. Winick, M. (Ed.). New York: Wiley.

Behrman, R. E., Seeds, E. A., Battaglia, F. C., Hellegers, A. E. & Bruns, P. D. (1964). The normal changes in mass and water content in fetal rhesus monkey and placenta throughout gestation. *J. Pediat.* **65**: 38—44.

Cheek, D. B. (1975). *Fetal and postnatal cellular growth*. New York: Wiley.

Dawes, G. S. (1968). The placenta and foetal growth. In *Foetal and neonatal physiology*: 42—59. Dawes, G. S. (Ed.). Chicago: Yearbook.

Dickerson, J. W. T. & Widdowson, E. M. (1960). Chemical changes in skeletal muscle during development. *Biochem. J.* **74**: 247—257.

Dobbing, J. & Sands, J. (1973). Quantitative growth and development of human brain. *Archs Dis. Childh.* **48**: 757—767.

Fomon, S. J. (1974). *Infant nutrition*. 2nd edn. Philadelphia: W. B. Saunders Co.

Jacobson, H. M. & Windle, W. F. (1960). Observations on mating, gestation, birth and postnatal development of *Macaca mulatta*. *Biol. Neonat.* **2**: 105—120.

Leitch, I., Hytten, F. E. & Billewicz, W. Z. (1959). The maternal and neonatal weights of some Mammalia. *Proc. zool. Soc. Lond.* **133**: 11—28.

McCallum, J. B. (1898). On the histogenesis of the striated muscle fibre and the growth of the human sartorius muscle. *Johns Hopkins Hosp. Bull.* **9**: 208—215.

McCance, R. A. & Widdowson, E. M. (1978). Glimpses of comparative growth and development. In *Human growth* I: 145—166. Falkner, F. & Tanner, J. M. (Eds). New York: Plenum Publishing Corp.

Minot, C. S. (1891). Senescence and rejuvenation. *J. Physiol., Lond.* **12**: 97—153.

Montgomery, R. D. (1962). Growth of human striated muscle. *Nature, Lond.* **195**: 194—195.

Portman, O. W., Behrman, R. E. & Soltys, P. (1969). Transfer of free fatty acids across the primate placenta. *Am. J. Physiol.* **216**: 143—147.

Southgate, D. A. T. & Hey, E. N. (1976). Chemical and biochemical development of the human foetus. In *The biology of human fetal growth*: 195—209. Roberts, D. F. & Thomson, A. M. (Eds). London: Taylor & Francis.

Widdowson, E. M., Crabb, D. E. & Milner, R. D. G. (1972). Cellular development of some human organs before birth. *Archs Dis. Childh.* **47**: 652—655.

Widdowson, E. M. & Dickerson, J. W. T. (1964). Chemical composition of the body. In *Mineral metabolism* 2A: 1—247. Comar, C. L. & Bronner, F. (Eds). New York: Academic Press.

Symp. zool. Soc. Lond. (1981) No. 46, 19–35

Size and Shape in Relation to Growth and Form

M. J. R. HEALY

London School of Hygiene and Tropical Medicine, London, UK

J. M. TANNER

Institute of Child Health, London University, London, UK

SYNOPSIS

The problem of distinguishing between size and shape differences is discussed. It is desirable to transform all the measurements defining the outline of an object or individual to logarithms before analysis; size can then be measured by the mean of the log measurements and shape differences are those remaining when the outlines have been adjusted to constant size. These residual differences can be summarized by a principal component analysis. Applying this technique to anthropometric data emphasizes the difficulties of choosing appropriate measurements and of making these with the necessary precision. Shape changes during growth can be described using the same basic notion, but call for more complex statistical methods.

INTRODUCTION

The distinction between size and shape is one of those intuitively obvious notions which become less and less clear the longer we look at them. When we look at a set of mature organisms of the same species (in this paper, man), it is immediately apparent that their bodily proportions differ and that some of them are, in some sense, bigger than others — they come, as the saying goes, in all shapes and sizes. Yet when we start to think about quantifying size and shape differences and consequently about distinguishing between the two, problems start to emerge. The same problems arise with added force as soon as we consider growing organisms. As a child grows, it increases in size — almost tautologically; but it also changes its shape in various more or less subtle ways, and it is not immediately apparent how these changes are to be described, particularly in quantitative terms.

There is of course a substantial literature on size and shape. In so far as the problem involves the interpretation of numerical measurements, it should be one that is of interest to statisticians. In fact, the contributions in the core statistical literature are quite meagre; very many of the papers have been written by biologists who enjoy mathematical topics rather than by mathematicians who have come to work on biological data. Some of this literature brings to bear upon the problem the standard techniques of multivariate analysis, such as principal component and canonical analysis; we shall try to make a small contribution of our own in this direction. Other papers have focused on the well-known allometry relationship

$$y = ax^b$$

which is empirically very successful in describing certain types of growth data. This relationship can work well in relating pairs of measurements, but there are a number of difficulties when we try to use it to describe sets of measurements in a more or less holistic fashion (Sprent, 1972).

Our own approach to the measurement of size and shape will be based on measurements of distances between defined landmarks on the body surface. This approach underlies most previous work, and indeed is usually taken for granted. It has been criticized by Bookstein (1978) in an important monograph. Bookstein points out that points located by means of the distances between them are consistent with quite different shape contours; he asserts that distance methods can only conceal the essential subtleties of biological shapes and shape changes. We feel that there is a good deal of force in this argument, but we intend to side-step it by making a distinction between shape in the large and in the small. In craniometry, for instance, fine details of the skull profile assume considerable importance and it may well be difficult to define sufficient unequivocal landmarks to describe these properly. In postcranial physical anthropometry, on the other hand, we are dealing with a much more complex outline and we can only hope to describe its shape in very general terms. For such a purpose — the description, let us say, of "build" — distances between landmarks should in our view be quite adequate.

THE DEFINITION OF SIZE

Our starting point has been to look more closely at what we mean by
saying that two outlines have the same shape and consequently differ
only in size. (We shall deal only with outlines of the human figure,
and need not concern ourselves with the problem of orientation
which presents certain difficulties in the context of craniometry or
of single bones such as the scapula.) Let us start with simple geometric figures and see where they lead us. We say that two triangles
have the same shape — are *similar*, in the technical language of
geometry — if the sides of one are proportional to those of the other
(Fig. 1A). In the same way, we say that two ellipses have the same

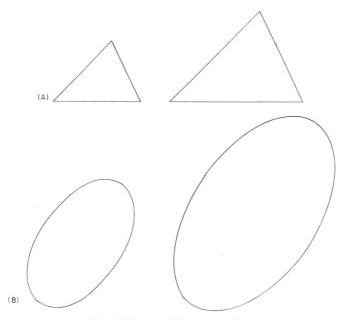

FIG. 1. Figures with the same shape.

shape if the ratio of the two major axes is the same as that of the two
minor axes (Fig. 1B). In both cases, one figure when suitably oriented
is simply a magnification of the other and each part of it can be
generated by multiplying the corresponding part of the other figure
by a constant factor. The important point is that, for differences to
be of pure size, they must be *proportional*, that is to say multiplicative rather than additive.

When we have similar figures, there is no difficulty about defining a measure of size. We simply take a particular figure as the unit and adopt as the size measure for any other figure the magnification factor which is needed to produce it starting from the unit. For some purposes, some monotonic function of the factor may be preferable.

Almost all the standard statistical techniques rely on addition rather than multiplication, dealing with differences rather than with ratios. This is especially true of multivariate analysis whose main task is optimizing in some appropriate sense the coefficients of a weighted sum. If therefore we wish to get a measure of size which will lend itself to statistical manipulation, we must change our multiplicative situation into an additive one, and this can be done by converting all the measurements into logarithms before analysis. It should be noted that the logarithmic transformation stems directly from the nature of the problem; it is not determined, as is sometimes the case in statistical work, by any considerations of Normality of distribution or of constancy of variance. We shall see, however, that it does produce certain extra advantages.

If we transform to logarithms a set of measurements taken on two similar figures, there is no difficulty in arriving at a measure of relative size; using for the purpose the logarithm of the constant magnification factor, we may take the difference between the log measurements of any two corresponding parts. With a whole set of measurements, which may be subject to measurement error, it is fairly natural to take the mean of all the differences, and simplicity suggests that the mean might as well be unweighted. By a suitable choice of the unit figure, we can thus use as our measure of size the unweighted mean of the logarithms of all the available measurements.

SHAPE DIFFERENCES

Once we have defined size in this way, differences in shape must include all those which are in some way *residual* to differences in size — those which remain when differences in size have been allowed for. Two methods of allowing for size differences suggest themselves. We need in effect the value of each measurement when the whole figure is adjusted to constant size. We can obtain this adjusted value by applying a suitable magnification to each measurement (adding a constant to all the logarithms); or we may calculate the regression of each log measurement on the size measure and obtain the adjusted value in this way. In practice, the two techniques seem to give almost identical results. Once we have the adjusted measurements, the next problem, that of the description of shape, is to determine ways in which they can be summarized.

One way of approaching this problem is to ask how many *dimensions* of shape there are — how many quantities we need to determine the shape of a figure once the size has been fixed. Statistically speaking, this is the problem that is usually tackled by principal component analysis. In our situation, this will involve calculating the dispersion matrix of the log measurements, partialling out the size measurement and evaluating the eigenvalues and eigenvectors. If a few of the eigenvalues are a good deal bigger than the remainder, we may suppose that most of the shape information is contained in the linear functions of the log measurements whose coefficients are given by the corresponding eigenvectors.

Principal component analysis is a less satisfactory statistical technique than is sometimes realized and it is our view that the shape problem is one of the relatively few instances in which its use is appropriate. In many applications, the results depend critically upon the scales of measurement, so that we get quite different answers when we change from pounds and feet to kilogrammes and centimetres — we are not impressed by suggestions that we should routinely express all measurements in terms of their own standard deviations, which is what is implied by the common advice to work with the correlation matrix. Here, however, differences of scale seem unrealistic (they correspond to changes in the base of the logarithms), and scale differences between the different original measurements are of no consequence. Note that the standard deviations of logarithmic measurements are closely related to coefficients of variation of the originals.

As an artificial example of the technique, let us consider a set of isosceles triangles on each of which we make four measurements — the lengths of the base and one side and of two perpendiculars from the vertices to the opposite sides. Figure 2 shows a collection of 10 such triangles. We can tabulate the logarithms of the measurements and the size measures given by the means of the logarithms, and from these it is straightforward to obtain the matrix of residual variances and covariances. Proceeding to the principal component analysis of this matrix, we find that there is automatically one eigenvalue which is exactly zero. In our case there are two more very small eigenvalues (Table I) suggesting that there is essentially only a single shape dimension. This is obviously correct; apart from their sizes, the triangles are completely defined by the angle at the base. We can calculate the value of the shape variable for each triangle and it can be seen from Fig. 3 that it is almost exactly equivalent to this angle. In particular, the two triangles which are similar to each other but of different sizes (numbers 3 and 8 in Fig. 2) have the same shape score.

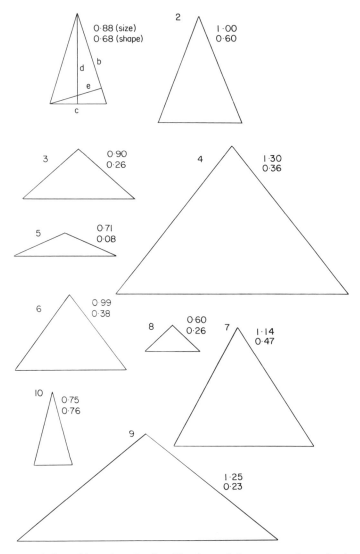

FIG. 2. A population of isosceles triangles. The size and shape scores for each triangle are indicated.

SIZE AND SHAPE IN ADULT HUMAN DATA

We have only just begun to investigate the applicability of this technique to anthropometric data, but we have already had some surprises. The data immediately available to us were far from ideal for the purpose. They relate to 114 young men and 112 young women,

TABLE I

Shape components of a set of triangles

	I	II	III	IV
Eigenvalues	0.4228	0.0192	0.0000	0.0000
Eigenvectors	−0.19	−0.68	0.54	0.41
	0.68	−0.19	−0.45	0.57
	−0.68	0.19	−0.45	0.57
	0.19	0.68	0.54	0.41

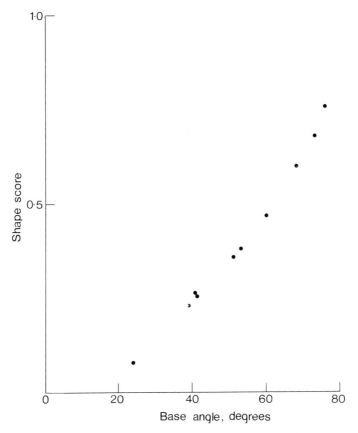

FIG. 3. Shape score and base angle for the triangles in Fig. 2.

and the measurements used are shown in Fig. 4. Four of these (subischial leg length, sitting height, bi-acromial and bi-iliac diameters) were taken on the body — subischial leg length is in fact a derived measurement calculated as the difference between stature and sitting

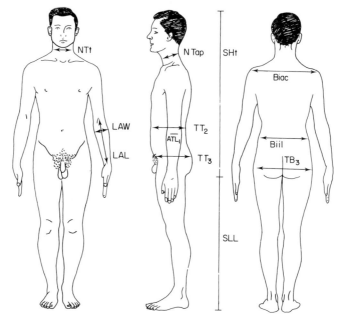

FIG. 4. Measurements for shape study.

height. The remainder were taken from standardized somatotype photographs. It is of some interest that we can ignore the difference in scale between the two sets of measurements, though we shall see that it has indirect consequences later on.

When we do the analysis, the technique appears to work rather well. The results for men and women are closely consistent; about half the total variance is accounted for by the first two principal components and a further 30% by the next three. Closer inspection, however, suggests a less than satisfactory situation. If we look at elements of the first principal component (Table II) (these add up to zero and have been standardized to unit sum of squares), we see that the linear compound is dominated by its last two coefficients and is essentially a contrast between the two transverse measurements taken on the lower arm. This is hardly what we expect to obtain as the best single description of shape.

A possible reason for this finding emerges if we go back to the residual dispersion matrix from which the principal components were derived (Table III). None of the correlations are very large; one of the four which are 0.5 or over is between the two arm-width measurements, and this brings to our notice the fact that these two measurements have by some way the largest standard deviations.

TABLE II

Shape components of anthropometric data

	I	II	III
Subisch.	0.13	0.51	−0.06
Sitt. ht.	0.10	0.10	0.17
Bi-ac.	0.08	0.16	0.13
Bi-il.	0.03	0.15	0.17
NTt	−0.11	−0.16	0.25
NTap	−0.07	−0.23	0.30
TT2	−0.09	−0.22	−0.71
TT3	−0.05	−0.25	−0.14
TB3	−0.11	−0.07	−0.39
LAL	−0.04	0.56	−0.10
ATL1	−0.60	−0.20	0.28
LAW	0.75	−0.36	0.10
% total variance	30	23	12

See Fig. 4 for key to abbreviations.

What we are looking at must, we think, be largely a posing artefact. Quite small changes in the angle at which the hand is carried relative to the body will increase one of our two measurements at the expense of the other and will produce the negative correlation that we observe. The large standard deviations (remember that these are analogous to coefficients of variation) are probably related to the fact that these measurements are quite small, in fact only a few millimetres on the photograph. The upshot is that the difference between the two measurements is an especially variable quantity and it is this that has been seized upon by the principal component analysis.

We have re-analysed the data omitting the two arm-width measurements, but before discussing the results we should spend a little more time on errors of measurement. Regarded as coefficients of variation, the standard deviations residual to size are really quite small, and it seems to us inevitable that a part of the observed variability must be associated with measuring errors, especially in the case of the photogrammetric measurements. If good estimates of measuring error were available it would be interesting to re-analyse the data after reducing the diagonal elements of the original dispersion matrix by appropriate amounts. We should perhaps stress that we are not criticizing the quality of the measurements; all were made by a single observer and since this observer was Mr R. H. Whitehouse we do not think that their quality could be greatly improved.

The first three principal components of the analysis omitting the arm widths are shown in Table IV. The pattern is not quite what we had expected. The first component links the longitudinal measure-

TABLE IIIA

Correlations residual to size

	Subisch.	Sitt. ht.	Bi-ac.	Bi-il.	NTt	NTap	TT2	TT3	TB3	LAL	ATL1	LAW
Subisch.	1											
Sitt. ht.	0.2	1										
Bi-ac.	0.4	0.4	1									
Bi-il.	0.1	0.1	0.1	1								
NTt	−0.4	0.0	0.0	−0.1	1							
NTap	−0.4	−0.1	−0.2	0.0	0.4	1						
TT2	−0.2	−0.3	−0.4	−0.3	−0.1	−0.1	1					
TT3	−0.5	−0.2	−0.3	−0.2	0.1	0.1	0.2	1				
TB3	−0.3	−0.4	−0.3	0.0	−0.1	−0.1	0.2	0.2	1			
LAL	0.6	0.1	0.2	0.0	−0.4	−0.5	−0.2	−0.2	−0.4	1		
ATL1	−0.4	−0.3	−0.3	−0.2	0.3	0.1	0.0	0.0	0.1	−0.1	1	
LAW	−0.1	0.1	0.0	−0.1	−0.2	−0.1	−0.1	−0.1	0.0	−0.3	−0.3	1

TABLE IIIB

Standard deviations (units of 100 $\log_e$)

4.6	2.7	2.7	3.8	3.2	3.8	4.9	3.5	4.3	4.9	6.4	7.3

See Fig. 4 for key to abbreviations.

TABLE IV

Shape components of anthropometric data

	I	II	III
Subisch.	0.27	0.33	0.32
Sitt. ht.	0.19	0.07	−0.20
Bi-ac.	0.25	0.06	−0.12
Bi-il	0.20	−0.29	−0.60
NTt	0.08	−0.15	−0.11
NTap	−0.07	−0.11	−0.10
TT2	−0.73	0.42	−0.12
TT3	−0.32	−0.04	0.02
TB3	−0.18	−0.66	0.56
LAL	0.33	0.38	0.36
% total variance	27	21	14

See Fig. 4 for key to abbreviations.

ments, subischial, sitting height and lower arm length, and contrasts them with the trunk widths and breadths taken from the photograph. However, the widths measured on the body, bi-acromial and bi-iliac, go with the lengths, not with the other transverse measurements. The largest loading by far is for TT2, an anteroposterior measurement at waist level; this reflects the large residual variance of this measurement. It would have been less dominant had we analysed the correlations instead of the variances and covariances.

We can claim very little in the way of originality for our analyses. The form of the allometry relationship suggests working on logarithmic scales, and Jolicoeur (1963) in particular, following Teissier (1960), has suggested using the first principal component of the dispersion matrix of the log measurements as a generalization of two-dimensional allometry, pointing out that it measures pure size when the coefficients happen to be equal in magnitude. Several previous workers used principal component analysis or factor analysis in attempts to quantify body build. Burt, to take an early example (Burt & Banks, 1947), factor analysed a set of body measurements, expecting apparently to find human physiques falling neatly into discrete Kretschmerian types; our Sheldonian background makes us less surprised than he was to find the factor scores making up a conventional unimodal distribution.

If we were to analyse the ordinary dispersion matrix of the measurements, whether with or without logarithmic transformation, we would undoubtedly obtain a first principal component all of whose coefficients were positive. Such an analysis has often been done and the resulting linear compound has been suggested as a measure of size. This we now think is incorrect. Principal component analysis is intended to emphasize all the differences between individuals, and this first component will in practice contain a mixture of size and shape. It is for this reason that we have found it necessary to define our size measure from external *a priori* considerations and to describe as shape differences everything residual to this. Among other things, this enables us to describe shape differences which are correlated with size differences. Such correlation can undoubtedly occur in artificial populations — we can picture a set of triangles falling into two categories, small fat ones and large thin ones, for example (Fig. 5) — and we would regard its presence or absence in human

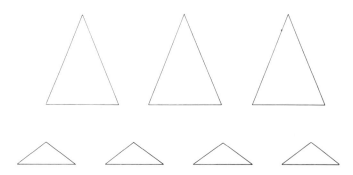

FIG. 5. Correlation between size and shape.

populations as a matter of fact, not of theory. In passing, we would apply the same comment to the choice of an index of build based on height and weight which seeks to find an index uncorrelated with height (Benn, 1971); we do not wish to be prevented from finding empirically that tall people are predominantly of slim build, should this in fact be the case.

CHANGES IN SHAPE DURING GROWTH

So far we have discussed the problem of defining size and shape differences in an adult population. The problems become more difficult and more interesting when we turn to growing organisms — specifically to children — and try to describe and quantify the shape changes that take place during the period of growth. We do not have any empirical findings to report on this occasion, but we can make some suggestions for the lines on which future work might be pursued.

The first obvious possibility is to use the linear compounds defining shape which have been obtained by the analysis of the adult measurements and simply to study the changes in these as growth proceeds. Such an analysis would be interesting but seems to us to miss the point of a real growth study; the phases of growth are of interest in themselves, not merely as stages on the path to maturity. We see no reason why a combination of measurements which best describes the different shapes of adults should be in any way optimal for children, nor do we consider that changes in the quantity so defined will be necessarily good descriptions of the changes in shape that take place during growth. The issue is in some way analogous to that involved in somatotyping children. We may on the one hand define a subject's somatotype from his adult build, and regard this as a built-in characteristic which gradually becomes apparent during the growth period; or we may on the other hand try to describe as well as possible the physiques of children as they are actually growing, either at a fixed age (be this chronological or maturity-based) or with reference in some way to a range of ages.

Given longitudinal data on a single child, we could adjust each set of data to a constant size and derive shape measures from the residuals as before. In itself this would not be of great interest, but it would become more so if we were to apply the analysis to a dispersion matrix obtained by pooling the data from a large group of children. Care would be necessary to make proper allowance for gaps in the data — genuine data are never purely longitudinal and unnecessary difficulties can be created if we do not use correct estimation techniques. Such techniques have been quite fully studied over the past decade or so, and it turns out that they form a fairly simple special case of the EM algorithm (Dempster, Laird & Rubin, 1977) which has permitted the unification of a large number of apparently unconnected statistical problems.

The weakness of this second approach is that it mixes up between-subject and within-subject variation in a rather unexplicit way. Its dependence upon longitudinal data is also a practical drawback since such data will always tend to be scarce and (what is more serious) unrepresentative. A further possibility would be to do a canonical analysis within and between age-groups so as to obtain linear compounds which distinguish between these groups as clearly as possible. Such an analysis would actually be better done on a large cross-sectional data set than on a longitudinal one. As before, we would do the analysis on residuals of logarithmic measurements after removing the effects of size.

All three methods we have considered share one feature. This is that the linear compound defining a particular aspect of shape has the same coefficients at all ages. We can envisage a more subtle definition of shape which tries to assess a single underlying component of build by means of a linear compound of measurements whose coefficients change as growth progresses. We have had a rather tantalizing glimpse of such a technique in our analysis of the remarkable longitudinal data on small children collected by Professor Low in Aberdeen between the wars and complemented by measurements on the same individuals as adults made by Professor Low's successors and ourselves (Tanner, Healy, Lockhart, Mackenzie & Whitehouse, 1956). Speaking informally, we found that there was almost no correlation in size between the birth and adult measurements, in that the correlations for single measurements were uniformly low, of the order of 0.2. However, a canonical correlation analysis of the two sets of measurements gave rise to two linear compounds, one for birth measurements and one for the same subjects as adults between which the correlation was as high as 0.8. This suggests to us the existence of some kind of almost invariant shape variable, measured by different linear compounds at the two ages. We have never published this analysis for various reasons, the main one being our total failure to interpret the two linear compounds; the patterns of the coefficients, even of their signs, seemed to us to make no sense at all. This may have been partly due to the rather small samples (canonical variable coefficients are even more badly behaved from the viewpoint of sampling error than are multiple regression coefficients, which in our opinion is saying a good deal) and partly to the rather odd set of measurements chosen by Professor Low some 50 years ago. We were also unable at the time to see how to extend our analysis so as to include the data at intermediate ages which were also available. It was by no means obvious to us how the two-group canonical correlation analysis (as distinct from the rather more familiar canonical variate analysis, which is a special case of it) should be extended to three or more sets of measurements — with two groups we aim to maximize a correlation coefficient, but what should we do with more? One possibility is to think of arranging the subjects in order of their shape values at each age, and to choose coefficients so that as far as possible the rank order of each subject stays constant from one age to the next. Some recent progress has been made in this direction by statisticians, including a comprehensive paper by Kettenring (1971) which does not seem so far to have achieved practical application.

In common with most others who have worked in this field, we have become increasingly impressed by the difficulty of expressing in proper quantitative terms the subtle changes of shape that take place during growth. To illustrate them, it is sufficient simply to superimpose two size-adjusted outlines of the same girl, the first at age $3\frac{1}{2}$ and the second at age 19 (Fig. 6). If we omit the head and

FIG. 6. Shape comparison at ages $3\frac{1}{2}$ (——) and 19 (. . . .).

arms, the main differences are those around the waist, and it is clear that we have no hope of describing these unless we have made detailed measurements in this area. This underlines the fundamental tenet of multivariate analysis, that the results depend upon what measurements were chosen in the first place. Always in quantitative biology of this kind, one of the most difficult problems on which all subsequent analysis depends is the choice of what measurements to make, and how we are to make them with sufficient precision. A fairly obvious difference between the two outlines in Fig. 6 is the angle between the legs; but this could be regarded simply as another posing artefact. Happily, modern equipment makes the acquisition of data from photographs a less overwhelming problem than it used to be, so that at least we can try out any new ideas we may have.

TRANSFORMATION GRIDS

With the title we have chosen, we cannot leave the study of size and shape in relation to growth and form without at least a mention of transformation grids. Ever since D'Arcy Thompson (1917) introduced them, these grids have haunted the literature and there have been repeated attempts to bring them within the reach of routine calculations. Our two outlines (Fig. 6) are taken from one of the latest papers in this field, by Goldstein & Johnson (1978) using methodology due to Sneath (1976), and we show in Fig. 7 the grids

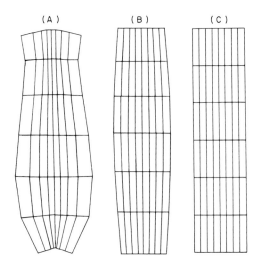

FIG. 7. Transformation grids between ages $3\frac{1}{2}$ and 19.

which they gave rise to. Clearly these have only captured the shape changes in very broad terms and much of the detail has escaped. One further very interesting development is described in Bookstein's 1978 monograph to which we have already referred. Bookstein points out that the standard approach starting with a rectangular grid for one outline is unnecessarily asymmetrical. Instead, one can derive from the two outlines a curvilinear orthogonal grid for each, the two being related by a simple magnification along the two grid directions at each interior point. He shows interesting applications of the method to outlines of primate skulls. Bookstein's methodology is, in his own phrase, not yet fully mature, but we feel it represents a real step in the direction of quantifying shape change; it may be more appropriate to fairly simple rigid objects such as the skull, perhaps less so to outlines of the whole body.

SIZE, SHAPE AND STATISTICS

It may be unnecessary to emphasize the contributions of statistics to the study of size and shape. Many of the standard statistical methods have been brought to bear upon the problems and some progress has been made with their elucidation. What we would like to stress is the importance of size and shape studies in particular, and human biology problems in general, in the development of multivariate statistical methods. Many of these methods have been initially worked out in a mathematical context and it is only through applying them to real data that their strengths and weaknesses become apparent and that directions for further research suggest themselves. The fact that statistical conclusions in the field of human biology can often be checked by everyday observations makes this field one of the most valuable for those whose business is the development of new statistical techniques.

REFERENCES

Benn, R. T. (1971). Some mathematical properties of weight-for-height indices used as measures of adiposity. *Br. J. prev. soc. Med.* **25**: 42–50.

Bookstein, F. L. (1978). *The measurement of biological shape and shape change.* Berlin: Springer-Verlag.

Burt, C. & Banks, C. (1947). A factor analysis of body measurements for British adult males. *Ann. Eugen.* **13**: 248–256.

Dempster, A. P., Laird, N. M. & Rubin, D. B. (1977). Maximum likelihood from incomplete data via the EM algorithm. *J. R. statist. Soc.* (B) **39**: 1–38.

Goldstein, H. & Johnson, F. E. (1978). A method for studying shape change in children. *Ann. Hum. Biol.* **5**: 33–39.

Jolicoeur, P. (1963). The multivariate generalization of the allometry equation. *Biometrics* **19**: 497–499.

Kettenring, J. R. (1971). Canonical analysis of several sets of variables. *Biometrika* **58**: 433–451.

Sneath, P. H. A. (1976). Trend-surface analysis of transformation grids. *J. Zool., Lond.* **151**: 65–122.

Sprent, P. (1972). The mathematics of size and shape. *Biometrics* **28**: 23–37.

Tanner, J. M., Healy, M. J. R., Lockhart, R. D., Mackenzie, J. D. & Whitehouse, R. H. (1956). Aberdeen growth study: I. *Arch. Dis. Childh.* **31**: 372–381.

Teissier, G. (1960). Relative growth. In *The physiology of Crustacea*: Ch. 16. Waterman, T. H. (Ed.). New York: Academic Press.

Thompson, D'Arcy W. (1917). *On growth and form*. Cambridge: University Press.

Symp. zool. Soc. Lond. (1981) No. 46, 37–62

Facial Growth in Primates with Special Reference to the Hominoidea

W. J. MOORE

Department of Anatomy, University of Leeds, UK

SYNOPSIS

From the relatively few species (principally man, the great apes, the baboon and the macaque monkey) that have been investigated, it appears that a common model of facial growth exists in Old World anthropoid primates. Such major departures from this model as occur are found in man. These comprise: (1) a relative reduction in the forward component of human facial growth achieved partly by a decrease in sutural growth and partly by the appearance of large resorptive areas on the anterior aspect of the jaws; (2) a tendency for the spheno-ethmoidal angle in man to decrease postnatally instead of increasing as in other primates; (3) growth of the human mandible along a more tightly whorled logarithmic spiral than occurs initially in great apes and possibly monkeys; and (4) a loosening of the co-ordination of growth between the different regions and subregions of the human facial skeleton as compared to that of the great apes.

The first three of these characteristics of the human mode of facial growth are clearly related to man's possession of an orthognathic facial outline. Although the fourth characteristic is probably of considerable clinical significance, in that it may underlie the very high prevalence of disorders of facial growth in modern man, its morphogenetic basis and evolutionary significance are not clear.

INTRODUCTION

The skull of man differs from that of all other primates — including his closest living relatives, the apes — in possessing a large braincase but a small and non-protrusive facial skeleton, contrasts which have been widely used in the assessment of the taxonomic affiliations of hominoid fossil crania. Knowledge of the morphogenetic processes underlying these differences in skull form is, however, extremely uneven. For the human skull there is available a vast, indeed almost overwhelming, amount of data relating to the postnatal growth of the skull, collected as part of the clinical endeavour to prevent and treat the common disorders of facial growth. When the need has arisen for experimental investigation, recourse has frequently been made to the rhesus macaque monkey on account of its ready availability as a laboratory animal, so that for this species, too, much information is available about postnatal skull growth. Interest has also focused on facial growth in the great apes because of its special significance in the interpretation of the evolutionary changes which have produced the characteristically shaped skull of man, but in this case investigation has been hampered by the problems involved in studying living apes with the result that practically all the data have been collected on dried skulls and are necessarily, therefore, of a cross-sectional nature with all the attendant disadvantages. With few exceptions, detailed information about postnatal growth in the remaining primate species is lacking. There appear to have been few, if any, comparative studies of prenatal skull growth in any primate species.

Despite this overall scarcity of information, the species that have been studied cover a sufficient range of Old World anthropoid primates to provide the basis for at least a tentative comparative appraisal of the general principles of postnatal facial growth in this group. It is convenient to make this comparison under three headings: (1) the pattern; (2) the sites; and (3) the control of facial growth.

PATTERN OF FACIAL GROWTH

Virtually all the early investigations were concerned with the pattern of facial growth and were, perforce, based on the study of the dried skull. Notable amongst these were the studies by Zuckerman (1926) of the baboon (*Papio porcarius*), by Krogman (1930a,b, 1931a,b,c) of the three great apes, and by Hellman (e.g. 1929) of man. From these studies, a general pattern of anthropoid skull growth can be discerned. The facial skeleton has been found to enlarge predominantly in a downward and forward direction. In general, the amount of forward growth (i.e. growth in anteroposterior depth) exceeds downward growth (i.e. growth in height) and increases in a superior to inferior gradient. Growth in width is less than that in either depth or height.

More recent studies, using the technique of radiographic cephalometry introduced by Broadbent in 1931 (see Moore & Lavelle, 1974, for a detailed account and full bibliography of this extensive body of work), have fully confirmed these general conclusions so far as man is concerned. Furthermore, since the technique allows the collection of longitudinal data, it has permitted investigation of changes in the rate of growth. It has emerged that enlargement of each of the dimensions of facial depth, height and width follows the somatic growth pattern, exhibiting a clear adolescent spurt which, in its timing, corresponds closely with the spurt in stature and other measures of overall body size. Whether or not an adolescent growth spurt occurs in apes, either in total body proportions or in the dimensions of the facial skeleton, is still uncertain because of the continuing lack of longitudinal data.

The early craniometric work established that the pattern of facial growth in apes contrasts with that in man principally in that it exhibits: (1) an overall greater amount of growth; (2) a more marked forward component of growth and superoinferior gradient; and (3) a tendency, especially in the gorilla and orang-utan, for the lower part of the facial skeleton to tilt upwards during the later stages of growth. These characteristics of pongid facial growth are accentuated in the males of each species. An essentially similar contrast between man and the baboon emerged from Zuckerman's (1926) craniometric study of skull growth in *Papio porcarius*.

The growth movements of the facial skeleton are associated with changes in the cranial base. As conventionally defined for cranio-metric investigation, this consists, in the median plane, of the basi-cranial axis, extending from the basion or endobasion to the pro-sphenion or pituitary point, and an anterior extension, reaching forwards from the anterior limit of the basicranial axis to the nasion (a posterior extension is also described but, since this is virtually identical with the anteroposterior diameter or foramen magnum, it has no significance in the present context). Essentially similar defi-nitions have been employed in radiographic investigations. Growth of the basicranial axis in man (Brodie, 1955; Zuckerman, 1955; Lewis & Roche, 1972), and in the great apes and macaque (Ashton, 1957) follows, like the facial skeleton itself, the somatic pattern with increments continuing until maturity and, in man at least, a marked acceleration during adolescence. The anterior extension as defined classically also follows the somatic pattern. However, if the anterior extension is measured to foramen caecum (i.e. the anterior edge of the cribriform plate) rather than to the nasion, its growth is found to correlate more closely with that of the brain and neurocranium (man — Ford, 1958; apes — Scott, 1958) being largely complete by the time the permanent dentition begins to erupt and exhibiting no adolescent spurt. It appears, therefore, that the region between the anterior limit of the cribriform plate and the nasion is one of adjust-ment between the growth of the braincase, following the neural pattern, and that of the facial skeleton, following the somatic pattern.

The spheno-ethmoidal angle (the angle between the basicranial axis and its anterior extension) increases in both apes and monkeys during postnatal growth, the amount of increase between the stage of the milk dentition and maturity (when the angle is about $160°$) being of the order of 5% (Ashton, 1957). In man, by contrast, the angle, after increasing prenatally to reach a value of about $140°$, decreases postnatally (by about 7%), particularly in the period to six years (Zuckerman, 1955). This angular growth change appears to be a unique human feature and clearly correlates with the reduced component of forward growth in the facial skeleton as compared with other anthropoid primates.

SITES OF FACIAL GROWTH

Bone tissue, with its rigidly mineralized matrix, is incapable of interstitial growth. Enlargement takes place solely by deposition of new tissue on to pre-existing surfaces which may be either external or internal. Such accretional growth is by itself, however, incapable of bringing about an increase in the size of a bone without progressive distortion of its shape. There is, for example, no way in which accretion alone could produce a femur of adult proportions from the corresponding bone of an infant. In order for proportions to be preserved it is necessary that accretion at some surfaces be accompanied by resorption at others — these differential activities being known collectively as remodelling (a not altogether satisfactory term in that it rather suggests that the process is concerned merely with small-scale changes whereas, in fact, it is fundamental mechanism of bone growth).

Studies of the sites of deposition and resorption in the facial skeleton have been made for even fewer primate species than have those dealing with the pattern of facial growth. The traditional method of defining the sites of these activities has been by means of intravitally administered dyes which mark growing bone. The most frequently used dye has been alizarin which stains mineralizing bone red. Administered by feeding madder root, this was in use as long ago as the 18th century to demonstrate bone growth in the skull (Hunter, 1771). A wide variety of other dyes and bone markers is now available, each with its own special uses and attendant advantages and disadvantages.

A rather different method of obtaining detailed information about remodelling changes has been employed by Enlow and his colleagues (Enlow, 1968). This consists of examining microscopically sections of bone, taken at regular intervals across the region under study, and identifying which surfaces were undergoing deposition and which resorption at the time of death. In addition, the architecture of the internal parts of the bone gives an indication of the mechanisms involved in its formation — bone formed periosteally consisting of densely packed circumferential lamellae, while bone formed endo-steally (i.e. opposite a resorbing periosteal surface) usually has a more irregular structure due to the filling in of cancellous tissue. Although laborious, this method has provided the bulk of the information about the detailed remodelling changes in the facial skeleton of man. It has also been used in the macaque monkey and the chimpanzee.

Cranial Base

The bones of the cranial base ossify in the cartilage of the prechordal and parachordal parts of the chondrocranium. The joints remaining between the definitive bones of the cranial base represent remnants of the chondrocranium and are, therefore, of the hyaline cartilage variety. In the majority of mammals, three such synchondroses remain after birth — the spheno-occipital (between basi-occipital and basisphenoid), midsphenoidal (between basisphenoid and presphenoid) and spheno-ethmoidal (in those groups lacking a mesethmoid ossification, notably ungulates and insectivores, the anterior part of the cranial base is ossified by forward extension of the presphenoid and consequently no spheno-ethmoidal synchondrosis is formed). The cranial base of man, the great apes and the macaque monkey possesses initially all three synchondroses, but in man and the apes the midsphenoidal synchondrosis fuses at about the time of birth, while in the macaque monkey this synchondrosis persists throughout much of the growth period (Michejda, 1971, 1972). The spheno-ethmoidal synchondrosis has fused by birth or early postnatally in the macaque and great apes (Scott, 1958) but in man the cartilage in this region becomes replaced by fibrous tissue shortly after birth and the joint then remains a growth site of some significance until about six to eight years (Baume, 1968).

Endochondral ossification at these synchondroses is responsible for much of the linear growth of the basicranial axis and its anterior extension. The early closure of the midsphenoidal and spheno-ethmoidal synchondroses correlates with the relatively early cessation of growth in the anterior extension (when measured to the anterior margin of the cribriform plate). The spheno-occipital synchondrosis, by contrast, remains an active ossification site for much of the growth period (certainly up to the age of adolescence in man — Latham, 1966, 1972) as would be expected of the principal site of enlargement of the basicranial axis, since this segment of the cranial

base follows the somatic growth pattern. Some additional growth of
the basicranial axis probably takes place by deposition at the basion
and, as already noted, remodelling at the nasion undoubtedly contributes to the growth of the anterior extension as traditionally defined,
especially in the period after the permanent dentition begins to erupt.

   Growth at the synchondroses may also play a part in bringing
about the angular changes observed in the cranial base. There is
considerable evidence for man (Hoyte, 1975), the macaque monkey
(Michejda, 1972) and for some non-primate species (Vilmann, 1971)
that growth increments at the synchondroses are not always symmetrical in rate or amount. The asymmetry may occur both across
and along a synchondrosis. Clearly such differential growth could
produce changes in angulation and there are indeed reported correlations between the timing of cessation of growth changes in the
spheno-ethmoidal angle and fusion of the associated synchondroses,
of which the best documented is that associated with midsphenoidal
closure in the macaque monkey. However, the early closure of the
midsphenoidal synchondrosis in man and apes precludes the possibility that asymmetrical growth at this joint could be involved in the
postnatal growth changes in the angle in these species. An alternative
mechanism must, therefore, be sought.

   One mechanism probably involved in the changes in the hominoid
angle of cranial flexion appears to be surface remodelling of the
cranial base. This phenomenon has been described in detail for the
human skull by Enlow (1968) using the technique of examination
of the internal bony architecture described above. The floor of the
braincase can be considered, at its simplest, as consisting of two
cortical plates (one endocranial and the other ectocranial) separated
by cancellous bone, although in some regions other structures, such
as air sinuses and the middle and inner ear, intervene. The predominant remodelling pattern in man consists of bone deposition on the
periosteal surface of the ectocranial cortical plate and resorption
from the corresponding surface of the endocranial plate (the endosteal surfaces undergoing opposite changes from those taking place
periosteally) (Fig. 1). In this way, the whole cranial floor is moved
relatively downwards. This type of growth, termed cortical drift,
produces the requisite amount of enlargement of the endocranial
cavity with less disruption of the spatial arrangements of the structures associated with the cranial base than could be achieved by
synchondrotic growth alone.

   Superimposed upon the overall remodelling pattern are a number
of localized variations. Of particular significance in the present
context is the area around the sella turcica where the periosteal

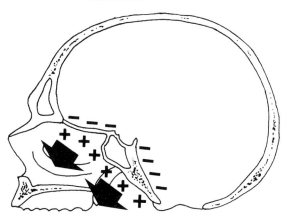

FIG. 1. Remodelling in the human cranial base. +, bone deposition; —, bone resorption; arrows indicate direction of cortical drift. Based on Enlow (1968).

surface of the endocranial plate is depository. As a result, the sella turcica increases its prominence relative to the surrounding resorptive areas. According to Enlow, the sella turcica tends to move in a relatively anterosuperior direction, whereas Latham (1972) has described it as moving posterosuperiorly — these discrepancies probably representing no more than individual variation. In either case, the upward movement of the sella turcica (together with the associated landmarks used to define the boundary between the basicranial axis and anterior extension) could well account for the postnatal decrease observed in the human spheno-ethmoidal angle.

In the cranial base of the macaque monkey, endocranial resorption is restricted to relatively small areas in the frontal and occipital regions and in the anteroventral part of the middle cranial fossa. The contrast between man and the macaque in the extent of cortical drift in the cranial base is probably related to the difference between the two species in the amount of brain growth. The relatively modest degree of cerebral expansion in the monkey can be accommodated by growth at the joints but the much greater expansion of the human cerebrum requires, in addition, a substantial amount of cortical drift in order to avoid undue disruption of the spatial relationships of the structures of the cranial base which would result if enlargement of this part of the skull were achieved exclusively, or even predominantly, by synchondrotic growth.

No equivalent information appears to be available for the remodelling changes in the pongid cranial base.

Upper Facial Skeleton

Most of the bones of the upper facial skeleton are of the dermal variety and the joints between them are, accordingly, of a sutural nature. Like the synchondroses of the cranial base, the sutures are important growth sites. The investigations by Pritchard and his colleagues (e.g. Pritchard, Scott & Girgis, 1956) have shown that the fully formed suture has five layers — namely, two cambial layers, each immediately adjacent to the margin of one of the two bones forming the suture and containing numerous osteoprogenitor cells and a layer of osteoblasts lining the bone; two capsular zones, of a more fibrous nature, one adjacent to each cambial layer; and a central loose cellular zone. The essential growth sites are the two cambial layers. The presence within a single suture of two such growth zones, separated by relatively inactive tissue, permits growth at each bony margin to be independent in direction and rate.

The sutural system of the primate facial skeleton can be subdivided into three components — the circummaxillary, craniofacial and sagittal groups of sutures — on the basis of the position and orientation of the joints (Scott, 1956). Thus, in the human facial skeleton, the constitution of the three groups is as follows: (1) circummaxillary — comprising the sutures between the maxillae, on the one hand, and the frontal, nasal, lacrimal, ethmoid, palatine, vomer and zygomatic bones and the pterygoid processes of the sphenoid, on the other; (2) craniofacial — the sutures separating the nasal, lacrimal, facial part of ethmoid, palatine, vomer and zygomatic bones from the frontal, perpendicular plate of ethmoid and sphenoid bones; (3) sagittal — the interfrontal, internasal, intermaxillary and median palatal sutures. The orientation of the circummaxillary and craniofacial groups is such that their growth contributes to the downward and forward growth of the facial skeleton relative to the cranial base while the sagittal sutures are so disposed that their growth leads to an increase in facial width. The facial sutures in other anthropoids have an essentially similar arrangement.

It has been shown repeatedly (e.g. Enlow & Bang, 1965; Latham, 1968) that one of the most active sites of bone deposition in the human facial skeleton is at the pterygomaxillary suture. Deposition at this site is even more marked in the macaque monkey (Enlow, 1966). Since the pterygoid processes of the sphenoid are firmly anchored to the cranial base, being in effect buttresses, the result of growth at the pterygomaxillary sutures is to impart a powerful component of forward movement to facial growth (McNamara, Riolo & Enlow, 1976). The greater growth activity at this site in the macaque monkey, as compared with man, correlates with the contrast in the degree of forward facial growth between these two species. Again, equivalent information for the apes is lacking but it seems reasonable to surmise that in these animals too the pterygomaxillary suture is a site of particularly rapid bone deposition, with the amount of growth at this suture exceeding that in man.

In the nasal region of the facial skeleton there is an extensive cartilaginous region representing the original nasal capsule. Much of this cartilage is replaced by bone during the growth period but a considerable part of the nasal septum remains cartilaginous and plays an important role in facial growth. In man, the vomer ossifies in the lower part of the septum at about the eighth week of intra-uterine life while the mesethmoid centre appears in the superior part of the septum at the time of birth. For the first two or three years of life the vomer and mesethmoid remain separated by a "sphenoidal" tail of septal cartilage but then ossification extends into this tail and the definitive septovomerine joint is formed.

An active endochondral growth site is present at the junction between the septal cartilage and the perpendicular plate of the ethmoid (mesethmoid) for much of the growth period in man, the macaque and many non-primate mammalian species (Baume, 1961, 1968). Its presence has led to the suggestion that the growth of the septal cartilage is a principal determinator of the degree and direction of facial enlargement in the sagittal plane, having a role rather like the growth of an epiphyseal cartilage in the extension in length of a long bone of the postcranial skeleton. However, attempts to demonstrate experimentally this suggested primacy of the septal cartilage have produced inconclusive results (see Moore & Lavelle, 1974, for detailed discussion).

The growth increments at the sutures and septal cartilage are accompanied by extensive remodelling of the external and internal surfaces of the bones of the upper facial skeleton. Detailed information about these remodelling changes has been provided for man by Enlow & Bang (1965) and Enlow (1968) and for the macaque monkey by Enlow (1966) on the basis of examination of the internal architecture of the facial bones. The patterns of remodelling in these two species, while having much in common, display a number of differences which are probably of considerable significance in determining the contrasts in facial shape between man and the monkey.

A large area of the external aspect of the human facial skeleton, including the periosteal surfaces of the premaxillae, of the maxillae as far posteriorly as the roots of the zygomatic processes, and of the zygomatic bones around the inferior and lateral orbital borders, is a site of resorption. The presence of this extensive resorptive area results in much of the buccal surface of the maxillary alveolar arch being a site of bone removal. At the roots of the zygomatic processes, the buccal surface turns from facing anterolaterally to face postero-laterally and here resorption gives way to deposition. The lingual aspect of the maxillary alveolar arch is entirely depository and the posteriorly facing tuberosity, as part of the pterygomaxillary suture, is a site of particularly rapid bone formation.

When seen in occlusal view (Fig. 2), the upper alveolar arch has

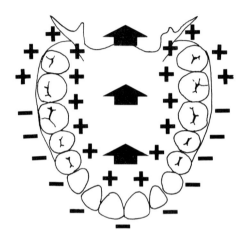

FIG. 2. Human maxillary arch seen in occlusal view. +, bone deposition; —, bone resorption; arrows indicate direction of intrinsic growth. Based on Enlow & Bang (1965).

the form of a rounded V and the remodelling just described is typical of that found in many other V-shaped regions (notably the meta-physes of the postcranial long bones). The combination of deposition on its inner surface and "free" ends and resorption from its outer surface results in the arch growing in a posterior direction (i.e. as in other V-shaped regions, the growth movement is towards the open end) and thus increasing in both its sagittal and transverse dimensions. This growth movement is, of course, intrinsic to the upper jaw com-plex; the whole complex is simultaneously being displaced down-wards and forwards by growth at the sutures and septal cartilage.

Viewed in coronal section (Fig. 3), the maxillary alveolar arch, together with the hard palate, can be seen to enter into a second V-shaped system, but in this case the V is inverted and somewhat truncated. Again the outer aspect of the V (i.e. the floor of the nasal cavity and maxillary paranasal sinuses and the buccal surface of the alveolar arch) is resorptive and the inner aspect (i.e. the oral surface of the hard palate and the lingual surfaces of the alveolar arch) is

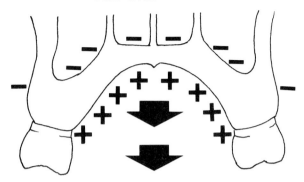

FIG. 3. Human maxillary arch seen in coronal section. +, bone deposition; −, bone resorption; arrows indicate direction of intrinsic growth. Based on Enlow & Bang (1965).

depository. Thus, the intrinsic growth movement of the region is in an inferior direction, adding to the downward component of movement produced by growth at the sutures and nasal septum.

Thus, two V-shaped systems exist in the maxillary complex. One, that formed by the alveolar arch, lies in the horizontal plane, while the other, that formed by the palate, occupies the frontal plane. Both undergo remodelling changes characteristic of V-shaped regions. The remodelling of the palatal V adds a further component of vertical growth to that produced by increments at the facial sutures and nasal septum while remodelling of the alveolar arch, by contrast, tends to offset the forward component of sutural growth. Coupled with the tendency in the human skull for the vertical component of growth at the sutures to exceed the forward component, these remodelling changes result in the face increasing rather more in height than in depth. Furthermore, remodelling of these two systems results in an increase in the height of the nasal cavity and maxillary sinuses and growth in length, width and height of the alveolar arches.

The principal difference between the pattern of facial remodelling in man and the macaque monkey is that in the latter the area of resorption over the anterior periosteal surface is lacking, this surface being entirely depository in nature. It appears that otherwise the growth of the maxillary arch in the macaque monkey is comparable in all essentials with that of the human arch.

It thus emerges that, in comparison to the macaque monkey, the characteristically flattened facial profile in man is associated with three interrelated morphogenetic features: (1) the decrease in the spheno-ethmoidal angle during postnatal growth: (2) the reduced velocity of forward growth contributed by increments at the sutures and nasal septum; and (3) the presence of an extensive area of resorption over much of the periosteal surface of the maxillary complex. While sufficient comparative data exist to make it possible to assert with some confidence that man is unique amongst primates in the possession of the first two of these features, no such confidence is possible so far as possession of the third is concerned. Nevertheless, consideration of the general pattern of facial growth and of the shape of the face achieved in the adult stage of apes and monkeys suggests that the distribution of facial remodelling in anthropoid primates generally is likely to be closer to that observed in the macaque monkey than that in man.

Remodelling in the orbit and in the region of the zygomatic arch is similar in both man and the macaque. The greater part of the periosteal surfaces facing into the orbit is depository in both species. This results in the orbit growing forwards and increasing in overall size, providing yet another example of the V-principle of bone growth within the facial skeleton. The anterior periosteal surface of the zygomatic arch is resorptive while the posterior surface (i.e. facing towards the temporal fossa) is depository. As a result, the attachment of the arch to the facial skeleton is relocated backwards so offsetting, to some extent, the lengthening of the arch produced by sutural growth.

The resorptive area on the front of the zygomatic arch is extended in man, but not in the macaque monkey, around the inferior and lateral margins and on to the anterior part of the lateral wall of the orbit. As a result, the orientation of the human orbital rim changes as growth proceeds, with the inferior border coming to lie in a more posterior plane than that of the superior border and the lateral border being displaced posteriorly relative to the medial border. In the macaque no such change of orientation occurs.

Mandible

Although the mandible is a dermal bone, as judged by both its ontogenetic and phylogenetic history, one of its principal growth sites is cartilaginous. During prenatal life, a number (varying between species) of cartilaginous areas appears in the developing mandible which, up to that point, has been composed entirely of intramembranously ossifying bone. The cartilages undergo rapid endochondral ossification and by birth have been completely replaced by bone except for that in the condylar region. This last persists throughout life and during the growth period is a site of continuous endochondral ossification. The actual direction of intrinsic growth produced by the condylar increments depends upon the precise shape of the condylar region, but is generally posterosuperior, thus contributing to increase in mandibular height and depth. Since the condyle abuts onto the cranial base at the temporomandibular joints this intrinsic growth movement results in the mandible being relocated antero-inferiorly.

The cartilages which form in the mandible differ in many important respects from those found in the epiphyses and synchondroses of the primary endoskeleton, and are frequently termed secondary to emphasize their distinctiveness. To take the condylar cartilage as an example: (1) it is formed by tissue which phylogenetically and ontogenetically appears to be of periosteal origin rather than perichondral; (2) it lies directly below the articular cartilage (composed of fibrocartilage) not separated from the joint surface by bone as occurs in epiphyseal cartilages; (3) unlike both epiphyseal and synchondrotic cartilage, condylar cartilage is composed of chondrocytes formed at an intermediate cell layer located between the condylar and articular cartilages, which do not undergo subsequent cell division (and do not therefore, form the columns of cells typical of more usual endochondral ossification sites); (4) the chondrocytes in the condylar cartilage remain vital and emerge from the ossification front capable of redifferentiating into osteogenic cells; in endochondral ossification sites in the endoskeleton the chondrocytes die and disintegrate subsequent to hypertrophy.

Despite these differences, the condylar cartilage is often considered to be the principal determinator of mandibular growth, being attributed, in this respect, a role similar to that of an epiphyseal cartilage in the growth of a long bone. As with the nasal septum, experimental evidence bearing on this point is inconclusive (see Moore & Lavelle, 1974, for detailed discussion). Nonetheless, there can be no doubt that the condylar cartilage is a major site of bone addition in the normally growing mandible, even if this should eventually prove to be of an adaptive or secondary nature.

Associated with the incremental growth at the condylar cartilage are a number of complex remodelling changes which serve to maintain the proportions of the mandible as it increases in size. These changes have been intensively studied since Hunter's (1771) pioneering investigation of mandibular growth in the pig which did so much to establish the fundamental principles of bone growth. Most subsequent investigators have followed Hunter in employing intravital dyes, but more recently the trend has switched to the use of metallic implant markers with serial radiography and to the study of internal bony architecture.

In each of the mammalian species that have been studied by intravital staining or implant marking (e.g. pig: Brash, 1934; Sarnat, 1968;

rat: Baer, 1954; macaque monkey: Turpin, 1968; McNamara & Graber, 1974; man: Bjork, 1968), the posterior rameal border has been found to be a site of especially rapid bone deposition and the anterior border of the coronoid process a site of resorption occurring at an only slightly slower rate. Together with the posterior component of growth contributed by increments at the condyles, this remodelling results in the mandible growing intrinsically in a predominantly posterior direction, and in both the ramus and body of the mandible increasing in anteroposterior depth.

Investigations based on the study of internal bone architecture in man (Enlow & Harris, 1964), the macaque monkey (Enlow, 1966) and the chimpanzee (Johnson, Atkinson & Moore, 1976) have added considerable detail to our knowledge of mandibular remodelling in primates. The overall pattern is remarkably similar in all three species, but certain local differences occur which can be readily related to regional variations in shape.

The species similarity is greatest in the ramus. In each case, the buccal periosteal surface of the coronoid process and the adjacent area below the sigmoid notch and extending onto the lateral surface of the condylar process is resorptive (Fig. 4). The opposed lingual surface (posterior to the temporal crest) is depository. Since the

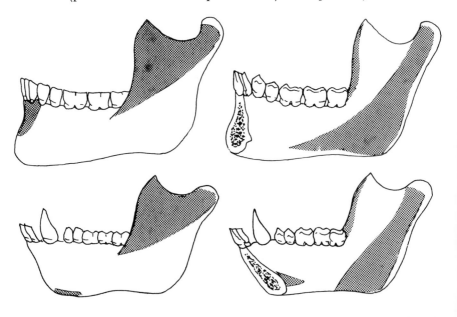

FIG. 4. Lateral (left) and medial (right) views of human (upper) and chimpanzee (lower) mandible to show surface remodelling. Bone deposition indicated in white, bone resorption in stipple. Based on Enlow & Harris (1964) and Johnson, Atkinson & Moore (1976).

latter surface faces superiorly and posteriorly as well as medially, the remodelling changes just described result in the coronoid process growing upwards, backwards and, following the V-principle, moving apart (the two coronoid processes forming, in frontal view, an inferiorly incomplete V). On both buccal and lingual surfaces of the ramus there is an obliquely running reversal line at which the direction of bone growth changes. On the buccal surface, the line passes from the posterior aspect of the condylar process in an antero-inferior direction to end at the base of the coronoid process, while on the lingual side the line runs from the anterior surface of the condyle again antero-inferiorly, to end below the molar teeth (the exact point of termination varying between species — see below). The ramus, in cross-section, tends to be somewhat convex buccally. As the ramus grows posteriorly by additions along its rear edge, bone regions formed here become progressively relocated in an anterior direction. Because of the convexity just described, this relocation also involves a lateral component which is brought about by the combination of buccal deposition and lingual resorption taking place posterior to the rameal reversal lines.

It will be noted that the position of the rameal reversal lines is such that the periosteal surfaces of the condylar process are resorptive. This arrangement is similar to that seen in the metaphysis of a long bone and has a similar function, serving to narrow down the condylar process in an inferior direction from the rather broad condylar cartilage to the narrower region where the process merges with the ramus.

Increase in the depth of the mandibular body and in the separation between the two sides is to a large extent the inevitable consequence of the growth and remodelling changes taking place in the ramus. There are, in addition, localized remodelling processes in the body itself, and it is in these that the principal species differences occur (Fig. 4). In the human mandible, the molar teeth are supported in alveolar bone which overhangs a quite deeply excavated sublingual fossa. The resorptive zone posterior to the reversal line on the lingual aspect of the ramus is continued forwards on to the lingual surface of the mandibular body as far as the first molar. Periosteal resorption and endosteal deposition in this region of the body hollow out the sublingual fossa and produce the overhang of the alveolar arch. In both the macaque monkey and the chimpanzee, by contrast, the area of resorption does not extend forward of the ramus and, in consequence, no sublingual fossa or alveolar overhang is produced.

The depository zone located posterior to the buccal reversal line of the ramus extends forwards on to the buccal surface of the body.

In the chimpanzee and macaque, the whole of this buccal surface, round to and including the symphyseal region, is depository in nature but in man a variable area of resorption is present below the incisor and canine teeth. It is this area of resorption which appears to be principally responsible for the formation of the characteristic human chin and its presence is presumably related, like that of the anterior resorptive zone in the upper facial skeleton, to the reduced forward growth of the human face.

LOGARITHMIC GROWTH OF THE MANDIBLE

The theoretical analyses of Thompson (1917) have demonstrated that the shape of structures growing by accretion, in which the increments of new material are incapable of further growth, can be predicted from a knowledge of the rates of accretion at the growth sites. If, for example, accretion is occurring predominantly at the base of a cone-shaped structure, equal increments of material at all points within the growth site, together with a constant ratio between the rates of accretion and extension in width, will maintain the conical shape. In fact, this mode of growth is rarely encountered. More usually growth gradients exist within a growth site so that the rates of accretion and width extension are greatest at one margin and least at the opposite margin. If the ratio of accretion and width extension between the two sides of the growth site remains constant, successive increments will be of the same form but of increasing bulk and the structure growing in this way will assume the form of a logarithmic or equiangular spiral (i.e. a spiral in which successive whorls cut the radius vector into lengths which increase in geometrical progression and in which the angle between the whorls and the radius remains constant − Fig. 5). Many biological structures have been found to grow in this way, including mollusc shells and horns (Thompson, 1917) and continuously growing rodent teeth (Herzberg & Schour, 1941).

As already described, growth in depth of the mandible is determined primarily by the amount of bone accretion taking place on the posterior border of the ramus (including the posterior component of growth at the condyle). Simultaneously, the height of this growth site is progressively extended by the vertical component of condylar growth. The differential rates of accretion at the posterior rameal border are not known but there is considerable indirect evidence, in man at least, that increments accrue more rapidly in its inferior than in its superior part (see Moore & Lavelle, 1974, for a detailed account

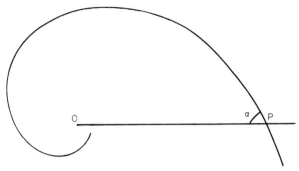

FIG. 5. A logarithmic spiral — successive whorls cut the radius vector (OP) into lengths which increase in geometrical progression and in which the angle (α) between the whorls and the radius vector remains constant.

of this evidence). Thus, some of the growth circumstances necessary to produce a logarithmic spiral form appear to exist in the human mandible.

The problem in testing whether or not such a spiral is actually produced lies in determining the essential shape of what is, at first sight, a rather irregularly shaped bone. A possible solution to this difficulty has been provided by Moss (e.g. Moss, 1960), who has suggested that the mandible comprises a number of separate regions — the skeletal units — each serving a different function and possessing considerable growth autonomy (Fig. 6). Much experimental

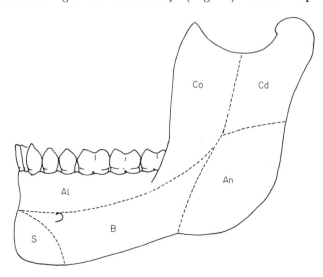

FIG. 6. Mandibular skeletal units. B = basal; Cd = condylar; Co = coronoid; An = angular; Al = alveolar; S = symphyseal.

evidence has now accumulated supporting this view (e.g. Moss, 1969; Moss & Meehan, 1970; Moore, 1973). The basal skeletal unit consists of the bone surrounding the inferior neurovascular bundle within the mandibular canal and, as its name implies, forms the basic "core" of the mandible to which all the other units (coronoid, angular, condylar, alveolar and symphyseal) are attached as superstructures. Its shape determines the essential form of the mandible and is the least modifiable by experimental procedures that interfere with bone growth.

Moss & Salentijn (1970) determined the growth of the basal skeletal units in a series of Amerindian skulls by recording the position, in the sagittal plane, of the mental, mandibular and oval foramina (i.e. the course of the inferior alveolar nerve) at different growth stages. They found that the records could be superimposed so that the positions of the three foramina lay, at each growth stage, along a common logarithmic spiral. Hildyard, Moore & Corbett (1976) have confirmed Moss & Salentijn's findings for skulls of mixed English stock and, moreover, found that the constants K (inversely related to the size of the angle between whorls and radius and therefore to the "tightness' of the spiral) and R (related to the position of the segment along the spiral) defining the segment of the spiral are very similar for the two groups of skulls (Amerindians: $K = 0.19$, $R = 3.73$; English: $K = 0.18$, $R = 4.08$). The fact that the specimens on which these studies are based are from widely different ethnic groups suggests that extension along a common logarithmic spiral (defined by values for the constants K and R of approximately 0.2 and 4 respectively) is a common feature of human mandibular growth (Fig. 7).

The similarity of the growth mechanisms in the lower jaw of the chimpanzee and macaque to those in man raises the question of whether or not mandibular growth follows a logarithmic spiral in non-human anthropoids. In order to test this possibility, Hildyard, Moore & Corbett (1976) extended their study to the skull of the chimpanzee and gorilla. It was found that in both sexes of the gorilla the value of K in the younger age groups was much greater (at between 0.6 and 0.9) than in man indicating that growth takes place initially along a less tightly whorled spiral than occurs in the human mandible. The value of K decreased but R increased with age suggesting that in this genus the mandible grows along a progressively more distal segment of a logarithmic spiral in which the angle between the whorls and the radius increases in value as growth proceeds (a combination referred to, rather inaptly, in our original paper as an unfolding of the logarithmic spiral). Similar but less marked trends were

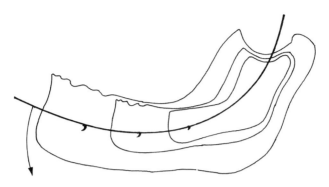

FIG. 7. Growth of the basal unit of the human mandible along a segment of a logarithmic spiral. In this figure enlargement is depicted as though occurring downwards and forwards relative to the cranial base which is the actual direction in which the *growth movement* occurs although, as described in the text, the *intrinsic growth* of the mandible is directed posteriorly; while the curve accurately depicts the growth movement, it does not, therefore, coincide exactly with the successive increments to the basal unit. The arrow indicates the rotation required to preserve the spatial relationships of the mandible.

found in the female chimpanzee while in the male K underwent no consistent age change but the value of R increased with growth, implying that in this group, too, growth takes place along a progressively more distal segment of the logarithmic spiral.

The functional significance of these contrasting growth patterns may be related to the differing degrees of forward growth of the facial skeleton in man and the African apes. Extensive growth along a tightly whorled logarithmic spiral would result in the mandible assuming a highly curved form inappropriate for the shape of the jaw in a large-faced, prognathous species, and in the necessity for a large amount of compensatory rotation of the mandible to preserve its spatial relationships with the remainder of the facial skeleton (Fig. 7). On such theoretical grounds, therefore, it might be expected that the "tightness" of the spiral would be inversely related to the adult size of the jaws. This expectation appears to be realized in the great apes where the value of K, in the younger age groups at least, appears to be related to the final size of the jaws, being much greater in the gorilla than in the chimpanzee.

Preliminary investigations indicate that a similar relationship between K and the adult size of the jaws exists in Old World monkeys.

CONTROL OF FACIAL GROWTH

Perhaps the most difficult aspect of growth to investigate is that of its control, largely because the factors regulating the underlying processes of cellular differentiation and proliferation are poorly understood. However, a useful first step is provided by an analysis of the associations between the dimensions of the region under consideration and between these and overall body proportions. Such an analysis will provide an insight into the manner in which growth is co-ordinated and will often suggest testable hypotheses about the regulation of this co-ordination. Furthermore, the collection and processing of the necessary data present no undue technical difficulties.

It is not surprising, therefore, that numerous studies have been made of associations within the skull, although these, almost without exception, relate to man (e.g. Pearson & Davin, 1924; Woo, 1931; Bjork & Palling, 1954; Koski, 1964). A particularly noteworthy study is that by Solow (1966) who has provided not only a comprehensive survey of the associations within the human facial skeleton but also a critical examination of the methodology involved. He distinguished topographical correlation, found between two variables sharing a common reference point, from non-topographical correlation where the variables do not share common reference points. A non-topographical correlation can be presumed to indicate a biologically determined association, while a correlation between two measurements made from a common landmark may indicate no more than the growth movement of that landmark which will, of course, affect both measurements.

Solow (1966) analysed the associations in the human facial skeleton (data collected from 102 adult male Danes), making use of principal components and factor analysis to interpret the large correlation matrices that he obtained. He was able to distinguish three separate groups of statistically significant non-topographical correlations.

(1) The first group comprised numerous associations between linear facial dimensions and overall body dimensions such as stature and length of limb segments.

(2) The second group was between measurements spanning common facial regions, reflecting co-ordinated variation within these regions.

(3) The third and largest group was made up of associations between dimensions of the jaws and dentition.

The only study of associations in the facial skeletons of primates other than man appears to be that of Moore (1977) in which the methods used by Solow (suitably modified to take account of species differences and more recent work on facial growth) were extended to the three great apes. Samples of male and female human skulls were also included. A total of 33 measures was taken of the

TABLE I

Correlation coefficients between facial dimensions.

Measurements	Homo		Gorilla		Pan		Pongo	
	M	F	M	F	M	F	M	F
Maxillary length (prosthion–pterygomaxillary fissure)/mandibular length (infradentale–condylion)	668	520	801	541	733	573	817	722
Maxillary inclination/mandibular inclination (both inclinations being the angles between the basal units of maxilla and mandible and the anterior extension of the cranial base)	455	352	656	691	662	500	411	545
Maxillary prognathism (infraorbital foramen–nasion)/mandibular prognathism (mental foramen–nasion)	192	066	745	816	701	492	814	403
Maxillary prognathism/upper arch length	-007	045	033	381	-014	-229	-012	280
Mandibular prognathism/lower arch length	067	134	195	353	-396	-327	-065	269
Spheno-ethmoidal angle/maxillary prognathism	-186	136	-511	-562	-096	-389	-299	-520
Spheno-ethmoidal angle/mandibular prognathism	-369	-148	-296	-686	-003	-671	-478	-472
Maxillary length/upper arch length	489	461	612	441	275	554	623	506
Mandibular length/lower arch length	292	471	529	132	-019	206	609	416
Mandibular prognathism/molar occlusion (distance between mesial surfaces of upper and lower left first molars)	314	607	156	069	101	-218	536	651
Mandibular prognathism/overject (distance between incisal edges of upper and lower left first incisors)	-573	-546	-305	020	395	-292	118	-027

Numbers of specimens in each group: *Homo*: M = 43, F = 26; *Gorilla*: M = 38, F = 37; *Pan*: M = 21, F = 50; *Pongo*: M = 16, F = 15.

proportions and position of the upper and lower jaws, of the size of
the teeth and dental arches and of the occlusal relationships. Some of
the more significant correlations between these measurements are
given in Table I.

The main conclusion to emerge from this study is that the pattern
of facial associations in the great apes is similar in essence to that in
man. The principal difference is the generally higher value of the coefficients in the apes, suggesting that pongid facial growth is more
tightly co-ordinated than is that of man. The very close associations
between measurements of maxillary and mandibular prognathism
and between maxillary and mandibular inclinations, for example,
indicate that variations in the position of the upper and lower jaws
are well co-ordinated and produce little occlusal disturbance. In man,
on the other hand, the association between upper and lower prognathism is weak and this, coupled with the absence of a mechanism
to compensate for discrepancies in the anteroposterior relationship
of the upper and lower jaws by variation in the size of the dental
arches (as indicated by low correlations between prognathisms and
arch and tooth size), frequently results in disturbed occlusal relationships.

Relatively high negative correlations were found between the
spheno-ethmoidal angle and the degree of mandibular prognathism,
as would be expected in view of the fact that the mandible articulates
at the temporomandibular joints with the lateral extensions of the
basicranial axis. Similar correlations are found between the spheno-
ethmoidal angle and maxillary prognathism in apes where there are
strong associations between upper and lower prognathism. In man,
however, the association between the prognathisms is low and the
correlation between the spheno-ethmoidal angle and maxillary prognathism does not follow that between the angle and mandibular
prognathism as closely as it does in apes. There appears to be no
powerful mechanism operating in man, through changes in jaw
lengths, to preserve the relationship of the upper and lower jaws in
the face of variation in cranial base flexion.

The patterns of correlations between measurements of the jaws
and dentition are remarkably similar in each of the hominoid genera.
Following Solow (1966), this group can be further divided into three
subgroups. In the first (associations between jaw, arch and tooth
dimensions — e.g. maxillary length/upper arch length and mandibular
length/lower arch length) and second (adaptation of the dental arches
to the relationship between upper and lower jaws — e.g. prognath-
isms/arch lengths) subgroups particularly there is fair consistency
between each of the genera. It was only in the third subgroup (corre-

lations between jaw relationships and occlusal estimates — i.e. molar occlusion and overjet) that any major difference was found between man and the apes, the associations in this group being generally much greater in man than in the apes because of the absence, already noted, of any apparent mechanism to compensate for discrepancies in the anteroposterior relationship of the upper and lower jaws in the human species.

CONCLUSIONS

Although the number of species so far investigated is clearly too small to allow more than tentative conclusions to be drawn, the many similarities in the pattern, sites and associations of facial growth in forms as far removed from each other as man, the great apes and the macaque monkey do suggest that the basic mechanisms of facial growth will be found to be the same in many, if not all, anthropoid primates. In the facial skeleton of each of the species that have been studied, a common fundamental plan is apparent in the direction of the major growth components, in the distribution of the growth sites, in the pattern of surface remodelling and in the co-ordination of growth. In most respects, man conforms to this plan but, in addition, displays several unique (amongst living forms) departures which are responsible for producing the characteristic shape of the human face. The principal departure is a reduction in the component of forward growth produced by a combination of a decrease during postnatal growth in the spheno-ethmoidal angle, a diminution in the amount of bone deposited at the sites contributing to forward growth, the appearance of areas of resorption on the anterior aspect of the upper facial skeleton and mandible and the tendency of the mandible to grow along a constant, tightly curved logarithmic spiral. As a result, the human facial skeleton is far less prognathous than that of all other primates. A departure possibly of rather less evolutionary importance, but of the greatest clinical significance, is the generally looser co-ordination of human facial growth, as compared with the great apes. There can be little doubt that this is one of the factors underlying the very high prevalence of facial dysharmony in modern man.

REFERENCES

Ashton, E. H. (1957). Age changes in the basicranial axis of the Anthropoidea. *Proc. zool. Soc. Lond.* 129: 61–74.

Baer, M. J. (1954). Patterns of growth of the skull as revealed by vital staining. *Hum. Biol.* **26**: 80—126.

Baume, L. J. (1961). The postnatal growth activity of the nasal cartilage septum. *Helv. ondont. Acta* **5**: 9—13.

Baume, L. J. (1968). Patterns of cephalofacial growth and development. A comparative study of the basicranial growth centres in rat and man. *Int. dent. J., Lond.* **18**: 489—513.

Bjork, A. (1968). The use of metallic implants in the study of facial growth in children: method and application. *Am. J. phys. Anthrop.* **29**: 243—254.

Bjork, A. & Palling, M. (1954). Adolescent age changes in sagittal jaw relation, alveolar prognathy, and incisal inclination. *Acta odont. scand.* **12**: 201—232.

Brash, J. C. (1934). Some problems in the growth and developmental mechanics of bone. *Edinb. med. J.* **41**: 363—387.

Broadbent, B. H. (1931). A new X-ray technique and its application to orthodontia. *Angle Orthod.* **1**: 45—66.

Brodie, A. G. (1955). The behaviour of the cranial base and its components as revealed by serial cephalometric roentgenograms. *Angle Orthod.* **25**: 148—160.

Enlow, D. H. (1966). A comparative study of facial growth in *Homo* and *Macaca*. *Am. J. phys. Anthrop.* **24**: 293—308.

Enlow, D. H. (1968). *The human face.* New York: Harper & Row.

Enlow, D. H. & Bang, S. (1965). Growth and remodelling of the human maxilla. *Am. J. Orthod.* **51**: 446—464.

Enlow, D. H. & Harris, D. B. (1964). A study of the postnatal growth of the human mandible. *Am. J. Orthod.* **50**: 25—50.

Ford, E. H. R. (1958). Growth of the human cranial base. *Am. J. Orthod.* **44**: 498—506.

Hellman, M. (1929). The face and teeth of man. A study of growth and position. *J. dent. Res.* **9**: 179—201.

Herzberg, F. & Schour, I. (1941). The pattern of appositional growth in the incisor of the rat. *Anat. Rec.* **80**: 497—506.

Hildyard, L. T., Moore, W. J. & Corbett, M. E. (1976). Logarithmic growth of the hominoid mandible. *Anat. Rec.* **186**: 405—412.

Hoyte, D. A. N. (1975). A critical analysis of the growth in length of the cranial base. *Birth defects* **11**: 255—282.

Hunter, J. (1771). *Natural history of the human teeth.* London: John Johnson.

Johnson, P. A., Atkinson, P. J. & Moore, W. J. (1976). The development and structure of the chimpanzee mandible. *J. Anat.* **122**: 467—477.

Koski, K. (1964). The Finnish female face in norma lateralis. *Trans. Eur. orthodont. Soc.* **40**: 463—469.

Krogman, W. M. (1930a). Studies in growth changes in the skull and face of anthropoids. I. The eruption of the teeth in anthropoids and Old World monkeys. *Am. J. Anat.* **46**: 303—313.

Krogman, W. M. (1930b). Studies in growth changes in the skull and face of anthropoids. II. Ectocranial and endocranial suture closure in anthropoids and Old World apes. *Am. J. Anat.* **46**: 315—353.

Krogman, W. M. (1931a). Studies in growth changes in the skull and face of anthropoids. III. Growth changes in the skull and face of the gorilla. *Am. J. Anat.* **47**: 89—115.

Krogman, W. M. (1931b). Studies in growth changes in the skull and face of anthropoids. IV. Growth changes in the skull and face of the chimpanzee. *Am. J. Anat.* **47**: 325—342.

Krogman, W. M. (1931c). Studies in growth changes in the skull and face of anthropoids. V. Growth changes in the skull and face of the orang-utan. *Am. J. Anat.* 47: 343—365.

Latham, R. A. (1966). Observations on the growth of the cranial base in the human skull. *J. Anat.* 100: 435.

Latham, R. A. (1968). A new concept of the early maxillary growth mechanism. *Trans. Eur. orthodont. Soc.* 44: 53—63.

Latham, R. A. (1972). The sella point and postnatal growth of the human cranial base. *Am. J. Orthod.* 61: 156—162.

Lewis, A. B. & Roche, A. F. (1972). Elongation of the cranial base in girls during pubescence. *Angle Orthod.* 42: 358—367.

McNamara, J. A. & Graber, L. W. (1974). Mandibular growth in the rhesus monkey (*Macaca mulatta*). *Am. J. Phys. Anthrop.* 42: 15—24.

McNamara, J. A., Riolo, M. L. & Enlow, D. H. (1976). Growth of the maxillary complex in the rhesus monkey (*Macaca mulatta*). *Am. J. phys. Anthrop.* 44: 15—26.

Michejda, M. (1971). Ontogenic changes of the cranial base in *Macaca mulatta*. *Int. Congr. Primatol.* 3(1): 215—225.

Michejda, M. (1972). The role of basicranial synchondroses in flexure processes and ontogenetic development of the skull base. *Am. J. phys. Anthrop.* 37: 143—150.

Moore, W. J. (1973). An experimental study of the functional components of growth in the rat mandible. *Acta anat.* 85: 378—385.

Moore, W. J. (1977). Associations in the hominoid facial skeleton. *J. Anat.* 123: 111—127.

Moore, W. J. & Lavelle, C. L. B. (1974). *Growth of the facial skeleton in the Hominoidea.* London: Academic Press.

Moss, M. L. (1960). Functional analysis of human mandibular growth. *J. prosth. Dent.* 10:1149—1159.

Moss, M. L. (1969). Functional cranial analysis of the mandibular angular cartilage in the rat. *Angle Orthod.* 39: 209—214.

Moss, M. L. & Meehan, M. A. (1970). Functional cranial analysis of the coronoid process in the rat. *Acta anat.* 77: 11—24.

Moss, M. L. & Salentijn, L. (1970). The logarithmic growth of the human mandible. *Acta anat.* 77: 341—360.

Pearson, K. & Davin, A. (1924). On the biometric constants of the human skull. *Biometrika* 16: 328—363.

Pritchard, J. J., Scott, J. H. & Girgis, F. G. (1956). The structure and development of cranial and facial sutures. *J. Anat.* 90: 73—86.

Sarnat, B. G. (1968). Growth of bones as revealed by implant markers in animals. *Am. J. phys. Anthrop.* 29: 255—286.

Scott, J. H. (1956). Growth at facial sutures. *Am. J. Orthod.* 42: 381—387.

Scott, J. H. (1958). The cranial base. *Am. J. phys. Anthrop.* 16: 319—348.

Solow, B. (1966). The pattern of craniofacial association. *Acta odont. scand.* 24 (Suppl. 46): 1—174.

Thompson, D'Arcy W. (1917). *On growth and form.* Cambridge: University Press.

Turpin, D. L. (1968). Growth and remodelling of the mandible in the *Macaca mulatta* monkey. *Am. J. Orthod.* 54: 251—271.

Vilmann, H. (1971). The growth of the cranial base in the Wistar albino rat studied by vital staining with alizarin reds. *Acta odont. scand.* 29(Suppl. 59): 1—44.

Woo, T. L. (1931). On the asymmetry of the human skull. *Biometrika* **22**: 324–352.

Zuckerman, S. (1926). Growth-changes in the skull of the baboon, *Papio porcarius. Proc. zool. Soc. Lond.* **1926**: 843–873.

Zuckerman, S. (1955). Age changes in the basicranial axis of the human skull. *Am. J. phys. Anthrop.* **13**: 521–539.

Primate Evolution, Biometrical
Morphology, Palaeontology

Chairman's Remarks

The intellectual appetite is whetted by the richness of the fare indicated on the menu of this session. Four main and attractive courses are set before us, each prepared by an anatomical *cordon-bleu*.

Our feast proves expectedly sustaining. Professor Ashton has manifested the fundamental importance of controlled statistical analysis in the elucidation of vexed morphological problems. Such mathematical approach has revolutionized investigation in a hitherto particularly contentious field, for too long bedevilled by subjectivity in the assessment of evidence. Henceforth such new evidence as the primate palaeontological record may reveal will automatically become subject to a necessary objectivity of assessment.

Professor Oxnard has likewise demonstrated the enlargement of understanding of biological problems and patterns likely to ensue from the judicious employment of biometrical techniques.

Professor Lewis has vividly justified the perennial importance of the traditional anatomical approach to problems of morphology and evolution and has shown convincingly how directly informative of functional activity is anatomical structure.

Dr Joysey has ingeniously adopted basic molecular evidence to the testing of accepted concepts of vertebrate phylogeny with some remarkable and unexpected results.

These several communications will duly provide, in desirable permanent form, information at once informative and thought-provoking.

A. J. E. CAVE

Symp. zool. Soc. Lond. (1981) No. 46, 67–126

The Australopithecinae: Their Biometrical Study

E. H. ASHTON

Department of Anatomy, University of Birmingham, UK

SYNOPSIS

From their initial discovery in 1924, the Australopithecinae have attracted attention because of apparent resemblances in certain features of the skull, teeth and postcranial skeleton to man. But such features overlie cranial proportions that are ape-like, and it is these that have been emphasized from the first by the dedicatee of the present symposium. Solly Zuckerman's early biometric appraisal of cranial capacity (amply confirmed in later and more extensive studies) showed that, in the fossil group, this was as in apes and not, as had been claimed, in excess of the ape range and approaching that of man.

Interest in the fossil group was renewed in the years after the Second World War, following the discovery and description of further specimens of the same group. Quantitative comparative studies, instituted and for many years led personally by Lord Zuckerman, have continued actively to the present day in the Department of Anatomy, University of Birmingham, while related studies are being pursued by his former students in various other parts of the world. These have now developed sufficiently to make apparent a coherent pattern of morphological similarity and contrast between the Australopithecinae, extant apes and man.

Superimposed upon the ape-like braincase and facial skeleton of the fossils are complexes of features, many of readily interpretable significance, in some respects ape-like, in others human. In combination, they are uniquely different from both extant groups, as distinct from lying in an intermediate position.

Thus, in their position on the skull base, the occipital condyles are as in apes, in angulation they are as in man. In overall proportions, as in certain more detailed aspects of their metrical morphology, the incisor and canine teeth of the australopithecine fossils are like those of man; the grinding teeth are as in apes. In certain features of the basicranial axis (and especially those relating to the face), the Australopithecinae are ape-like, in others they tend to be like man; in combination, they are unique. In basic configuration, the articular surface of the temporal bone is similar in man and apes but its proportions differ quantitatively. In this region the Australopithecinae, although varying from specimen to specimen, again show overall a unique pattern of similarity to and contrast with extant types.

Correspondingly unique patterns of similarity and difference are found in characters whose functional significance, although less obvious, can often be elucidated by comparative studies of related anatomical structures. Thus, the infraorbital foramen of the Australopithecinae, although single as in man, in lying well down the facial skeleton as in apes correlates with the prognathism of these types rather than with any fundamental difference in pattern of branching of the infraorbital nerve.

An extended series of studies of the pectoral girdle and arm — a biomechanically simple region, because linkage with the axial skeleton is almost entirely by muscular attachment — has provided a vehicle for enquiring first into methods of determining metrically definable features functionally related to the impressed force pattern (and thus to the habitual use of the region), and secondly into the use of multivariate statistics as a method of describing biological uniqueness and of distinguishing it from intermediacy.

The translation of these concepts to the study of the innominate bone — a region biomechanically more complex because of being subjected to forces due to bearing weight as well as to those resulting from muscular pull, but represented by an almost perfect fossil specimen — has shown that in groups of five features relating to muscle disposition and of four characters relating to the transmission of weight, there is again a unique pattern of similarity and contrast between the fossil Australopithecinae, extant apes and man.

Recently, multivariate combination of 25 biomechanically significant features of the primate innominate bone has produced a division of extant primates into groups correlating closely with the habitual use of the hindlimb in locomotion. It has also demonstrated an individuality of the Australopithecinae more striking than has emerged from any previous study. If the Australopithecinae were bipeds, as had been repeatedly claimed on the basis of qualitative appraisal of the morphology of the innominate bone, their bipedalism must thus have been quite different from that of *Homo sapiens*.

Such demonstrations of the uniqueness of the Australopithecinae are supported by similar results from related studies of the talus, the distal end of the humerus, part of the hand, and the proximal and distal ends of the femur.

As yet the overall functional significance of these findings is uncertain. Their bearing upon assessment of the phylogenetic position of the fossil Australopithecinae could well remain, in perpetuity, indeterminate.

INTRODUCTION

In this opening contribution to the second session of the present symposium, it is my pleasure to deal with an aspect of the study of human evolution that has intrigued the man whom we honour today from his earliest days as a research worker. Solly Zuckerman's interest in the quantitative study of hominoid fossils developed very shortly after the discovery, in 1924, of the first specimen of the Australopithecinae. It was an immature individual with only milk teeth and first permanent molars in place. But it comprised not only the teeth and jaws, but also much of the facial skeleton, part of the cranial vault, and an almost complete endocranial cast.

Notwithstanding the almost overwhelming resemblance of this fossil skull (*Australopithecus africanus*) to that of chimpanzees of comparable dental age, the first description of the remains (Dart, 1925) immediately pointed to a number of features — notably the dolichocephalic cranium, the shortened face, the lack of supraorbital ridges, the steep slope of the forehead, the small nasal aperture and short nasal bones, together with the small size of the canine teeth and the supposedly forward location of the foramen magnum — in which it was claimed that the fossil deviated from the living apes and approached extant man. In fact, Dart went so far as to submit that the specimen represented a group ". . . well advanced beyond modern anthropoids in just those characters, facial and cerebral, which are to be anticipated in an extinct link between man and his simian ancestor". Even if Dart's anatomical statements had been correct — and there was considerable doubt about this, partly because his comparative studies gave no indication that reasonable account had been taken of natural variability — it would now seem almost inconceivable that any worker could interpret such a complex of features — many of a minor character grafted on to a specimen of overwhelmingly ape-like proportions — as necessarily betokening a human ancestor. But matters were different in the mid-1920s. It was a mere 65 years since the beginning of controversy surrounding the view put forward by Darwin (1871), Huxley (1863), Mivart (1873), and others, that man was related, in an evolutionary sense, to the anthropoid apes.

It would be futile to attempt to review the sterile oratory of these and succeeding years, echoes of which can be heard even today. But, once generated, the popular misconception that a human ancestor would have been intermediate in morphological form between present day man and apes lingered, and found expression in the early descriptions of the australopithecine fossils.

Dart was by no means alone in this outlook. Only 12 years earlier an assemblage of remains comprising a human skull with an ape-like jaw had been reputedly discovered in deposits in the Thames valley and given the name of *Eoanthropus dawsoni*. It was soon to become popularly known as the Piltdown man and to remain an anthropological enigma until the mid-1950s when modern technology exposed the whole as having been fraudulently assembled (e.g. Weiner, Oakley & Le Gros Clark, 1953; Weiner, 1955). At that time (1912), the fossil record of the higher primates was sparse. Limited remains of early apes from the Lower Oligocene (*Parapithecus* and *Propliopithecus*) had been described, together with considerable numbers of specimens from the Middle and Upper Miocene extending into the Pliocene. Most of these latter had been assigned to the genus *Dryopithecus* and to a variety of its species almost corresponding with the number of specimens available.

Homo sapiens in a fossilized form was known by relatively abundant remains from the Upper Palaeolithic, while an appreciable number of specimens of Neanderthal man had also come to light, principally from Western Europe. *Pithecanthropus* (now *Homo erectus*) was known from Javanese specimens although the Chinese variant (initially styled *Sinanthropus*) was scarcely defined.

Controversy raged even more violently than at the present day as to when the line of evolution leading to man diverged from that leading to the anthropoid apes. At the one extreme, W. K. Gregory (e.g. 1920) upheld and meticulously documented a view that remains of the ape-like creature *Dryopithecus* represented, or lay close to, the common ancestor of man and the living apes — a view implying a separation of these groups not earlier than the Miocene. At the other extreme, Frederic Wood Jones (e.g. 1919) was putting forward — possibly with deliberate provocation — an admittedly extreme view that man had descended directly from some fossil tarsioid without the intervention of an ape-like phase — a view implying a separation of the human line of descent certainly as early as the Oligocene.

It was against this background of speculation and often heated controversy that the work of Professor Sir Grafton Elliot Smith flourished at University College, London. It was concerned with the probable pattern of evolutionary change that had occurred in the primate brain during the emergence of *Homo sapiens* (Smith, 1924). In the same College the Biometric School, led by Professor Karl Pearson, pursued most actively work started by Sir Francis Galton in the metrical definition of biological variation, with special reference to that of the human skeleton, together with its summary and comparison by means of statistical techniques.

It was into this general anthropological scene that Solly Zuckerman entered not very long after the discovery of the Taungs ape, as the original specimen of *Australopithecus africanus* had popularly become known. It was a time when the general climate was hostile to Dart's anatomical assessment and interpretation of this fossil, a group of senior anthropologists of the day (Keith, Smith, Woodward & Duckworth, 1925) joining in a single paper to express a view that the creature was essentially ape-like and although possibly differing from the living apes in certain minor respects, was a member of the ape radiation rather than the human lineage.

In the joint environment of University College, and the Zoological Society of London, Solly Zuckerman's interest in this fossil, first aroused in South Africa, found active expression in a quantitative enquiry into whether or not the brain volume of the specimen was, as had been hinted by its describer, in fact in excess of that in living apes. This enquiry into age changes in the skull of the chimpanzee, with special reference to growth of the brain, formed the subject of a major scientific paper, published in 1928 in the *Proceedings of the Zoological Society of London.* It was actually the third publication from his pen. The study was based on measurements taken from 112 chimpanzees of varying ages. The ages were judged in immature groups by the stages of dental eruption, and in adults by such factors as the extent of suture closure and of tooth wear. An overall conclusion was that after the eruption of the first permanent molars (these being the first teeth of the permanent dentition to erupt) the brain grows by only some 8%. A much greater percentage change, of course, occurs in the facial skeleton in phase with the eruption of the remainder of the permanent teeth. Projecting from this to the fossil group, it was estimated that had the Taungs ape lived to maturity, it is improbable that its brain volume would have exceeded about 550 cm³ — a figure that is well within the range of adult apes and much below the minimum for any type of man.

This study did not involve statistical techniques and thus, from a strictly purist viewpoint, could not be regarded as a complete biometric analysis. In fact probably the very first application of a full spectrum of statistical as well as mensural techniques to the analysis of remains of fossil subhuman primates was by Pearson & Bell (1919). Here, as part of an extensive metrical study of the long bones of the human skeleton, comparison was made with a femur attributed to the fossil ape genus *Dryopithecus*. But Solly Zuckerman's study can, I think, properly be regarded as the very first attempt to accumulate purposefully appreciable quantitative data on living apes as a basis for comparison with fossil types. The lack of statistical treatment was inevitable as only restricted samples of ape skulls were available at that time, and statistical techniques had not developed beyond the large sample methods of Pearson and his colleagues. Small sample techniques as fully developed by Sir Ronald Fisher and his School were still a decade away. It would thus have been virtually impossible for any worker to carry out a full biometrical study in the sense that we understand the term today.

In 1933, a little-quoted but often misunderstood paper: "*Sinanthropus* and other fossil men" (Zuckerman, 1933) appeared in the *Eugenics Review*, stimulated by recent finds in China of archaic fossil men (*Sinanthropus* in the current terminology) — of manifest similarity to those of *Pithecanthropus* from Java. The significance of quantitative methods in the study and classification of fossil men had already been underlined by the publication of biometrical studies by Morant on skulls from the Upper Palaeolithic (1926, 1930), and on various types of Neanderthal remains (1927, 1928). While the extent of metrical analysis that could be undertaken on the basis of the published dimensions of the two incomplete skulls of *Sinanthropus*

known at that time was limited, it did lead to two significant formulations. In the first place, the system of classification of the Homi-
nidae was shown to be inconsistent, there being no apparent grounds
for the generic separation of *Sinanthropus* from *Pithecanthropus*.
Secondly, although the braincase of Neanderthal man was much
larger than in this fossil group, several cranial features — for instance
the flattening of the frontal arc — were similar in the two groups
and contrasted with *Homo sapiens*. Such similarity could, it was
agreed, warrant the inclusion of all these archaic types in a single
taxonomic category, distinct from that appropriate to all groups
of *Homo sapiens* — living or fossil.

Also in this paper there was put forward a view, subsequently to
be much elaborated, that it could well be unjustified to place special
emphasis upon selected morphological features as a basis for classification. As a corollary, it was submitted that measures of similarity
and divergence should be based upon a combined assessment of all
morphological features.

Meanwhile, other remains of fossil creatures akin to *Australopithecus africanus* were coming to light. Robert Broom was one of those
who had, from the first, given active support to Dart's views about
the human-like affinities of *Australopithecus africanus*. In 1936 and
subsequently, he recovered from cave deposits in the Transvaal
remains (soon to be assigned to two new genera, *Plesianthropus* and
Paranthropus, but in recent years again regrouped along with many
related specimens in the single genus *Australopithecus*) that were
manifestly adult representatives of the same type of creature. From
the first Broom emphasized even more enthusiastically than had
Dart, what he regarded as human features in these fossils, and especially in their dentition. Equally boldly, these fossils too were
hailed as probable human ancestors. In this view he was, to no small
extent, supported by such workers as Weidenreich (e.g. 1937)
together with Gregory & Hellman (e.g. 1939).

The whole situation was summarized by Le Gros Clark (1940) in
an extensive review of the fossil evidence for man's ancestry. His
view, as then expressed, was cautious: the Australopithecinae (a
subfamily into which the three genera of South African fossils had
been placed by Gregory & Hellman (1939)), although deviating in
some morphological features from the living great apes and showing,
in certain respects, an approximation to living man, were members
of the ape radiation.

During the Second World War work of this type was much reduced, although in 1943 Franz Weidenreich produced an extensive monograph on the skull of archaic human remains from China, then continuing to be styled *Sinanthropus*.

It was in the immediate post-war era that activity developed in the metrical analysis of the australopithecine remains, more and more of which came to light, principally from sites in South and East Africa. Without in any way detracting from the significance of studies carried out by workers such as W. L. Straus (e.g. 1948), it can fairly be said that impetus in this field originated from the work of Solly Zuckerman in what was then his rapidly developing department in the University of Birmingham. His early fascination by the South African fossils was resuscitated by two events. The first was the publication by Broom & Schepers in 1946 of an extensive monograph, describing all the then known australopithecine material, and asserting with even greater enthusiasm — and again wholly on the basis of non-quantitative appraisal, often with only a poor appreciation of inherent variability — that many features of the group deviated from the living great apes and resembled man. The fossils were thus claimed to be ancesters of *Homo*. The second was that, at about this time, Le Gros Clark visited South Africa, where he was able to examine in detail the original fossils. He returned practically convinced of the validity of Broom's anatomical assessment (e.g. Le Gros Clark, 1947a & b).

Solly Zuckerman's approach to a quantitative appraisal of fossils remained as had been foreshadowed in his paper of 1933. The assessment of overall similarities and differences must rest upon a cumulative appraisal of individual features (taking into account, of course, their interrelationships). Equally, account must be taken of the variation of each feature in living and, whenever possible, in fossil groups. The definition of such features can be given precision by using metrical techniques which, coupled with statistical analysis, provide a method for summarizing inherent variability.

Such ideas formed the basis of work that has continued purposefully and progressively to the present day, not only in the Anatomy Department of Birmingham University, but also by other workers trained in that School and now holding senior appointments elsewhere: for instance W. J. Moore at the University of Leeds and C. E. Oxnard at the University of Southern California.

It is my intention to outline the principal pathways along which this research has developed. In avoiding any attempt at a strict chronological sequence, I shall try to indicate the stage that each facet of this research has now reached, and to present what appears to be the most likely interpretation of the findings.

BIOMETRICAL STUDY

Selection of Characters

It was as a final year undergraduate in the immediate post-war period that I was introduced to this work. For the next 15 years, first as a post-graduate scholar and then as a member of staff, I had the honour of being responsible to Professor Zuckerman, who was then personally involved in leading the work. My own responsibility was to collect and process the large amounts of mensural data that were involved. Initially I was deputed to comb relevant literature (e.g. Broom & Schepers, 1946; Le Gros Clark, 1947a, b; Broom, Robinson & Schepers, 1950; Le Gros Clark, 1949) and list morphological features upon which emphasis had been placed in published descriptions of the South African fossils. Most of these had been emphasized as deviating from the condition seen in the anthropoid apes and reputedly resembling or approaching that typical of extant man. They included the endocranial capacity, the position of the foramen magnum and occipital condyles (previously stressed in early descriptions (e.g. Dart, 1925) of *Australopithecus africanus*), the shape of the palate, the size of the incisor and canine teeth and the cusp morphology of the molar teeth, the fact that the infraorbital foramen appeared to be consistently single in the fossils, the slope of the frontal bone and the configuration of the occipital region. My brief was to define such features metrically, to establish their variance in the living apes and living man and to effect comparison with the australopithecine fossils. The first step involved defining and testing new techniques of measurement, following the general principles (although not the dimensions – general or specific) established by Professor Karl Pearson and the Biometric School.

A practical difficulty was that, in many cases, corresponding data had not been published for the fossils. Descriptions were mainly verbal although in some cases supplemented by certain overall dimensions. At that time few good casts of the fossils were available upon which our own series of dimensions could be reliably taken. But there was, in the published literature, a reasonably complete set of overall dimensions of virtually all of the deciduous and permanent teeth of the fossil specimens. These were the dimensions customarily used by odontologists to define general dental proportions and were, in the context of these descriptions, often used to support statements about the supposedly human nature of the fossil teeth. However, comparison with living types in the published studies had normally been based on corresponding measures of, at the most, one or two specimens from the living apes and living men in a procedure that gave little indication of variation in living (or fossil) types and which thus failed to show whether or not any supposed contrasts could be reasonably attributed to chance sampling errors. It was thus that in our early studies emphasis was given to overall dental dimensions, and this work took a decade to reach maturity. Even now, analysis of all the data collected during this period is not yet complete, although it is hoped that in the future there will be presented an overall spectrum of basic and comparative results.

This early odontometric study dealt with only limited groups of interrelated dimensions (numbering up to four or five per tooth), but posed considerable problems. It was, in fact, our first encounter with the concepts of multivariate statistics. Although these can now be applied by well-established statistical routines, the development of their proper use required several years' work in collaboration with professional statisticians of the highest calibre. Here one pays most grateful thanks to Dr Frank Yates and Mr (now Professor) Michael Healy for their ever generous and imaginative help, representing a continuation of a fruitful wartime friendship and collaboration between Dr Yates and Professor Zuckerman.

From an early stage other studies ran in parallel with such work on overall dental proportions. Some proved to be much more simple from a purely biometrical viewpoint. Others, although biometrically uncomplicated, proved to be difficult in morphological interpretation, and necessitated preliminary, and at times extensive, studies of the related "soft-part" anatomy.

These studies fall into a number of groups. While not forming a chronological sequence of execution, they demonstrate a coherent development of the principal biological problems involved.

Univariate Features of Readily Interpretable Significance

Endocranial capacity

In descriptions of the fossil remains stress was placed from the first upon apparent deviations of the endocranial capacities of the Australopithecinae from those of extant apes. This was because endocranial capacities are closely related to brain size (a feature in which living apes contrast markedly with all types of man, living or fossil), and because the expansion of the brain during human evolution is correlated with the development of man's unique mental abilities.

Quite early in Professor Zuckerman's programme, the not inconsiderable body of published data relating to endocranial capacity in the living great apes were collated and analysed as a basis for com-

parison with published values for the Australopithecinae. The brain
size of several adult specimens had been estimated, as also had that
of the juvenile specimen of *Australopithecus africanus*. The data for
the living great apes were limited to the chimpanzee and gorilla. They
were by no means as standardized as might have been desired, but it
was possible to get fair estimates of the mean value for endocranial
capacity together with its variance in both sexes of each of these
great apes. Also, values were calculated from the published data for
the fossil hominid *Pithecanthropus* (now *Homo erectus*). Although
some of the estimates of endocranial volume of the Australopithecinae
were imprecise, comparison indicated that, contrary to many of the
claims already advanced, the endocranial volume of these creatures
was quite within the range of that characteristic of the great apes
(and especially the gorilla). It was consistently below the lower 95%
fiducial limit for *Pithecanthropus* which is, in its turn, smaller in
brain size than is any form of living *Homo sapiens* (Ashton, 1950).

Some years later, this work was extended by Mr Spence and myself
(Ashton & Spence, 1958). In this study, using a carefully standardized
technique, we were able to measure the endocranial volumes together
with the dimensions of the foramen magnum, of much bigger series
of specimens of each of the extant great apes than had hitherto been
possible. Corresponding determinations were also made on similar
series from the major geographical subgroups of extant man. We were
also able to include reasonable series of immature specimens which
could be separated into groups of comparable developmental age on
the basis of the numbers and kinds of teeth that had erupted.

Our comparisons were between these newly collected data for
extant apes and men, and estimates for australopithecine brain size,
not only those used in the earlier study (Ashton, 1950), but also
those of several additional specimens that had been described during
the intervening years. The earlier conclusions stood: in endocranial
volume, all australopithecine specimens are significantly smaller than
man (living or fossil) and do not deviate from one or other of the
living great apes.

In this study we were also able to confirm the finding of the study
carried out 30 years earlier (Zuckerman, 1928), and limited of necessity to the chimpanzee, that by the time the first permanent molars
have erupted, the brain has attained more than 90% of its adult size.
The new study also extended this finding by establishing not only
that the overall growth pattern of the endocranial cavity is common
to the apes and man, but also that it is characteristic of dimensions
of the foramen magnum. The size of the latter did in fact correlate so
well with endocranial volume throughout the various phases of growth

that it proved possible to predict from it the likely adult endocranial volume of an immature specimen of *Australopithecus* (then styled *Paranthropus crassidens*), of which only the base of the skull had been preserved. This too proved to be completely ape-like and much below the values for corresponding age-groups in man.

At a later stage still (Ashton, 1972) it was possible to bring together even more data for the Australopithecinae that had been published in a carefully documented review by Tobias (1971). Such extended data permitted a statistical presentation of endocranial volume not only in the living apes and man (as derived from combined data from our earlier studies), but also in the australopithecine fossils. There was good separation of the apes, *Homo erectus* and *Homo sapiens*, but the Australopithecinae lay completely within the band spanned by the chimpanzee, gorilla and orang-utan.

The pattern of growth change in the endocranial cavity continued to fascinate us and a recent report (Ashton, Moore & Spence, 1976) gave the results of a study of age changes in endocranial volume that extended our earlier studies to include many genera of Old World monkeys. This showed that the growth pattern already established as common within the Hominoidea (Ashton & Spence, 1958) also extends throughout the Cercopithecoidea.

Thus, in endocranial volume, the Australopithecinae are ape-like and quite different from living or fossil man. It is the ape-like proportions of the braincase, coupled with a big and prognathous face, that result in the essentially ape-like appearance of the sagittal section of the australopithecine skull. Their functional interpretation is manifest, but in stressing the similarity of this feature of the fossils to the living apes one is conscious that only absolute size has been considered. Deviation from the extant apes might well emerge if it were possible to enquire into allometric relationships with overall size, or to study differential development of the several regions of the brain.

Position of foramen magnum and occipital condyles

Another cranial feature emphasized in descriptions of the Australopithecinae, definable metrically by a single ratio and of readily ascribable mechanical significance, is the position of the foramen magnum and occipital condyles relative to the anterior and posterior extremities of the skull.

In man, where the occipital region of the skull is prominently developed and where the facial skeleton is reduced and sited under the anterior part of the braincase, the foramen magnum and occipital condyles lie relatively far forward, in fact close behind the centre of gravity of the head. Thus, only a relatively small force applied by the nuchal muscles (attached to the planum nuchale of the occipital bone) is sufficient to counterpoise. But in apes, the occipital region of the skull is not nearly so prominent. The facial skeleton is not only big, relatively and absolutely, but also set more in front of the braincase. Thus the foramen magnum and occipital condyles lie relatively far back behind the centre of gravity. A correspondingly bigger force applied by the nuchal muscles is therefore necessary to counterbalance such face-heaviness, and this is reflected in their prominent development.

Such major contrast in position of the foramen magnum may well be functionally associated with man's assumption of an upright posture and gait. But there is no reason to believe that minor differences in position of the foramen magnum such as exist between many species of quadrupedal monkey, or between monkeys and apes, can be correlated with differences in posture and locomotion or in the way in which the head is poised on the vertebral column. In young apes, for instance, because the braincase attains a high proportion of its adult size at a relatively early stage of postnatal development, while the skeleton of the face remains relatively small, the foramen magnum lies further forward than in adults, although not nearly so far forward as in man. Yet there is no evidence that young apes carry their heads in any way different from that characteristic of adults (Ashton & Zuckerman, 1952a).

Nevertheless, in even the earliest descriptions of the Australopithecinae, emphasis was placed upon the position of the foramen magnum. It was presumed to lie relatively further forward than in the present-day great apes, in this respect approximating in some degree to the condition typical of man. This factor had received mention by Dart (e.g. 1925) and by Broom & Schepers (1946), but came into prominence in 1949 when Le Gros Clark, in the fifth William Smith Lecture to the Geological Society of London, defined the position of the occipital condyles by an index relating the segments of the skull behind and in front of the condyles as projected on to the Frankfurt horizontal. As comparative data he provided information based upon measurement, but without statistical analysis, of a considerable series of adult chimpanzees and gorillas, although not of living men. It was found that in the one adult australopithecine skull available at that time in which the index could be computed accurately, the occipital condyles, and thus the foramen magnum, lay further forward than on average in present-day great apes. In this respect they were claimed to make some approximation to the human condition. This observation was used in support of a thesis that these creatures had already attained some form of bipedal posture and gait.

A more extensive examination of this feature (including a simple univariate statistical analysis) not only in living apes but also in various subgroups of living man together with a variety of monkeys, showed that, although lying somewhat further forward than the average for apes and monkeys, the occipital condyles in the fossil Australopithecinae were much further back than in any form of *Homo*. They were, in fact, far more ape-like than human in position, (Ashton & Zuckerman, 1951a). This result was confirmed in a study carried out much later by Adams & Moore (1975), and also by Moore, Adams & Lavelle (1973), using the more extensive fossil material that had become available in the meantime.

In a second study (Ashton & Zuckerman, 1952a), an examination was made of the pattern of growth change in the position on the skull base of the occipital condyles. It showed that whereas the backward movement relative to the anterior and posterior extremities of the skull is less in man than in the living apes during postnatal development, adult australopithecine fossils lie consistently within the corresponding ape range and outside the human range. But values for fossil man (e.g. Neanderthal specimens) lie, equally consistently, within the range for extant man.

Such evidence as could be gleaned from a total of two adult and two immature specimens of the Australopithecinae showed that the growth change in the position of the occipital condyles was, in scale and pattern, much more like that characteristic of the living apes than that which obtains in living man (Ashton & Zuckerman, 1956a). Thus, irrespective of any question of posture, gait or poise of the head, the Australopithecinae were, so far as concerns position of the occipital condyles on the skull base, like the living great apes and quite different from extant man.

Several years later it fell to my colleague, Professor W. J. Moore (another of Lord Zuckerman's former pupils), to lead a comparative study of the angulation of the occipital condyles to the Frankfurt horizontal (Adams & Moore, 1975; Moore, Adams & Lavelle, 1973). In man, where the condyles point almost directly downwards, their angulation is relatively small, whereas in apes they tend to point backwards, and the angulation is correspondingly increased. In this respect the Australopithecinae proved to be as in living man, and differed from extant apes.

It is not easy to make a critical appraisal of the biomechanical significance of the contrast in these two features of the occipital condyles between the Australopithecinae, extant man and apes in relation to posture and gait. But the anatomical facts are indisputable: in respect to their position on the skull base, the occipital condyles and foramen magnum of these fossils are ape-like; in respect to angulation to the Frankfurt horizontal they are like man. In neither respect are the fossils intermediate between extant groups. Thus, in combination, they are different from both, i.e. they are unique.

Multivariate Features of Readily Interpretable Significance

Overall proportions of teeth

The human dentition contrasts with that of subhuman Primates in that the incisor and canine teeth are reduced in size. The human incisor and canine teeth are subequal, whereas in apes and especially in males, they are bigger, the upper canine articulating with a sectorial face on the lower first premolar to form a shear. Such contrasts correlate with man's omnivorous diet on the one hand and with the frugivorous diet of the apes, where the well-developed canine teeth tear through the outer rinds of fruits, on the other.

In the earliest descriptions of the Australopithecinae (e.g. Dart, 1925) emphasis was given to what appeared to be a reduction in size of the incisor and canine teeth of the milk dentition, together with many supposedly man-like as distinct from ape-like features of the cusp pattern and proportions of the grinding teeth. Such emphasis was given special prominence in later accounts of the permanent teeth (e.g. Broom & Schepers, 1946; Le Gros Clark, 1947a & b, 1949). As overall dimensions were published in support of statements about supposed differences in proportion of the fossil teeth from those of living apes and presumed resemblance to modern man, an obvious initial study was to examine the variation of these features in samples as adequate as could be obtained of living apes and living man and to effect statistical comparisons with the fossils.

At most three or four dimensions of each tooth had been published, or could, in a few cases, be obtained from casts of verifiable accuracy. We could neither expect nor hope that these measurements would give more than an impression of overall proportions as distinct from describing detailed anatomical structure. Even so, the problem of interrelationship of measurements, morphological and statistical, was present even if comparison were to be made only tooth by tooth rather than between groups of teeth. It obtruded itself upon us from the first. But at that time, even though the basic theory of handling multiple correlated measurements had been worked out (Fisher, 1936a; Hotelling, 1936; Mahalanobis, 1936), the application of methods appropriate to the problem with which we were faced was not yet feasible because of the nature and extent of the computations involved. Initially, therefore, comparison had to be effected by univariate techniques.

The first stage was to define measurements describing the overall proportions of each of the permanent and deciduous teeth, these being, so far as possible, comparable with those in the published literature. They were defined so as to ensure repeatability of technique, the series covering the complete tooth row. The next stage was to take these dimensions on such series of chimpanzees, gorillas and orang-utans as could then be obtained, and also on groups from certain of the major geographical variants of extant man. To the best of our knowledge these were by far the biggest samples of ape and human teeth that had been measured at that time, although the loss of teeth post mortem (and especially incisors and canines) resulted in the numbers of observations that could be made for certain teeth being limited. Nevertheless, the precision of the statistical estimates was generally good. Corresponding dimensions were taken from both the right and left sides and as these proved to be highly correlated, they were treated in pairs rather than singly. The mean and variance of each dimension of each fossil tooth was then compared by elementary statistical methods individually with the corresponding dimension in each group of living apes and men (Ashton & Zuckerman, 1950a & b).

The result was that a much greater proportion of dimensions of the fossil Australopithecinae — and especially of the cheek teeth — agreed with the corresponding dimensions in living apes than in living man. In this respect the results failed to substantiate many statements that had been made, on the basis of comparison with at the most one or two specimens of apes or men, about supposed dimensional differences between the australopithecine fossils and the living apes. Equally they failed to substantiate corresponding assertions about resemblances to man. A similar result emerged when the differences were considered tooth by tooth rather than dimension by dimension.

These results and the underlying methodology were understandably challenged by those who adhered to a view that the dental morphology of the Australopithecinae was essentially human, such critics almost invariably failing to appreciate that our conclusions referred only to overall dental proportions such as were defined by the measurements that we had taken. It would be inappropriate and indeed unprofitable to review the seemingly endless controversies and exchanges that took place in the columns of *Nature*, *Man* and elsewhere during these years. In retrospect, one is conscious of how difficult it was to convince the anatomical and anthropological world of the real purpose of these studies. At times it seems that even now one continues to be misunderstood. Nevertheless, the relatively great numbers of morphometric studies that appear in current literature indicate that acceptance of such an approach is now appreciably more widespread.

The wish to analyse such data by multivariate techniques led our statistical advisors (Dr Frank Yates and Mr (now Professor) Michael Healy) to examine our data. Their interest was possibly never more welcome than on this occasion for, when examining our statistical estimates of means and standard errors, they discovered that when computing the basic statistics, I had made a systematic error that resulted in all the standard deviations being too big by a factor approximately equal to the square root of 2. The means were unaffected and therefore the overall picture of similarity and difference between the fossils, extant apes and man remained valid, although, if a stated level of significance (say 5%) continued to be accepted, some comparisons that previously had failed to give a significant difference from any group, now revealed one. Nevertheless, the overall pattern of similarity and difference remained (Ashton & Zuckerman, 1952b).

The correction of this error in computation took some little time, and in a sense we were forestalled in the application of multivariate techniques to the analysis of dental dimensions of living and fossil hominoids.

Bronowski & Long were among those who had manifestly failed to appreciate the objectives of the biometrical study of the teeth of the Australopithecinae, apes and man that Professor Zuckerman had instituted. In 1951 they undertook a more detailed metrical examination of the lower milk canine of the australopithecine fossils with a view, not so much to establish overall dimensional resemblances and differences with extant man and apes, as firstly to define the proportions of the tooth more accurately by means of extra measurements and secondly to enquire whether or not, in these more detailed measures of shape and form, the fossil australopithecines were more like man or the living apes. In their analysis of this more detailed data, Bronowski & Long (1951) used a simple type of discriminant function devised by Fisher (1936a). Such a function enables a specimen, for which basic data have been compounded and which is known to belong to one or other of the groups included in the initial analysis, to be assigned with a definable degree of certainty to its appropriate category.

Bronowski & Long found that in the features they had examined the fossil tooth was more like that of man than like that of the living apes — as had been earlier claimed on the basis of visual inspection (e.g. Broom & Schepers, 1946; Le Gros Clark, 1947a & b). But as had already been pointed out, it was not on this topic that we had committed ourselves, and when an analysis, corresponding to that used by Bronowski & Long, was made of the dimensions that we had defined, it emerged that the apparent difference in the result was due not so much to the difference in method of analysis as to the difference in the series of dimensions initially chosen (Yates & Healy, 1951).

Yates & Healy also drew our attention to the fact that the type of discriminant function initially used by Bronowski & Long was inappropriate to the study of the Australopithecinae, as these could well belong to a group of hominoids distinct from living men or apes. In later publications Bronowski & Long (1952, 1953) attempted to overcome this defect by the application of alternative multivariate techniques. But a fully professional appraisal and application of discriminant functions as a tool in descriptive morphology was still some years away.

We had always been conscious of the restricted objectives of our earlier study and were as anxious as were our critics to make more informative quantitative comparisons of the teeth of the Australopithecinae, living apes and living men. Accordingly in 1953 work was started on accumulating appreciably greater series of dimensions of certain teeth and especially the incisors, canines and lower first premolars with a view to giving a more precise definition of their overall shape and proportions than had been obtained in our first analysis. Meantime, Dr Yates and Mr Healy, with whom we became ever more closely associated, were actively enquiring into which variety of multivariate analysis would be most appropriate to our studies.

From a purely theoretical viewpoint, it might be argued that the statistic of generalized distance (Mahalanobis, 1936) is the best overall measure of divergence between groups. But being a concept applying to multidimensional space, the results are not always easy to visualize. The closely related, and indeed derivable, technique of canonical analysis (Hotelling, 1936) appeared to offer a potentially good descriptive method in that it permitted much of the information relating to an array of groups positioned in multidimensional space to be represented in a limited number of orthogonal axes. From these, much of the essential information could be presented relatively easily in the form of a small number of simple two-dimensional graphs. Such methods make no assumptions about an unknown fossil group necessarily being a member of any of the extant groups forming the basic framework of the analysis.

Using the electronic computing facilities that had been developed in the Department of Statistics of Rothamsted Experimental Station

— limited by present-day standards but revolutionary compared with the desk calculating machines that they were superseding — it was possible to carry out a number of canonical analyses on certain of the teeth for which the basic dimensions had been accumulated. The results of this study were published shortly afterwards (Ashton, Healy & Lipton, 1957) and showed that in the proportions of their incisor, canine and lower first premolar teeth, the Australopithecinae did in fact show a considerable resemblance to man. But in the proportions of the remaining premolar teeth, as in those of the molar teeth, they were overwhelmingly like the great apes and contrasted with man.

Such results even now remain incomplete. However the study was taken sufficiently far 23 years ago to give a clear indication that, in being in some regions like man, and in others like the extant apes, the dentition of the Australopithecinae is, in its total proportions, uniquely different from both.

Basicranial axis

The basicranial axis extends from the anterior limit of the foramen magnum and broadly follows the skull base to the prosphenion in the anterior cranial fossa (or, more practically, the pituitary point on the anterior edge of the sella turcica). An anterior limb extends to the nasion, while a posterior arm traverses the foramen magnum.

The anterior extension of the axis lies broadly along the roof of the nasal cavity forming the upper boundary of the face, while the angle that it makes with the axis itself (the spheno-ethmoidal angle) gives a measure of the set of the face relative to the braincase. The posterior limb, in representing the sagittal diameter of the foramen magnum, gives an overall reflection of its size and thus of the lower end of the medulla oblongata, while the angle that it makes with the axis itself (foraminobasal angle) reflects the expansion of the posterior part of the brain.

In man, where the face is small and set under the front of the braincase, not only is the anterior limb of the axis relatively short, but the spheno-ethmoidal angle is reduced. Also, in correlation with the expansion of the brain — and especially of its posterior part — the posterior limb in the human skull is relatively big, the foraminobasal angle also being increased.

The basicranial axis has special significance in the present symposium in that growth changes in the axis, its two extensions and the two associated angles, were prominent in the first communication that Solly Zuckerman presented to the Zoological Society of London just over 53 years ago. This early study (1926) by Lord Zuckerman dealt with growth changes in the skull of the South African baboon (*Papio porcarius*). Almost 30 years later it was followed up by two enquiries into the corresponding pattern of growth changes in the human basicranial axis (Grossman & Zuckerman, 1955; Zuckerman, 1955). A year later I was invited to continue these analyses with a study of the corresponding growth changes in the great apes and in certain Old World monkeys (Ashton, 1957).

These studies of comparative growth were self-contained, but they also provided good basic data for each of the five features (three lengths and two angles) in adult living man, the great apes, and certain Old World monkeys for comparison with fossil skulls. At that time no good sagittal sections had been published of those australo-pithecine skulls that were sufficiently complete to permit measures of the axis to be taken. Such information did not in fact become available until relatively recently, making possible both univariate and multivariate comparisons of these features in the australopithecine fossils, apes, man and certain Old World monkeys (Ashton, Flinn & Moore, 1975). After appropriate adjustment of the data to compensate for contrasts in overall skull size in several extant genera, it emerged that man differed both in individual features of the basicranial axis and in their multivariate compound from the living apes which in turn deviated from the living monkeys. The australopithecine fossils resembled man in some features, linear and angular; in others they differed from man and resembled monkeys or apes. These groups also differed from each other. The two available australopithecine fossils, although lying a little closer to man than to other groups as judged by the generalized distance statistic, differed from *Homo sapiens* to an extent equal to that which separated man from apes and monkeys. The two specimens (one from the so-called robust type of australopithecine and the other from the so-called gracile group) appeared to differ from each other by an approximately similar amount. Each was thus unique.

Once again, multivariate analysis had proved to be an appropriate technique for summarizing overall contrasts, and especially appropriate in summarizing seemingly conflicting results of univariate study. It also illustrated that features in a fossil species, some of which are like man and others like monkeys and apes, in combination produce a complex, not intermediate between man and apes, but distinct from both — i.e. one that is unique.

Articular surface of the temporal bone
Detailed descriptions of the australopithecine fossils repeatedly drew attention to the contours of the part of the inferior surface of the

squamous temporal bone that articulates with the mandibular condyle to form the temporomandibular joint. Emphasis was given to
the presence of an articular eminence disposed anterior to the articular fossa, it being claimed that in this feature the Australopithecinae
were similar to man, and contrasted with the apes. It was inferred
that the existence of an articular eminence implied a rotatory pattern
of jaw movement, as is characteristic of man. It was correspondingly
claimed that in the apes an articular eminence is not present — or at
any rate is not prominently developed — and that, correlating with
the big interlocking canine teeth, the pattern of jaw movement in
apes and monkeys is primarily in a vertical direction.

A preliminary examination of the pattern of wear on the teeth of
men and apes (unpublished observations reported in Zuckerman,
1954) indicated a basic similarity between living man and extant
great apes, while even desultory observation of captive animals confirmed that in monkeys and apes rotatory patterns of jaw movement
do occur. Thus if any differences exist in this respect between apes
and man, they are of degree rather than kind.

It appeared likely that although the presence of an articular eminence was almost as characteristic of the living apes as of living man,
certain differences in proportions of the joint may well obtain between extant apes and man. A study was accordingly undertaken as
long ago as 1953 (Ashton & Zuckerman, 1954). This was based
largely upon data collected earlier by my colleague, W. J. Moore,
when a final year undergraduate student. This study gave a metrical
description of such features of the articular surface of the temporal
bone as the projection of the postglenoid tubercle, the depth of the
articular fossa, the projection of the articular eminence and the extent of antero-posterior "compression" of the joint area.

At that time it was possible to analyse these metrical data only by
univariate statistical techniques. Nevertheless, a number of significant
findings emerged. Firstly, it was confirmed that an articular eminence
is a normal feature of the joint of living apes — both immature and
adult — although in apes, as in man, certain age changes occur in its
proportions. Secondly, the human and ape temporomandibular
joints, although basically similar in structure, differ quantitatively in
that the human joint is more "compressed" in the anteroposterior
direction.

Any biomechanical significance that may attach to such differences, both between different age groups within genera and between
different extant members of the Hominoidea, was not apparent, Nor
has it subsequently become so in relation to pattern of movement of
the temporomandibular joint. In fact, the basic observation that jaw

movements and pattern of tooth wear appear to be essentially similar in living apes and man is supported by the observation that in living men, apes and monkeys the contours of the articular surface of the temporal bone are, to a large extent, smoothed out by the articular disc.

At that time (1954) it was still not feasible to make extensive comparison with the australopithecine fossils, partly because few casts of proven accuracy were available upon which dimensions comparable with our own could be taken. Such limited comparisons as were attempted gave a confused pattern of resemblance to and difference from each of the extant types. But some 20 years later (Ashton, Flinn & Moore, 1976) it was possible to undertake multivariate analyses of the original basic data, and to compare with dimensions obtained from a total of seven australopithecine specimens. The opportunity was also taken to compare with a total of five specimens of Neanderthal man (*Homo sapiens neanderthalensis*) and with a more limited representation of *Homo erectus*.

The results were once again significant. *Australopithecus africanus*, although closest to both *Gorilla* and to Neanderthal man, was not coplanar with these or other groups. In fact, it lay uniquely in multivariate space. *Australopithecus robustus* was about equidistant from Neanderthal man, gorilla and extant man, but its relationships with these types in multivariate space were such that the group also lay uniquely. No australopithecine fossil lay in a position intermediate between extant apes and man. *Homo erectus* was also isolated, although *Homo neanderthalensis*, in lying close to *Gorilla*, was less noticeably exceptional than were any of the other fossil groups.

In that variation in the contours of the articular surface of the temporal bone do not appear to be directly related mechanically to any observable difference in patterns of jaw movement or of tooth wear, it is just conceivable that they could be regarded as having some taxonomic significance. But irrespective of this or any other possible interpretation, one fact emerged strongly from this analysis: in yet another skeletal region, the australopithecines, in resembling living apes in some respects and living men in others, differ *in toto* from both, not being intermediate in form but contrasting uniquely.

Characters of Functional Significance Definable by "Soft-Part" Studies

Many similar contrasts in combinations of osteological characters were to emerge in later studies. In some cases possible functional significance was even less obvious than in the temporomandibular joint. A possible basis for interpretation was by comparative study of the associated soft parts. A good early example was provided by the infraorbital foramen.

Infraorbital foramen

Although the observation that the infraorbital foramen is single in the australopithecine fossils was yet another that had been prominent from an early stage of their description, it was not until a considerable number of specimens had been found and descriptions published that it could be asserted with some assurance that a single infraorbital foramen on both left and right sides is a regular characteristic of these fossils. The foramen transmits the terminal branches of the maxillary division of the trigeminal nerve as it emerges on to the face, as the infraorbital nerve, to supply sensory fibres to the maxillary region. It is single on both left and right sides in a high proportion of human skulls. But cursory observation indicated that, at any rate unilaterally, it may also be single in, for instance, the gorilla. Accordingly an enquiry was undertaken in which the numbers of external apertures of the infraorbital foramen were documented in long series for each of the great apes, the gibbon and in certain geographical varieties of extant man. The result of this simple analysis was clearcut. Although the foramen is single on both sides in the majority of human skulls, it is not infrequently single, even bilaterally, in the gibbon. In the gorilla it is occasionally single bilaterally, and more frequently single unilaterally. A single foramen on both left and right sides occurs much more frequently in man than in any monkey or ape (Ashton & Zuckerman, 1958).

The biological significance (if any) of the reduplication of the infraorbital foramen was not immediately apparent, and it was in the context of enquiring into the significance of multiple foramina that Charles Oxnard first entered the field in 1955. He was then reading for the degree of B.Sc. in Anatomical Science. In a most excellent piece of work, with whose later stages I had the pleasure of being associated (Ashton & Oxnard, 1958a, b), the maxillary nerve was dissected out and documented in a wide variety of mammals, special reference being made to its form and distribution in different types of primate. In this study we were greatly helped by the maturely critical advice of Professor A. J. E. Cave (then of St Bartholomew's Hospital Medical College, now of the Zoological Society of London). It marked the beginning of close association between Professor Cave and Professor Zuckerman's junior colleagues that has continued to the present day. We all record with grateful humility the beneficial imprint upon each of our studies that has resulted from scrutiny by this last of the great comparative anatomists.

Our study showed consistency in the basic pattern of branching of the maxillary division of the trigeminal nerve throughout the mammals. A similar consistency obtained in the pattern of division of its terminal branch: the infraorbital nerve. This splits at a variable point into nasal and labial divisions which traverse the infraorbital canal and emerge on to the face through the infraorbital foramen (foramina). The nasal division supplies an area of the side of the nose, together with the nasal vestibule or rhinarium (if present). It also supplies some upwardly directed palpebral twigs to the skin of the lower eyelid.

The labial division supplies the skin of the cheek and upper lip together with the mucous membrane on its inner aspect. Some twigs of the labial division become fasciculated with branches of the seventh cranial nerve supplying motor fibres to the facial muscles, and probably provide a proprioceptive pathway from these. The relative development of the two principal branches of the infraorbital nerve can be readily correlated with the development of the various structures that they supply. In the primates, the nasal division becomes progressively reduced and the labial division more prominent, in accordance with the elimination of the rhinarium (present in some prosimians but not in monkeys, apes or man) and with the progressive elaboration of the muscles of facial expression that characterizes the Order, reaching its greatest degree of complexity in man.

Despite the consistency of these findings, it appeared that there was no correlation between the numbers and arrangement of infraorbital foramina and either the basic pattern of branching of the infraorbital nerve or variations in the relative prominence and distribution of its two principal subdivisions. As far as I know the biological significance of the multiplication of the infraorbital foramen in subhuman primates remains undefined, and we were thus left with a simple empirical observation that with respect to frequency of foramina, the australopithecine fossils are more like man than like the living apes.

As an extension of this study, an assessment was made of the position of the foramen on the face relative to the inferior orbital margin and to the alveolar margin — the latter distance being measured both parallel to the sagittal plane and perpendicular to the alveolar margin. The position of the foramen, if single, or the centre of the cluster of foramina, if multiple, was expressed by means of a simple ratio. It was found that the fossil Australopithecinae were far more like the apes than like man in this respect — a finding that appeared to correlate with the prominently developed and prognathous facial skeleton of these fossil forms (Ashton & Zuckerman, 1958).

Once more, there had emerged findings relating to a single morphological complex in which the fossils were, in one respect, like man,
and in another, like the living apes. In combination they were again
unique.

Shoulder and arm

In the late 1950s, the Birmingham team was greatly strengthened by
the addition on a full-time permanent basis of Dr (now Professor)
Charles Oxnard who, having completed his training in clinical medicine, had opted for a career in Anatomy. His continued association
with this work to the present day, coupled with a similarly close
co-operative relationship with Dr (now Professor) W. J. Moore, has
given to the whole field an imaginative impetus, now manifest world-
wide, that would have otherwise been lacking following Lord Zuckerman's gradual withdrawal during the 1960s from active day-to-day
leadership. At about this time another member of the Department
— Mr Tom Spence — also became much involved. Through the years,
he likewise has contributed much to the development of the work by
his assiduous collection of osteometric data.

Charles Oxnard was initially presented, as a major long-term
research project, with a study of the biomechanics of the innominate
bone in primates. His objective was to enquire whether or not major
inferences about locomotion and gait in the Australopithecinae could
be made from an almost complete innominate bone of *Australopithecus* (formerly *Plesianthropus*). It was immediately realised that the
full metrical definition of the form and proportions of a shape as
complex as the innominate bone would be a virtual impossibility.
Even if this could have ultimately been achieved by a process of geometrical triangulation, it would have been impossible with the computing facilities available at that time (and even now barely possible)
to combine all the information in a single multivariate analysis.

It appeared that an initial step in solving the problem was to define
features of the innominate bone directly related to, and statistically
correlated with, variations in the force pattern that is habitually
impressed upon the pelvic girdle, and thus with variation in locomotor use of the hindlimb.

An early appreciation of the problem made us realize that, in the
case of the pelvic girdle, the impressed force pattern is complex. It
results from: (1) the transmission of weight via the vertebral column
and sacro-iliac joints, downwards through the hip joints to the legs;
and (2) the pull of the principal attached blocks of muscles. Such
complexity of the force pattern imposed upon the pelvic region
made it appropriate to enquire, in other biomechanically less involved

skeletal regions, into the extent to which defined aspects of morphological form correlate functionally with impressed force patterns.
The pectoral girdle appeared suitable because there the only bony
attachment to the axial skeleton is through the sternoclavicular joint.
Virtually the entire complex of impressed forces results from the pull
of attached muscle blocks, this being correlated with locomotor use
of the forelimb.

Thus began a further long series of enquiries which continues with
full vigour to the present day, based in the Anatomy Department of
Birmingham University. It is a programme in which Lord Zuckerman
has maintained a consistent interest, remaining associated, even after
his retirement, with certain of the more significant publications
resulting from this work (e.g. Zuckerman, Ashton, Flinn, Oxnard &
Spence, 1973). This programme has received from him not only
inception, but constant stimulation and personal interest throughout
the years.

As a first step of what was to be manifestly a biomechanical analysis of the pectoral region and forelimb, Charles Oxnard and I set out
to assess, and so far as possible to classify, the habitual pattern of use
of the forelimb and shoulder girdle in each primate genus with a view
to being able to infer the force pattern normally impressed. In an
initial publication (Ashton & Oxnard, 1964a) we summarized much
published literature on primate locomotion, mostly scattered
throughout monographic studies of primate taxonomy and in
naturalistic accounts of primate behaviour. At the same time we
attempted to make a corresponding analysis of hindlimb function.

A major result of this work was to show that there is often little
correlation between the pattern of use of fore- and hindlimbs. Thus
unless there is accepted a locomotor classification with a large number of subdivisions, and one whose utility in comparative biomechanical problems is thus reduced, it is virtually impossible to derive
a scheme that is applicable jointly to both fore- and hindlimbs, and
thus to the animal as a whole. For instance there are several genera of
monkeys of both the New and Old World (e.g. *Macaca, Cercocebus,
Erythrocebus*) that are habitually quadrupedal, moving on the larger
branches of the middle canopy of the tropical rain forest or on the
ground, and in which both fore- and hindlimbs, and thus both pectoral
and pelvic girdles, are subject to forces of compression. There are
others (e.g. *Presbytis, Colobus*) in which the animal not infrequently
suspends itself from its forelimb, which is thus subjected to forces of
tension, while the hindlimb is the propulsive agent in extensive
leaping movements, its habitual compressive forces thus becoming
impulsive. In other primate types, in which the forelimb may corre-

spondingly be used to an appreciable extent for suspension (e.g. *Lagothrix, Perodicticus*), the hindlimb may seldom or never take part in leaping movements and may be used to suspend the animal from branches of trees and thus be subjected to tensile forces.

As our principal interest at that time centred on studies of the pectoral girdle and forelimb, and as it seemed that there was more information available in the literature about the use of this region in locomotion than applied to the pelvic girdle and hindlimb, it was possibly understandable that in this early study attention was mainly directed to an analysis and classification of forelimb use. It was in this context that we grouped genera in the subhuman Anthropoidea into quadrupeds, semibrachiators and brachiators, each group comprising types in which the locomotor use of the forelimb was, despite quantitative variations, broadly similar. The series formed a progression from, at the one extreme, Primates in which the force pattern imposed upon the forelimb was essentially compressive to, at the other, ones in which it was predominantly tensile. They were parallelled by the groups quadrupeds and hangers in the Prosimii. Man was set alone in a separate category, his forelimb not participating in locomotor activity.

Despite such emphasis upon the locomotor use of the forelimb, this early review contained sections dealing with the hindlimb and with inferences therefrom relating to the habitual pattern of impressed forces. But partly because of our preoccupation at that time with the forelimb and pectoral girdle, and partly because significantly less information appeared to be available about hindlimb use, the locomotor classification of this region that we proposed did not, subsequently, prove to be nearly so significant in relation to biometrical analysis as did the forelimb classification. It was based upon an attempt to grade the various primate genera according to the extent to which the hindlimb takes part in leaping, when it is subjected to compressive forces of an impulsive nature.

This classification was effective in singling out certain groups: for instance those in which leaping is practised to an extreme extent (e.g. *Galago* in the Prosimii and *Presbytis* in the Anthropoidea). But it failed to give prominence to others. These included the prosimian lorisines, that habitually hang from all four extremities; types of monkey (e.g. *Lagothrix*) that may, again not infrequently, suspend themselves from the hindlimbs and prehensile tail whilst the forelimbs are used for foraging; the apes (e.g. *Pongo*), where the hindlimb participates in diverse acrobatic activities.

The suggested forelimb classification did appear to give a reasonable working basis for quantitative studies of that region, and it was

very much to our regret that groups of workers (e.g. Napier, 1963;
Napier & Napier, 1967; Ripley, 1967) apparently misunderstood, to
some extent, our intention, Napier & Napier attempting to use, and
Ripley to criticize such terms as "semibrachiation" and "semibrachi-
ator" as an overall pattern of locomotion. Of course such extension
from the use of the forelimb cannot be valid because of the low
correlation between fore- and hindlimb use. Even after a lapse of
15 years, it would seem necessary to reiterate that the purpose of
these classifications — as also with their later refinements (e.g.
Oxnard, 1974) — was to provide a descriptive pattern of the habitual
use of a circumscribed skeletal region (in this case the forelimb and
pectoral girdle) that would enable living primates to be assigned to a
relatively small number of groups, the members of each sharing a
similar pattern of forces habitually impressed upon that region.

The next stage of the investigation of the pectoral girdle comprised
a detailed analysis of the form and disposition of the muscles of the
shoulder region as revealed by dissection of 52 primates representing
22 genera from the Anthropoidea and six genera from the Prosimii
(Ashton & Oxnard, 1963). In this study note was made not only of
such features as the detailed bony attachments of the muscles, but
also of their relative weights and the orientation of their fibres. It
emerged that a number of significant contrasts in form and disposition,
both of individual muscles and of functional muscle groups, could be
readily correlated with defined differences in the use of the forelimb
in locomotion and with the resultant pattern of impressed forces.
Thus, for instance, in brachiators which habitually suspend and
propel themselves by the forelimb and where the principal movements are thus in the upper quadrant, the muscles responsible for
raising the arm (e.g. m. deltoideus providing the principal power in
abduction, together with m. trapezius and the radiating fibres of m.
serratus magnus, that provide the components of a force couple to
rotate the scapula during the later phases of arm raising) are well
developed. Again, in quadrupeds where the principal movements are
in the craniocaudal direction and in the lower quadrant, mm. pec-
torales are correspondingly directed in a cranial direction. In contrast, in brachiators the pectoral muscles are tranversely directed,
thus contributing to adduction of the arm. M. latissimus dorsi, only
weakly developed in quadrupeds, is more powerful and directed
more cranially in brachiators, thus enabling the forelimb to be re-
tracted efficiently when in the raised position typical of this group.
It thereby contributes to the power stroke in swinging and climbing.
Together with the reorientation of the pectoral muscles, such orientation of m. latissimus dorsi provides for a greater range of diversity

of arm movement in brachiators than characterizes quadrupedal types where movement is principally in the craniocaudal direction. Other muscular features of this region also show contrasts that can be readily correlated with differences in pattern of arm movement and use.

In semibrachiators, where the pattern of use of the forelimb, and hence the overall pattern of impressed forces, is sometimes as in quadrupeds and sometimes as in brachiators, the characteristics of these several muscle groups are intermediate between those typical of brachiators on the one hand and of quadrupeds on the other.

In the Prosimii, there is a close parallel. The muscular configuration in prosimian quadrupeds is similar to that characteristic of quadrupedal members of the Anthropoidea, while in the prosimian hangers (where the forelimb and pectoral girdle are frequently subjected to tensile forces) the configuration is similar to that found in groups from the Anthropoidea (brachiators and semibrachiators) where tensile forces form a significant component of the total pattern impressed upon the forelimb.

In man, the situation is different. In some features of the shoulder muscles (e.g. the transversely orientated pectoral muscles and the powerful m. deltoideus) the similarity is with brachiators. But in others (e.g. the size and disposition of both m. latissimus dorsi and the radiating digitations of m. serratus magnus), the correspondence is with quadrupeds. The resulting arrangement is thus unique and appears to contribute to a more balanced overall mobility of the shoulder than in monkeys and apes. The arrangement is also consistent with one in which the principal movements of the forelimb, although extensive in all directions of space, take place mainly in the lower quadrant.

Even though it was not possible to carry out comparative electromyographic studies of living types and to synchronize with high speed cinematography, the contrasts were so big that it was relatively easy to make biomechanical inferences about muscle function directly from a study of dissected parts.

The next question to which we had to address ourselves was crucial to the analysis of fossil remains. It was whether or not bony features could be defined which were directly related, morphologically and biomechanically, to the functionally significant locomotor contrasts in muscles and muscle blocks, and which varied in phase with these and in such a way as to enhance the mechanical significance of the established locomotor contrasts in muscular form and disposition. As a first step (Ashton & Oxnard, 1964b) a series of six angles and indices of the scapula were designed to be related to the mechanism of arm-raising. They included features such as the medial extent of insertion of m. trapezius, the angulation of insertion of this muscle, the caudal prolongation of insertion of m. serratus magnus, together with the relative distance between these two insertions — this last quantity representing the length of the arm of the biomechanical couple that they form. Corresponding measurement was also made of the distal extent of insertion on the humerus of m. deltoideus.

Three additional quantities were measured as being of mechanical significance in relation to the orientation of the shoulder joint. They described the direction in which the glenoid cavity points, the torsion of the clavicle, and the lateral projection of the shoulder.

Following the customary biometric tests to define the accuracy of measuring techniques, these quantities were measured in a total of 17 genera from the Anthropoidea and ten from the Prosimii, comprising 398 specimens in all. After the calculation of basic statistical data, corresponding measures for the several representatives of each locomotor group (brachiators, semibrachiators and quadrupeds from the Anthropoidea (together with hangers and quadrupeds from the Prosimii) were compared both with each other and with the corresponding measures in man. At this stage, before extensive computing facilities were readily available, each angle or index was treated separately. Despite the accepted limitations of such univariate study, a number of well-defined conclusions emerged. For instance, in brachiators from the Anthropoidea together with hangers from the Prosimii, the bony features associated with the insertions of m. trapezius and m. serratus magnus together with those relating to m. deltoideus were, in all cases, such as to enhance the mechanical effect of the greater degree of development of these muscles in these locomotor groups. Thus, they further facilitated arm-raising. Again, the angulation and disposition of the shoulder joint in brachiators were such as to facilitate not only arm-raising, but free usage of the arm in the upper quadrant. In quadrupeds they were consonant with more restricted movement, and in the lower quadrant.

In semibrachiators most features are intermediate, thus correlating with the observation that, in this group, the forelimb is sometimes used as in quadrupeds and sometimes as in brachiators.

In man some characters are as in brachiators (e.g. those associated with arm-raising), while others are as in quadrupeds (e.g. the orientation of the glenoid cavity). Raising of the arm is thus facilitated, but its greatest freedom of movement is, in contrast to brachiators, in the lower quadrant.

In the Prosimii, the contrast between quadrupeds and hangers parallels, in muscular form and disposition, that between, on the one hand, quadrupeds and, on the other, semibrachiators and brachiators in the Anthropoidea.

It thus appeared that in the shoulder region osteometric features could be defined that were closely correlated with the habitual pattern of use of the forelimb, and thus with the spectrum of impressed forces. This finding has proved fundamental to all subsequent studies in the series.

Comparison with corresponding quantities in fossil types (presuming such information could be obtained) might be expected to give some indication of the probable force pattern to which the region had been subjected, and thus of the pattern of habitual use of the forelimb and shoulder girdle. However, fossil remains of this region are few and fragmentary, the thin blade of the scapula seldom, if ever, being preserved. Nevertheless, some time later, my colleague Professor Oxnard (1968a, b) was able to make certain tentative comparisons with a fragmentary scapula of australopithecine origin and with a less than complete clavicle. Again in such features of these fossils as could be examined, the groups, although manifestly hominoid, differed uniquely from both living man and apes.

The primary purpose of studies of the primate shoulder was to establish a method of defining functionally significant metrical features. Having apparently succeeded, it was felt that the next logical step would be to enquire whether or not a further group of dimensions could be defined that were not obviously correlated with features such as muscular form and disposition that had a manifest correlation with locomotor pattern, and which might thus form a basis for taxonomic separation.

In the next study (Ashton, Oxnard & Spence, 1965) eight ratios and angular measures were defined in this way, the series of specimens being mainly those used in the preceding study of locomotor dimensions. It was again possible to carry out only a simple univariate analysis, and when the intergeneric contrasts in certain of these features were examined it appeared that some of the angles and indices gave a reasonable separation between certain of the established taxonomic groupings. Such separation occurred at various systematic levels from the genus to the superfamily. These dimensions were styled "residual" and it was felt that an examination of such features in fossil remains might contribute to taxonomic placing. It was noted that the overlap between taxonomic groups was so big for each dimension that it was unlikely that an unknown specimen could have been placed on this basis with any degree of certainty.

Although anxious to be able to return to studies of the pelvic

girdle — especially because of the excellent fossilized australopithecine specimen of which accurate plaster casts were available to us —
it was realized that further development of methodology was desirable. Thus with the collaboration of Professor Michael Healy we were
able to carry out a multivariate analysis compounding the nine established locomotor dimensions used in the first of our studies of the
pectoral girdle. The results of the study were presented as plots of
pairs of canonical variates (Ashton, Healy, Oxnard & Spence, 1965).

Each individual dimension had shown a clear-cut pattern of contrast between quadrupeds, semibrachiators and brachiators from the
Anthropoidea, and also between hangers and quadrupeds from the
Prosimii. It was thus not surprising that the combined dimensions
revealed a clear-cut contrast between different groups of primates
in accordance with the force pattern to which the forelimb and
shoulder girdle were subjected. Thus, in the first canonical axis,
quadrupeds from both the Prosimii and Anthropoidea were clearly
separated from brachiators of the Anthropoidea. Semibrachiators
occupied an intermediate position, with the prosimian hangers broadly
overlapping. In contrast to what emerged when most (although not
all (Oxnard, 1963)) of the dimensions were viewed singly, the multivariate compound revealed a graded pattern of difference between
genera within each locomotor group. This too correlated with the
extent to which the forelimb and shoulder girdle are habitually
subjected to forces of compression or tension. In the second direction of multivariate space, there was a further gradation in each
locomotor group, extending from types that are completely arboreal
to those that are, to a greater or lesser extent, terrestrial.

In this analysis, the majority of information relating to all groups
of subhuman primates was contained in the first two directions of
canonical space, most contrasts between genera in individual higher
axes being little more than random fluctuations. However, man projected markedly from all subhuman primate groups in the third
canonical axis and was thus well differentiated from all other primate
groups. This finding, illustrating man's uniqueness, correlated with
the observation, derived from univariate study, that in certain features
man is like quadrupeds, while in the remainder he is like brachiators.
It was thus that there emerged a method of depicting the concept of
overall uniqueness. It also demonstrated this as distinct from the
type of intermediacy that characterizes semibrachiators in both
individual features and in their multivariate compound.

By this time, facilities for carrying out multivariate studies had
greatly improved with the advent of the second generation of electronic computers in the major universities. The work at Birmingham

was also much facilitated by the involvement, from this time on-wards, of Dr R. M. Flinn. In collaboration with Professor Healy and his colleagues, he has consistently contributed much to the application of multivariate techniques in these studies.

The next development in such studies was the multivariate combi-nation of the eight "residual" dimensions, for whose contrasts uni-variate analysis had apparently shown some correlation with the taxonomic divisions of the primates. Somewhat to our surprise, it emerged that when these eight "residual" dimensions were thus compounded, again using the now well-established descriptive tech-nique of canonical analysis (Ashton, Flinn, Oxnard & Spence, 1971), there was effectively no correlation in any direction or combination of directions of multivariate space with established taxonomic group-ings of any grade. Such correlation as did emerge appeared to be the locomotor groupings, although the separation of the principal loco-motor categories was less clear than had emerged from the corre-sponding multivariate study of nine "locomotor" dimensions, each specifically chosen as defining features of biomechanical significance in relation to the force pattern habitually impressed upon the region (Ashton, Healy, Oxnard & Spence, 1965).

The two groups of measurements thus reinforced each other, and when all 17 dimenions were compounded in a single analysis (Ashton, Flinn, Oxnard & Spence, 1971) a spectrum of locomotor separation emerged, even more clear-cut than that produced by multivariate analysis of the nine locomotor dimensions.

In this last analysis, the first three directions of multivariate space gave significant separation between groups of subhuman primates, man being uniquely separated from all other groups in the fourth canonical axis. Within each locomotor group of both the Anthro-poidea and Prosimii, individual genera were separated even more clearly in accordance with individual variations in locomotor pattern (and hence in force pattern impressed upon the shoulder region) than they had been by the earlier multivariate combination of the nine locomotor dimensions.

Thus, it appeared that the form of the primate scapula is more closely associated with the mechanics of the region and thus with its impressed force pattern than with systematics of the relevant genera.

As a next stage in the development of methodology, it was felt appropriate to extend these analyses to the long bones of the arm, even though, as in the case of the shoulder girdle, we were well aware that it was most unlikely that even reasonably complete fossil speci-mens would become available for comparison.

As with earlier studies of the shoulder girdle, the first stage was to analyse quantitatively the form, proportions and disposition of the principal muscle blocks of the arm (Ashton, Flinn, Oxnard & Spence, 1976). This was done in a total of 145 specimens representing six genera from the Prosimii and 21 from the Anthropoidea. Simple univariate statistical analysis showed a number of well-defined contrasts. For instance, the flexors of the elbow are proportionately bigger in the more acrobatic types in which flexion plays a bigger part in locomotion, whereas extensors of the elbow are proportionately bigger in quadrupedal forms where extension of the elbow contributes markedly to the power stroke in locomotion. Again, for instance, in the Anthropoidea supinators of the forearm are relatively smaller in quadrupedal species, correlating with the habitually prone position of the forearm during locomotion.

On the basis of such established muscular contrasts, and of related features, 19 dimensions were defined as being of possible biomechanical relevance (Ashton, Flinn, Oxnard & Spence, 1976). These were taken on 525 sets of matching arm bones (humerus, radius and ulna) representing 39 genera of extant primates. Preliminary univariate study established that in eight of these there was a significant contrast between the different forelimb locomotor categories (brachiators, semibrachiators and quadrupeds from the Anthropoidea together with hangers and quadrupeds from the Prosimii). These dimensions were thus regarded as of locomotor significance. The remaining 11 dimensions did not show such clear-cut contrasts and were designated "potentially residual".

Multivariate combination of the locomotor dimensions of the brachium and antebrachium resulted in a separation of the subhuman primates that accorded clearly (although appreciably less so than in the corresponding dimensions of the pectoral girdle) with the extent to which the forelimb is subjected to forces of tension or compression. Once again man emerged as uniquely separated from other primate types, although less obviously so than had emerged in the analysis of dimensions of the shoulder girdle.

In multivariate combination, the potentially residual dimensions gave, in contrast to what had emerged from studies of the scapula, a good separation of the major taxonomic groups within both Prosimii and Anthropoidea. Combination of locomotor and residual dimensions of the arm resulted in a separation that was basically taxonomic, although overlain by a subsidiary division in accordance with locomotor function. In the arm, therefore, it appeared that although locomotor features could be dimensionally defined, the majority of features varied more closely in phase with taxonomic groupings.

As bones from many of the individuals used in the earlier studies of the shoulder girdle had featured in this later analysis of the arm, it was possible to combine data from the two regions in a total of 241 individuals representing 37 primate genera. In this combination, shoulder dimensions were all practically related to locomotor function, while those from the arm were split between a measure of relation to locomotor categories and a more clearly defined taxonomic correlation. Overall they gave a clear-cut locomotor spectrum within each of the major primate taxonomic groups. These were themselves well separated, man lying distinct from all subhuman primates.

In all these multivariate studies we had relied heavily upon the technique of canonical analysis. In the case of tooth dimensions (Ashton, Healy & Lipton, 1957) and of locomotor dimensions of the scapula (Ashton, Healy, Oxnard & Spence, 1965), this had proved to be a highly informative method of summarizing multivariate data by representation as two-dimensional plots. However, our success had depended upon the fact that in these studies the great majority of biologically relevant information happened to be contained in the first two, three, or at the most four planes of canonical space, individual remaining axes contributing little more than random fluctuations. In later studies (e.g. of the arm), in which greater numbers of dimensions were often compounded, it became apparent that, although the mathematical derivation of the technique meant that each individual higher canonical axis contributed progressively less to the overall separation, the total information contained in the higher axes was by no means insignificant. However it could not readily be represented in bivariate plots. We had thus to turn to more detailed analysis of the matrix of squared generalized distances (Mahalanobis, 1936) between all pairs of groups, each squared generalized distance reflecting the sum total of separation in all directions of canonical space. In order to elucidate groupings within such a matrix, a single link cluster analysis was used (e.g. Blackith & Reyment, 1971). The results were presented either as a dendogram (e.g. Ashton, Flinn, Oxnard & Spence, 1971) or as a minimum spanning tree, stylized in two dimensions (e.g. Ashton, Flinn, Oxnard & Spence, 1976). The latter, while depicting accurately the main links between groups and their association into, for instance, locomotor or taxonomic categories, does not necessarily portray faithfully their relative placings in multivariate space.

In other instances (e.g. as summarized in Oxnard, 1975), it proved informative to superimpose such a tree upon the first two canonical axes. As Professor Oxnard has demonstrated in yet other contexts (e.g. Oxnard, 1975), high dimensional plots of canonical variates as outlined by Andrews (1972, 1973) can give much significant information. No single method is universally or completely satisfactory, and the choice of the appropriate one often depends upon the type of data to be analysed and the numerical results obtained.

Such experience in presenting multivariate data was, of course, only a logical continuation of development of biometric technique that had taken place continually from the time of the earliest and most simple morphometric studies. Under the guidance of our statistical advisers, we had from the beginning defined rigorously the technique of measurement. By means of repeated measurements on short series of specimens we had also tested whether or not the variance due to unavoidable inconsistency in measurement by one worker, or due to minor variation in technique between different members of the group, was significant compared with that existing between specimens. Normally it was not, but when necessary, definition and technique were refined until such criteria had been met. Again, before any computation was carried out, it was habitual to enquire — usually by bivariate plots of correlated dimensions — whether or not any recorded dimensions were grossly abnormal due to mismeasurement or to the accidental inclusion of a highly unusual specimen.

When the work progressed to the multivariate field, we were similarly concerned with problems such as the equalization of variance between groups, the effects of differences in group size, and the possible distortion of the relative arrangements of groups in individual planes of multivariate space (although not of the generalized distances between them) by the introduction of groups with only small numbers of specimens. Again we have been, and shall continue to be, much concerned with selecting the best method of presenting the results of multivariate analysis, whether by plots of pairs of canonical axes, by cluster analysis of the squared generalized distance matrix, by high dimensional plots (Andrews, 1972, 1973) of the canonical variates, or by other techniques still to be developed.

As seems inevitable in morphometric studies of higher primates, even these aspects of our work have often attracted ill-informed criticism. Most recently Corruccini (1978), himself an avid user of morphometric techniques, made a number of observations, mainly concerned with studies carried out by Professor Oxnard (e.g. as summarized in Oxnard, 1975) as extensions of the Birmingham work with which he himself had been associated. Corruccini's review contains much cautious morphometric wisdom, but makes scant mention of the basic studies reviewed in the present paper, and in his criticism shows little awareness of or sensitivity to the rigorous procedures that have, through the years, been developed and consistently applied. Nor does he appear alive to our philosophy of selecting dimensions to have a definable functional significance.

Amidst continued, and in many ways welcome attention of this type, the results of our work on the pectoral girdle and forelimb give a clear indication that metrical analysis of individual bony regions can provide a basis for enquiry into both functional and taxonomic affinities. It was on this broad basis that an analysis of the pelvic girdle was developed.

Innominate bone

An almost perfect innominate bone attributed originally to *Plesianthropus transvaalensis* (now *Australopithecus africanus*) was described in preliminary announcements (e.g. Broom & Robinson, 1947) and then in more detail by Broom, Robinson & Schepers (1950). The bone had an expanded iliac blade which, at a first glance, gave a strong impression of being like man and of contrasting with the monkeys and apes, where the iliac blade is narrow transversely and elongated craniocaudally. Such apparent resemblances to man were stressed in the descriptions of the bone and were used, along with several other features (e.g. as listed in detail by Le Gros Clark, 1949, 1955, 1962), as further evidence that the Australopithecinae were bipeds.

In his own description of the bone, Broom (Broom *et al.*, 1950) drew attention to other features in which he felt the bone differed from man and was possibly more in line with the conditions obtaining in monkeys and apes. Such features characterized mainly the pubis and ischium, but any significance that might attach to them appeared to be largely overlooked. In subsequent descriptions (e.g. Le Gros Clark, 1949, 1955, 1962) it was the supposedly man-like features of the bone that were consistently emphasized in support of the thesis that the fossil creatures had been bipeds.

An early quantitative enquiry into a number of overall pelvic dimensions (Williams — unpublished observations recorded in Zuckerman, 1954) gave, within the limited confines of the type of analysis that was possible in the early 1950s, some indication that overall resemblances might not be nearly so much to man as to certain apes. A second study, carried out by Dr (now Professor) S. R. K. Chopra in Professor Zuckerman's laboratory at Birmingham (recorded in Zuckerman, 1966), indicated that the angulation between the plane of the iliac blade and that of the ischiopubic rami is also more ape-like than human.

Real advance in the study of the australopithecine innominate bone was possible only after Charles Oxnard had joined the group permanently and after much of the preliminary work on the pectoral girdle and arm had been carried out in order to establish a viable methodology.

In this study of the pelvic girdle, and in contrast to that of the shoulder and arm, we were fortunate in that the available specimen of a fossil australopithecine innominate bone was remarkably complete. The bone had few imperfections, and such as were evident (e.g. the lack of an anterior superior iliac spine) could be readily reconstructed from fragments that were recovered from the contralateral bone of the same individual. Again, reconstruction of the ischiopubic region to correct manifest distortion and minor erosion during fossilization proved relatively easy, and in the first major study of the pelvic girdle (Zuckerman, Ashton, Flinn et al., 1973) a single "average" reconstruction of this type was used.

The first stage in analysis consisted of developing a locomotor classification of the hindlimb that would adequately reflect variation in the force pattern to which it is subjected during locomotion. It was already apparent that our earlier classification (Ashton & Oxnard, 1964a), based upon the extent to which leaping occurs, was manifestly incomplete. The question was one that had been actively pursued for some time by Professor Oxnard following his emigration to the United States of America in 1966, and the locomotor subdivisions ultimately selected for use in. our joint study of the pelvic girdle (Zuckerman, Ashton, Flinn, et al., 1973) were elaborated in some of his own publications (e.g. Oxnard, 1973, 1974, 1975, 1976). The classification was based upon the general principle that the pattern of forces to which the hindlimb is subjected in different primate genera does not, in contrast to that imposed upon the forelimb, form a spectral arrangement progressing from forces predominantly of compression to those of tension. It is, in fact, better represented as a number of specialized functional units branching from a central generalized core of types that are basically quadrupedal. Close to this core of "regular" quadrupeds (including such genera as Macaca, Cercocebus and Papio) there are others (e.g. Lemur, Cheirogaleus) in which quadrupedalism is associated with running and a measure of climbing. Peripheral to these are more extreme groups such as certain lorisines (including such genera as Nycticebus and Perodicticus), in which the hindlimb is used for suspending the animal from the underside of branches and in which it thus participates in acrobatic activities. In certain anthropoid genera (e.g. Alouatta, Lagothrix, Pongo) the hindlimb participates to an extreme extent in such acrobatic movements.

At another pole, a few genera of Prosimii (e.g. *Propithecus*) are typified by a locomotor pattern in which the hindlimb produces the propulsive force in leaping movements and in which it also participates along with the forelimb in clinging — often vertically — to branches upon which the animal lands. In yet another prosimian group, comprising such genera as *Galago* and *Euoticus*, the basic mode of progression is quadrupedal with a considerable amount of leaping superimposed. The hindlimb again participates in pronounced leaping movements of yet another type in certain genera of monkeys (e.g. *Presbytis* and *Colobus*). Other genera of monkeys (e.g. *Aotus* and *Saimiri*), although again basically quadrupedal, leap on occasion and may be styled "facultative leapers".

Man is unique in that his hindlimb is the principal functional unit in a striding form of bipedal gait.

From such a classification, it is relatively easy to derive the general pattern of forces — compressive (smooth or impulsive), tensile or both in combination — to which the hindlimb is habitually subjected.

The next stage comprised an attempt to define features of the pelvic girdle — more specifically of the innominate bone — that are mechanically related to such a pattern of impressed forces. Certain of these forces are imposed upon the pelvic girdle by the pull of the attached blocks of muscle — a mechanical system that our earlier studies of the shoulder girdle and arm had shown to be likely to exhibit contrasts of definable biomechanical significance between different locomotor groups. Others result from the downward transmission of weight through the sacro-iliac and hip joints to the bones of the lower extremity.

As a basis for study of the first group of features, an analysis was made of quantitative variations in form, disposition and proportions of the several muscle blocks attached to the pelvic girdle and acting upon the hip joint. These comprise: firstly, the flexors (principally the iliopsoas); secondly, the extensors of the hip (including primarily the hamstring muscles, supplemented in man by the powerful but, in many ways, exceptional m. gluteus maximus); thirdly, the adductors lying on the medial aspect of the thigh but orientated in quadrupeds so as to contribute appreciably to the movements of flexion and extension; fourthly, the abductors of the hip. In man these comprise the two lesser gluteal muscles (m. gluteus medius and m. gluteus minimus), both of which lie lateral to the hip joint owing to the lateral orientation of the iliac blade. But in sub-human primates, because of the dorsal alignment of this bony element, the lesser gluteal muscles have less effect as abductors, the main power of abduction apparently coming from the few fibres of

m. gluteus maximus which, in these types and in contrast to man, pass lateral to the hip joint. A fifth group comprises the short muscles, spanning the hip joint and contributing to the maintenance of its stability.

The results of this part of the analysis were clear-cut. Although the flexors constitute approximately the same proportion of the total hip musculature in both man and subhuman primates, the extensors of the hip are more prominent in subhuman primates while the adductors are relatively less well developed in these types than in man. The biggest contrast occurs in the abductors of the hip, which are more prominent in man than in the subhuman primates.

A corresponding group of bony features — five in all — were thus selected for examination. One related to the attachment of the extensor muscles, defining the relative length of the lever-arm to which they are attached in relation to the hip joint. Others related to the orientation of the iliac blade and to the disposition of the lesser gluteal muscles, a lateral orientation indicating that these were likely to have been abductors of the hip as in man. This, in turn, is a biomechanical prerequisite for a human type of striding gait.

The definition by biometrical methods of osteological features directly related to the transmission of weight through the innominate bone had, of necessity, to be approached on a more empirical basis. It had been realized (e.g. Oxnard, 1967, 1973) that several engineering techniques (e.g. photostress and photoelastic analysis) might well give clues as to the type of overall force pattern that a particular bony shape is best able to withstand and transmit. At the time of conception of this study, however, it had not been possible to carry out such analyses, and morphological features associated with this part of the total force pattern to which the pelvic girdle is subjected had, of necessity, to be judged by simple mechanical appreciation. Nevertheless, it appeared that the effectiveness of weight transmission through the innominate bone would, in all likelihood, be related to the disposition and orientation of the two principal joints (sacro-iliac and hip joints) in relation firstly to each other and secondly to the boundaries of the innominate bone. Four dimensions were devised to represent these quantities.

The total of nine dimensions (linear and angular) was taken on 430 specimens representing 13 extant genera from the Prosimii together with 28 from the Anthropoidea (including man).

In the subsequent statistical analyses (both univariate and multivariate), attempts were made to compensate for the wide difference in overall size between different primate genera, firstly by the use of simple indices (ratios — logarithmically transformed), and secondly by adjustment of each value (again after logarithmic transformation) to the midpoint of a regression line computed for each dimension in turn, against a standard. That chosen was the dorsoventral dimension of the innominate bone projected on to the standard plane of orientation, tests having shown this to be highly correlated with overall body size. The indices, as is now well recognized, give an overall representation of aspects of shape as it exists in each specimen. When compounded the dimensions adjusted by the regression technique give a representation of overall shape, excluding the component that is dependent upon differences in overall size — i.e. upon allometric factors. In the event, both approaches pointed to quite similar conclusions, although the logged indices, in containing an extra component of information, naturally demonstrated a somewhat greater scale of contrast.

The first stage in analysis comprised the calculation of basic statistical data (means, standard deviations and standard errors of the means) from which were plotted, for each genus, simple bar graphs giving the means and 90% fiducial limits for each of the nine variates examined. From an inspection of each of these nine compilations it was possible to see major overall contrasts and to make a rough assessment of the groups to which the fossil specimen of *Australopithecus* appeared to approximate more closely.

Although in several instances a number of groups of subhuman primates were distinguished from the others (the hanging lorisines often featuring prominently in this context), the major contrast was between man and the subhuman primates as a whole. In each of the five dimensions relating to the disposition of muscle blocks, the fossil innominate bone tended to resemble the subhuman primates more closely than man. However, in the four quantities relating to the relative positions of the sacro-iliac and hip joints, the fossil tended to be like man and to contrast with the several groups of subhuman primates (Zuckerman, Aston, Oxnard & Spence, 1967).

Each of these conclusions was reinforced when first the five quantities relating to muscular orientation, and second the four measures relating to joint disposition, were combined by multivariate analysis (Zuckerman, Ashton, Flinn *et al.*, 1973). In each instance, a clear contrast emerged between man and all subhuman primates. In the combination of four variates relating to joint position such differentiation of man was fully revealed in the first two canonical axes. In the analysis of the five dimensions relating to muscular disposition, it was in the third direction of canonical space that man was mainly separated. In the first analysis (relating to joint position) *Australopithecus* lay within the range of extant man. In the second (relating to muscular disposition) the fossil lay in the range for the subhuman primates.

These two multivariate analyses did little more than confirm and reinforce the results of univariate study.

When all nine variates were combined in a single canonical analysis, a well-marked contrast emerged between man and all groups of sub-human primates, among which various locomotor categories were distinguished to a varying extent. Understandably, in view of the close similarity to man in certain features and to subhuman primates in others, *Australopithecus* emerged as distinct from both. Contrasts between man, subhuman primates, and *Australopithecus* were in fact spread throughout the first three dimensions of the multivariate space, both *Australopithecus* and man being differentiated in the second canonical axis and *Australopithecus* being, in its turn, separated from man in the first and third canonical axes. Nevertheless, the extent of its distinction as illustrated by bivariate plots of the lower canonical axes was less than, for instance, that applying to man in a multivariate compound of the locomotor dimensions of the shoulder girdle of extant primates.

We were all aware that the descriptive result produced by this method, as also from a cluster analysis of the squared generalized distances between all genera — a technique which again emphasized the individual distinction of both *Australopithecus* and of man — did not demonstrate the uniqueness of the fossil genus as clearly as might have been anticipated from the contrasting and relatively clear results that had emerged from both univariate and multivariate analysis, firstly of characters relating to joint position and secondly of those relating to muscular disposition.

Appraisal of the possible biomechanical significance of individual contrasts, as set in the framework of the collective result from multi-variate study, led us to believe that if *Australopithecus* had been bipedal, its bipedalism must have been quite different from that characteristic of *Homo sapiens*. This was based, possibly above all else, on the observation that the dorsal orientation of the iliac blade would have implied that the lesser gluteal muscles were orientated as in subhuman primates. They were thus unlikely to have provided a powerful source of abducting the hip and hence a means for stabilizing the pelvic girdle such as is a prerequisite of the stepping-off phase of a striding type of bipedal gait.

In addition to such difficulties of interpretation, we were conscious of a number of shortcomings in the study. Some of these would be very difficult to rectify although others were susceptible to critical examination. For instance, although adequate samples of bones were available for many primate genera (e.g. *Homo, Presbytis, Pan, Lemur,*) in other cases (e.g. *Daubentonia, Propithecus, Nasalis*) the available samples were minimal, sometimes comprising only a single specimen. To have increased these to a size comparable with those available for certain of the better represented groups would have involved unwarrantable administrative effort and travelling. In any case, in certain instances the total numbers available in the museums of the world could well have been insufficient to form a representative sample. The precision of the mean and variance of each dimension is of course reduced in such small samples. But after definition of variance in groups represented by adequate numbers of specimens coupled with mathematical transformation where necessary, it was possible to equalize intergeneric variance and to base the multivariate analysis upon a uniform variance between primate genera. It is a not unreasonable assumption that such quantities would also obtain in the genera (possibly even including the fossil species) that are represented by only small numbers of specimens.

Again, the cast of only a single fossil specimen had been available to us for study, and even this needed a certain amount of reconstruction. The version used for comparison in this initial multivariate study was a "median" one. But again there is no certainty that this is necessarily correct, nor did we, on this occasion, enquire fully into the effects of various methods of reconstruction upon the final results.

Above all we were aware that extensive though this analysis had been, we had been able to study only nine biomechanically significant quantities and these could scarcely give a comprehensive representation of the overall functional significance of a structure as complex as the pelvic girdle. Professor Oxnard, Dr Flinn, Mr Spence and I therefore felt that we ought to define an extended series of functionally significant measurements that could be added to the first and thus give a basis for a better overall biomechanical appraisal of the significance of observed contrasts in pelvic morphology. We have now been able to extend the original series of nine angles and indices to a total of 25. We have also been able, as a result of assiduous and persistent work by Mr Spence, to take the additional 16 dimensions on a very high percentage of the same bones that were used in the earlier study of the pelvic girdle. The fact that such a large part of the original material could thus be traced and used again is a tribute to the efficient systems of Museum and University Department curators in this country. In a few instances, where extra specimens of poorly represented genera had become available, both the original series of nine dimensions and the extra series of 16 were taken.

As in the first study, each dimension was chosen with a specific biological aim. Some, as with the first nine, were chosen because of obvious association with such biomechanical features as muscular attachment or weight transmission. Others related to other functional aspects of pelvic morphology while a few were deliberately chosen as not having obvious functional significance but as contributing to the overall general description of the bone.

A full analysis of this new material is now nearing completion, and results have already been processed from a total of seven indices and nine angles from the new series in multivariate combination with the nine angles and indices used in the earlier study. The multivariate technique originally used was that of canonical analysis. However preliminary examination of the results showed that in this study very considerable amounts of information were contained in the higher canonical axes, and that simple bivariate plots could, therefore, scarcely present any comprehensive picture of the pattern of intergeneric contrast. Recourse was thus had to clustering (by means of a single link technique) of the matrix of squared generalized distances. This was summarized as a minimum spanning tree whose major features proved to be reproduceable in two dimensions. A number of significant findings emerged. Firstly, in both Prosimii and Anthropoidea there was a good separation of locomotor groups as earlier defined on the basis of hindlimb function. Secondly, there was a good separation between the locomotor groups within the Prosimii and between those within the Anthropoidea. The two suborders were connected by relatively few links. Within each suborder specialized groups (e.g. leapers, hangers, acrobats) radiated from a centralized area occupied by the more generalized quadrupedal forms. Man linked with the generalized quadrupeds through the group of hindlimb acrobats represented by the extant apes. Closely associated with the several geographical subgroups of living man (all very similar to each other) was a Neanderthal innominate bone (Skhūl IV) of which an accurate cast had become available and which was also included in the general analysis. *Australopithecus* also linked with the living great apes, but along a line quite different from that by which man was attached.

In this analysis we used a number of reconstructions of the innominate bone of *Australopithecus*. Some of these were deliberately made so as to be as extreme as possible in either the human or apelike direction without, of course, being morphologically grotesque. Although the five reconstructions were thus made deliberately to be as different as possible from each other, all formed, in the multi-variate analysis, a single cluster attaching to that comprising the living great apes. Moreover, it lay in multivariate space separated as far from man as both the human and australopithecine clusters were individually separated from the cluster comprising the living apes.

This result thus provided a demonstration, possibly more convincing than any that had emerged previously, of the uniqueness of the Australopithecinae.

DISCUSSION AND CONCLUSION

It would be wrong to claim that in this summary I have attempted to cover, even cursorily, large areas of the now vast literature that has accumulated about the australopithecine fossils. More especially, it has been possible to make mention of only a sample of the biometric studies that have been carried out on this material.

In the preparation of any such eclectic review, there is always a danger of bias. It may be claimed — possibly with justification — that undue emphasis has been given to those studies that point to the uniqueness of the Australopithecinae.

I have, for instance, made no mention of the extensive analyses that were carried out some 25 years ago in Lord Zuckerman's laboratory at Birmingham and under his personal direction, of the morpho-genetic development of muscular crests on the primate skull (Ashton & Zuckerman, 1956b). The occipital crest, extending from mastoid to mastoid, gives attachment on its undersurface to the extensive nuchal musculature appropriate to balancing the face-heavy skull of the great apes (and especially adult males). Its upper surface gives attachment to the horizontally disposed posterior fibres of the temporal muscle — this again being big in accordance with the large jaws and facial skeleton of subhuman hominoids. On occasion, the two temporal muscles approximate in the mid-line of the skull resulting, in adult males, in the formation of a sagittal crest extending forward from the occipital crest along the line of the sagittal suture.

Such crests, of course, never appear in man, the formation of both crests being correlated in apes not only with the large size of the facial skeleton and the dorsal position of the foramen magnum (resulting in "face-heaviness" of the skull on the vertebral column) but also with the relatively small cranium in apes. This correlates with their relatively small cerebral development and provides, on its outer surface, only a limited area for muscular attachment. Study of available australopithecine remains did in fact show that contrary to a previously held belief (e.g. Broom & Robinson, 1952) the disposition of muscular crests and hence presumably the morphogenetic processes underlying their formation, were identical in the fossil group with those that obtain in the living apes: the nuchal crest forms first and is followed later, if the temporal muscles approximate to the mid-line, by the formation of a sagittal crest extending forward from the inion. The Australopithecinae are, as judged from several well-preserved specimens, in this respect quite like the extant apes and contrast with man.

But such results emphasize those of the work that I have selectively described and point to a clear overall conclusion. In the Australopithecinae, superimposed upon an overall resemblance in cranial and facial proportions to the living apes, there is, in each of the morphological complexes examined, in certain respects a resemblance to man, and in others, to the living apes. In scarcely any instance has there emerged a situation in which a feature or group of features of the fossils is intermediate between these living groups. In combination, ape-like features, associated with some of a human-like nature, produce a pattern that differs from both man and the living apes and is thus unique.

It might be claimed — and again with justification — that in this account I have given disproportionate emphasis to work that was not only directed by Lord Zuckerman personally, but which has been carried out wholly or mainly in the University of Birmingham. However, several of his former pupils and colleagues have themselves been productive in this area, their work producing results consonant with those to which I have principally drawn attention. One such study relates to the talus, and was carried out with characteristic thoroughness by Dr (now Professor) F. P. Lisowski. The work was initiated and the data collected when Dr Lisowski was a member of Lord Zuckerman's staff at Birmingham. But final analysis of the results awaited his assumption of the Chair of Anatomy in the University of Hong Kong. With the collaboration of Professor Oxnard and his colleagues (then in Chicago), Professor Lisowski has been able to show not only a contrasting morphological pattern of this bone in living men and apes, but also that the australopithecine fossils, in showing a resemblance to man in some respects and to apes in others, are uniquely different from both (Lisowski, Albrecht & Oxnard, 1974).

Such a conclusion has also emerged from studies by other workers, in some of which there has been a less conscious attempt than in the case of, for instance, our own study of the pelvic girdle, to define metrical features in accordance with the functional pattern of the region. Thus, Rightmire (1972) has described a multivariate study of the first metacarpal bone of 105 specimens of man and the living apes, together with a fossil specimen attributed to *Paranthropus* (now *Australopithecus*) *robustus*. The fragment had previously been described by Napier (1959) and also by Le Gros Clark (1967), these workers having found a combination of ape-like and human features leading to a conclusion that it was essentially primitive in its morphological structure.

The overall finding, from multivariate analysis — extended somewhat by Professor Oxnard beyond that published by Rightmire — was simple: the fossil specimen, in the complex of features examined, differed from both man and the living apes — and to an extent greater than that which separates any of these extant groups. In other words, it is unique (Oxnard, 1975).

Again I have not mentioned the equally significant study of McHenry (1973) of the distal part of the humerus. This analysis, based upon a series of 18 dimensions taken on 221 humeri of the chimpanzee, gorilla and man, also led to a conclusion that the distal end of the fossil humerus is unique among living hominoids — a view that was reinforced by the submission of these data in Professor Oxnard's own laboratory (Oxnard, 1975) to analysis by multivariate methods even more sophisticated than those deployed by McHenry.

Yet another illustration, possibly even clearer, of the unique morphological nature of the Australopithecinae derived from a study by McHenry & Corruccini (1976) of the proximal part of the femur. Using a series of ten carefully chosen measurements, taken in good samples of living man and apes, and analysed by the canonical technique, they found: (a) that whereas man and the extant apes were co-planar in the first two canonical axes, man was well separated in the first axis; (b) whereas in the third axis there was some separation of the chimpanzee from other extant groups, in the fourth axis, the Australopithecinae were conspicuously different from all living types, occupying a unique position in the multidimensional space.

In sum total, the picture is consistent: superimposed upon a basically and often overwhelmingly ape-like structure, the characteristics of the australopithecine fossils are, in a wide variety of circumscribed regions, such that in some respects they bear resemblance to man, in others to apes while in yet others they may differ from both. In combination of characters the fossils are, in each instance, quite different from both extant groups. This concept is in contrast to views that they are morphologically intermediate.

It is possibly ironical that this concept of uniqueness as distinct from intermediacy emerged, at any rate in basic principle, from what was, to the best of my belief, the first analysis ever to be published of the australopithecine remains using modern metrical and (univariate) statistical techniques. Its publication in 1948 actually antedated the first of the Birmingham studies. It was carried out by Dr W. L. Straus and centred upon the one distal end of an australopithecine humerus that was then known — a specimen attributed to *Paranthropus* (now *Australopithecus*) *robustus*. Straus's conclusion was interesting: he found the fossil to be no more human (as had been claimed by certain earlier workers — e.g. Broom & Schepers, 1946) than it was ape-like. In some features it resembled man, in others it agreed with the living apes. In other words, it is ". . . in no manner intermediate between the humeri of the anthropoid apes and that of man . . ." (Straus, 1948). It showed exactly the unique combination of human and ape-like features that has emerged in many later studies.

Again, Kern & Straus (1949), in a corresponding study of the distal end of a femur attributed to *Plesianthropus* (now *Australopithecus*) found that despite certain resemblances to man in some features, other features deviated and in these, while differing from the apes, the fossil agreed with extant monkeys. In total complex, so far as could be judged from univariate analysis (which was all that was practicable at that time), the fossil was no more human than it was monkey-like.

Here, then, was the germ of the concept that developed in the studies that have emanated through the years from members of the Birmingham and ex-Birmingham School, as also, more recently, from certain others working independently.

It has not yet been possible, and it may be a long time before it becomes feasible, to evaluate the biomechanical significance of overall patterns of similarity, contrast, and uniqueness such as have emerged from these studies. But any evaluation of the significance of such unique combinations of morphological features in relation to the possible phylogenetic position of these fossils is even more difficult and may well prove to be, for all time, impossible. Such speculative evaluation has, however, never formed a part of the studies carried out at Birmingham, where the objective has been limited to quantitative morphological comparison and, where possible, cautious biomechanical inference.

Another criticism of the work that I have summarized could be that an element of bias might have been introduced because, to no small extent, attention was first directed to morphological features upon which emphasis was placed in the initial descriptions of the fossils, and these were often claimed to have a human-like configuration. However subsequently the biometric studies embraced much wider aspects both of the general morphology of the fossils and of their individual regions.

Again, it might, with equal justice, be claimed that I have given too little prominence to a view repeatedly put forward by scholars of distinction (e.g. Le Gros Clark, 1949, 1955, 1962, 1967) that notwithstanding the ape-like proportions of the australopithecine skull, the apparently man-like nature of certain selected features indicated a creature which, if not a human ancestor, was of a morphological type that might have been postulated, on theoretical grounds, as filling this position. In reply it must be reiterated that the Birmingham studies have had a much more limited objective than to attempt what some, including myself, would regard as the almost intractable problem of deciding the phylogenetic position of a fossil group. It has been their purpose, from the first, to develop quantitative techniques to give precision to morphological description and comparison, and hence to move towards an assessment of overall degrees of similarity and difference. Concurrently it has been the objective, realizing that it is and will continue to be virtually impossible ever to examine all definable morphological features, to develop methods for selecting those that are of functional significance in the context of a definite problem — e.g. an enquiry into whether or not a given bony configuration might provide a mechanical system better adapted to a bipedal posture and gait than to a quadrupedal one.

This approach has been basic throughout, but as many have apparently failed to appreciate the true objectives of our work, it has, as with many other branches of the study of human evolution, attracted more than its fair share of controversy. It would be sterile to review the public exchanges that have taken place during the past 30 years or the detailed criticisms formerly advanced by those who were sceptical of any significance that might attach to quantitative metrical study as a means of giving precision to morphological observation. Naturally it fell almost exclusively in the earlier years to Professor Zuckerman to take the brunt of this type of criticism and to put foward the general philosophy and objectives of the work. This was done in a series of reviews and lectures extending from an address to the British Association in the summer of 1947 (subsequently published in 1950 in *Biological Reviews* under the title "Taxonomy and Human Evolution") through addresses to the Royal Anthropological Institute in 1951 and to the Linnean Society in the same year, through a masterly review in a volume presented to Sir Julian Huxley on the occasion of his 65th birthday in 1954 and culminating in an address (Struthers Lecture) to the Royal College of Surgeons of Edinburgh in 1965 (Zuckerman, 1950, 1951a, b, 1954, 1966). In all these communications, as in the first paper on the topic published in 1928, the underlying theme was the basically ape-like nature of the australopithecines — a view often projected in the context of comparison between mid-sagittal sections of skulls of the living apes, of the australopithecine fossils, of *Pithecanthropus* (now *Homo erectus*) and extant *Homo sapiens*. The outline of the australopithecine skull is remarkably similar to that of living apes, despite certain differences such as some shortening of the front of the face correlating with the reduced size of the incisor and canine teeth in the fossils, and in the configuration of the occipital region of the skull. Certainly the overwhelming impression is of a much greater similarity to the living apes than to any type of man — living or extinct. But, ironically, it has not yet proved feasible to quantify the extent to which, in sagittal outline, fossil skulls of this group differ from each of the living types of ape and man and also from other fossil species. In studies of the Australopithecinae, this might, in fact, be seen as scarcely necessary, so great is the ape-like resemblance. But if analysis is extended to other fossil groups (e.g. as represented by the skull "1470" recovered in 1972 from East Rudolf (Leakey, 1973)), the extent of similarity and difference is not so clear-cut and quantification becomes desirable.

Such a problem may well be regarded as basic — especially if translated into three-dimensional space to encompass the entire skull. But the technology of comparison of complex variable shapes (such as the sagittal outlines of hominoid skulls) is so involved that only very recently has its application become feasible. In basic principle, this problem is not dissimilar from the one approached by Professor Karl Pearson and his workers in the Biometric School who, in the course of many assiduous studies ranging over the first 35 years of the 20th century, attempted, by means of rigorous techniques of measurement, to define the overall proportions of the human skull by a process of geometrical triangulation (e.g. Buxton & Morant, 1933; Morant, 1936). The variability of each dimension (linear, curvilinear or angular), and of derived ratios, computed to express aspects of shape, was assessed in as extensive series of human crania as could be accumulated. Then by means of a statistical technique designed to compound differences between groups as exhibited by individual measurements, some overall assessment of morphological contrasts was derived. In the studies of Pearson and his pupils, comparison was by the "Coefficient of Racial Likeness" (Pearson, 1926).

Such a basic overall approach to the comparison of complex variable shapes, although modified by the use of such multivariate statistical concepts as the "Generalised Distance" (Mahalanobis, 1936) to replace the Coefficient of Racial Likeness (shown by Fisher (1936a & b) to have so many theoretical defects as to be effectively invalid) remains good to the present day. It has, in recent years, been applied with considerable success by, for instance, Howells (1973) to an analysis of human crania from the principal geographical subgroups of living man.

A number of theoretical drawbacks to the approach remain. For instance, even with the relatively large numbers of dimensions that were designed and taken by Pearson and his school, the overall picture provided of even a sagittal outline of the skull is no more than approximate. It is even less complete when extended into three dimensions. A possibly more serious objection stems from the observation that the fixed points between which dimensions are taken (often the point of junction of cranial sutures) may not necessarily be biologically equivalent (homologous) in different species and genera. Possibly such an objection may not weigh heavily when comparison is restricted to human groups, but uncertainty increases if comparison is made between extant *Homo sapiens* and subhuman primate types.

A possible alternative is the technique of coordinate transformation put forward as long ago as 1917 by the late Sir D'Arcy Wentworth Thompson. In this the skull (or other part) is placed in a rectilinear co-ordinate grid which is then systematically deformed until the presumably related shape is derived. The extent of deformation provides a measure of overall difference. But the geometrical illustration of such techniques is not easy and, as shown by the pioneering work of Medawar (1945, 1950), the mathematical definition of such transformation presents many problems. These have not, to the best of my knowledge, been overcome even at the present day. Related techniques (e.g. Sneath, 1967) have involved the superimposition of outlines, their scaling to eliminate differences in overall size and their rotation so as to give the best degree of fit as assessed by statistics derived from the distances between related points. But these techniques, as also does the related technique of resistant fitting in which areas apparently broadly similar are fitted in the best possible way irrespective of imperfections of the fit in other less similar regions (e.g. Andrews, Bickel, Hampel, Huber, Rogers & Tukey, 1972; Siegel & Benson, 1979), continue to depend upon the identification of points that appear to correspond in each of the groups under comparison.

Certain mathematical techniques are capable of overcoming such defects by treating the entire outline as a complex mathematical curve. But their application is involved and it is conceivable that greater promise may attach to certain simple factors defining individual aspects of shape, and based upon such quantities as overall area, perimeter, greatest diameters, etc. (e.g. Exner, 1978). Such work, even at the present day, remains relatively in its infancy, but critical studies carried out collaboratively in the Universities of Birmingham and Leeds have not only emphasized the ape-like nature of the Australopithecinae, but have also shown degrees of similarity and contrast between extant apes, man and several groups of fossil hominids (e.g. *Homo erectus* and *Homo neanderthalensis*) that correlate well with their taxonomic placings.

This most recent finding relating to the Australopithecinae further emphasizes the outcome of a total of 50 years' investigation of the group: superimposed upon a basically ape-like creature, there is a complex of features, in some respects ape-like, in others human, and in sum total, uniquely different from both.

In giving emphasis to the metrical study of the australopithecines, carried out to enquire into aspects of quantitative and functional anatomy, I have not been able to mention related quantitative studies of primate evolution that were instituted and initially led personally by Solly Zuckerman, to be subsequently pursued by his former pupils as parts of derived research programmes.

One of these (Ashton, 1960; Ashton, Flinn, Griffiths & Moore, 1979; Ashton & Zuckerman, 1950c, 1951b, c & d) has given information about the extent of morphological divergence that can develop in a primate population, geographically isolated for a defined time. It relates to a population of green monkeys (*Cercopithecus aethiops sabaeus*) whose progenitors were transported, some 300 years (100 generations) ago from their natural habitat in West Africa to the West Indian island of St Kitts. An early study of the St Kitts green monkey was carried out by the late Sir Frank Colyer, who, in 1938, was given almost 100 skulls collected from the West Indian island. Colyer's interest was in dental abnormality (e.g. the presence of extra teeth in the tooth row; the occasional deletion of certain teeth; gross malpositioning etc.) which, he established, now occurs more frequently in the island population than in the present-day descendants of the parent population from West Africa (Colyer, 1948).

In 1947 I was, on Professor Zuckerman's behalf, able to take extensive measurements of the skulls and teeth of this series and to obtain corresponding comparative data, not only from a good sample of the present-day descendants of the parent African stock, but also from two related African subspecies (*Cercopithecus aethiops aethiops* — the grivet monkey from Ethiopia; and *Cercopithecus aethiops pygerythrus* — the vervet monkey from South Africa) together with two additional full species of *Cercopithecus* from the West African mainland: *Cercopithecus nictitans nictitans* (the white-nosed monkey) and *Cercopithecus cephus cephus* (the moustached monkey).

A number of univariate analyses of these data were carried out between 1949 and 1960 and, within the well-known and acknowledged limitations of this type of approach, gave clear indication that in addition to the increase in variability of meristic dental features as described by Colyer (1948), the skulls and teeth of the island monkey are now bigger and less variable in their dimensional characters than are those of the present-day descendants of the African stock (Ashton & Zuckerman, 1950c, 1951b). In addition, the variation between corresponding dental and cranial structures from the left and right sides is now greater in the island monkey than in the present-day West African green monkey — i.e. the St Kitts monkey is less symmetrical than are the present-day descendants of its parent stock (Ashton & Zuckerman, 1951c). The extent of the difference (Ashton, 1960; Ashton & Zuckerman, 1951d) appeared to be as great as that which now exists between the present-day African green monkey and certain of the other mainland groups to which distinct subspecific or specific status is accorded.

The development during the 1960s and early 1970s of the necessary hardware and expertise to apply techniques of multivariate statistics, thus making accurate allowance for the correlation between dimensions and producing adequate representation of overall morphological "distance" between different groups, has made it possible in recent years to extend this study. The results have been published recently (Ashton, Flinn, Griffiths & Moore, 1979). By an extensive application of multivariate techniques, it has been possible not only to confirm the general conclusions of the earlier univariate studies, but also to establish that the overall quantitative difference in cranial and dental characters between the island and mainland groups of green monkeys is as big as that which now typically, in these features, separates the African green monkey from the other two groups (*Cercopithecus nictitans nictitans* and *Cercopithecus cephus cephus*) to which full distinct specific status is normally accorded.

The total pattern of contrast in size and variance would appear to indicate a genetic basis to such differences as now exist, although the remote possibility cannot be excluded that these have resulted purely from the effects of environment upon an unchanged genetic constitution. But there would appear to be little likelihood of such rapid change continuing even under intense selective pressures, and the results could scarcely be extrapolated into the context of morphological change of an extent seen in the fossil record of the primates.

This study of the St Kitts green monkey, along with analysis of the functional morphology of the pelvic girdle, have special significance in showing how concepts, dating from as long as 30 years ago and explored in the immediately succeeding period within the limitations of the technology then available, have formed a basis for progressive lines of development, which today continue to bear fruit in several centres of the world. Such lines in being complete, ongoing and viable programmes of work, cannot be, in any way, regarded as mere fragments that remain of research programmes that were in operation 30 years ago. Equally there is no question, in the minds of any of us who have the pleasure of remaining deeply involved in this field, firstly that the studies we are undertaking, both in concept and in inspiration, originate in the work which was initiated more than 50 years ago by Solly Zuckerman, and secondly that the present impetus of the work stems ultimately from his interest and personal leadership during the 15 years or so following the Second World War.

His concepts remain viable and continue to be developed by former pupils in different centres of the world. One is gratified that a group of them, including two of the other speakers in this symposium (C. E. Oxnard and W. J. Moore) have not only developed independent and highly distinguished research lines described in their own contributions, but have also remained in close association, continuing to publish joint studies of biometrical morphology. Such productive continuity is an incontestable demonstration that in this symposium we honour one of the truly great biologists of the present century.

SUMMARY

(1) From the earliest discovery, in 1924, of a specimen of the Australopithecinae, emphasis has been placed in published descriptions upon features of the skull, teeth and postcranial skeleton which, on the basis of qualitative examination, are reputed to differ from extant apes and to resemble man.

(2) But any such characteristics are superimposed upon an overwhelmingly ape-like resemblance in general proportions of the braincase and facial skeleton.

(3) Enquiry by quantitative methods (used to obviate the uncertainties of visual comparison and to take account of variability) into

patterns of similarity and contrast between the australopithecine fossils and extant types, was initiated in 1928 by the dedicatee of the present symposium. It was much developed by him following the Second World War.

(4) An early quantitative study (Zuckerman, 1928) showed that in contrast to certain expressed beliefs, the endocranial volume of the first australopithecine fossil (*Australopithecus africanus*) was wholly like that of apes of a corresponding degree of maturity and quite dissimilar from that in man. This result was subsequently fully confirmed by studies of additional fossil material and with more extensive comparative data.

(5) Many studies carried out in the past 35 years have shown that, superimposed upon their ape-like cranial profile, the Australopithecinae display combinations of characters in some instances like apes, in others like man. *In toto* the fossil group is uniquely different from both, as distinct from being intermediate in form.

(6) The occipital condyles are, in position, as in apes; in angulation, as in man.

(7) In their proportions, the biting teeth of the Australopithecinae are as in man; the grinding teeth are as in apes.

(8) In such features of the basicranial axis as relate principally to the size and set of the face, the Australopithecinae are ape-like. In others they show resemblance to man. In combination they are again different from both.

(9) The articular surface of the temporal bone, although basically similar in man, living apes and the Australopithecinae, varies quantitatively in its proportions. The three groups contrast uniquely with each other.

(10) In being consistently single, the infraorbital foramen of the australopithecine fossils is as in man. In lying relatively far down the prognathous facial skeleton, it is as in apes.

(11) An extensive series of studies of the pectoral girdle and forelimb has provided a basis for enquiry into methods of selecting osteological features, metrically definable and related biomechanically to the pattern of forces impressed upon the region during locomotion. This work has also provided a vehicle for developing the use of multivariate statistical techniques and especially in differentiating between uniqueness, as opposed to intermediacy between extant types.

(12) These principles have been translated to the pelvic girdle (an almost perfect australopithecine innominate bone being available). This region is biomechanically more complex than the pectoral girdle because, in addition to forces produced by muscular pull, there are others that result from weight-bearing.

(13) A total of 25 metrically defined features have been selected so as to be biomechanically significant in relation to muscular pull, weight-bearing and other aspects of pelvic function and form. From their multivariate combination, extant primates separate into groups that correlate well with the use of the hindlimb in locomotion.

(14) In combination of these features, the Australopithecinae have proved quite distinct from extant apes and man, lying isolated in multivariate space as opposed to being in an intermediate position. The conclusion has not been affected by variation in necessary reconstruction of the fossil bone.

(15) This analysis has provided a demonstration, probably more convincing than any predecessor, of the uniqueness of the Australopithecinae.

(16) Such results have been fully supported by related quantitative analyses of the talus, the humerus, and a part of the hand together with the proximal and distal ends of the femur.

ACKNOWLEDGEMENT

Through the years the work reviewed in the present paper has depended greatly upon the efforts of technical and clerical assistants. The preparation of the manuscript has been much facilitated by the support of Miss Cathy Sallis (University of Birmingham) and Mrs Barbara Whitehead (University of Leeds).

REFERENCES

Adams, L. M. & Moore, W. J. (1975). A biomechanical appraisal of some skeletal features associated with head balance and posture in the Hominoidea. *Acta anat.* 92: 580–594.

Andrews, D. F. (1972). Plots of high dimensional data. *Biometrics* 28: 125–136.

Andrews, D. F. (1973). Graphical techniques for high dimensional data. In *Discriminant analysis and applications*: 37–59. Cacoullos, T (Ed.). New York: Academic Press.

Andrews, D. F., Bickel, P. J., Hampel, F. R., Huber, P. J., Rogers, W. H. & Tukey, J. W. (1972). *Robust estimates of location: survey and advances*. Princeton: University Press.

Ashton, E. H. (1950). The endocranial capacities of the Australopithecinae. *Proc. zool. Soc. Lond.* 120: 715–721.

Ashton, E. H. (1957). Age changes in the basicranial axis of the Anthropoidea. *Proc. zool. Soc. Lond.* 129: 61–74.

Ashton, E. H. (1960). The influence of geographic isolation on the skull of the green monkey (*Cercopithecus aethiops sabaeus*). V. The degree and pattern of differentiation in the cranial dimensions of the St Kitts green monkey. *Proc. R. Soc.* (B) 151: 563–583.

Ashton, E. H. (1972). *Know thine anatomy*. Inaugural lecture published by

University of Birmingham: 1—48.

Ashton, E. H., Flinn, R. M., Griffiths, R. K. & Moore, W. J. (1979). The results of geographic isolation on the teeth and skull of the Green monkey (*Cercopithecus aethiops sabaeus*) in St Kitts — a multivariate retrospect. *J. Zool., Lond.* **188**: 533—555.

Ashton, E. H., Flinn, R. M. & Moore, W. J. (1975). The basicranial axis in certain fossil hominoids. *J. Zool., Lond.* **176**: 577—591.

Ashton, E. H., Flinn, R. M. & Moore, W. J. (1976). The articular surface of the temporal bone in certain fossil hominoids. *J. Zool., Lond.* **179**: 561—578.

Ashton, E. H., Flinn, R. M., Oxnard, C. E. & Spence, T. F. (1971). The functional and classificatory significance of combined metrical features of the primate shoulder girdle. *J. Zool., Lond.* **163**: 319—350.

Ashton, E. H., Flinn, R. M., Oxnard, C. E. & Spence, T. F. (1976). The adaptive and classificatory significance of certain quantitative features of the forelimb in primates. *J. Zool., Lond.* **179**: 515—556.

Ashton, E. H., Healy, M. J. R. & Lipton, S. (1957). The descriptive use of discriminant functions in physical anthropology. *Proc. R. Soc.* (B) **146**: 552—572.

Ashton, E. H., Healy, M. J. R., Oxnard, C. E. & Spence, T. F. (1965). The combination of locomotor features of the primate shoulder girdle by canonical analysis. *J. Zool., Lond.* **147**: 406—429.

Ashton, E. H., Moore, W. J. & Spence, T. F. (1976). Growth changes in endocranial capacity in the Cercopithecoidea and Hominoidea. *J. Zool., Lond.* **180**: 355—365.

Ashton, E. H. & Oxnard, C. E. (1958a). Some variations in the maxillary nerve of Primates. *Proc. zool. Soc. Lond.* **131**: 457—470.

Ashton, E. H. & Oxnard, C. E. (1958b). Variation in the maxillary nerve of certain mammals. *Proc. zool. Soc. Lond.* **131**: 607—625.

Ashton, E. H. & Oxnard, C. E. (1963). The musculature of the primate shoulder. *Trans. zool. Soc. Lond.* **29**: 553—650.

Ashton, E. H. & Oxnard, C. E. (1964a). Locomotor patterns in primates. *Proc. zool. Soc. Lond.* **142**: 1—28.

Ashton, E. H. & Oxnard, C. E. (1964b). Functional adaptations in the primate shoulder girdle. *Proc. zool. Soc. Lond.* **142**: 49—66.

Ashton, E. H., Oxnard, C. E. & Spence, T. F. (1965). Scapular shape and primate classification. *Proc. zool. Soc. Lond.* **145**: 125—142.

Ashton, E. H. & Spence, T. F. (1958). Age changes in the cranial capacity and foramen magnum of hominoids. *Proc. zool. Soc. Lond.* **130**: 169—181.

Ashton, E. H. & Zuckerman, S. (1950a). Some quantitative dental characteristics of the chimpanzee, gorilla and orang-utan. *Phil. Trans. R. Soc.* (B) **234**: 471—484.

Ashton, E. H. & Zuckerman, S. (1950b). Some quantitative dental characters of fossil anthropoids. *Phil. Trans. R. Soc.* (B) **234**: 485—520.

Ashton, E. H. & Zuckerman, S. (1950c). The influence of geographic isolation on the skull of the green monkey (*Cercopithecus aethiops sabaeus*) 1. A comparison between the teeth of the St Kitts and the African green monkey. *Proc. R. Soc.* (B) **137**: 212—238.

Ashton, E. H. & Zuckerman, S. (1951a). Some cranial indices of *Plesianthropus* and other primates. *Am. J. phys. Anthrop.* (n.s.) **9**: 283—296.

Ashton, E. H. & Zuckerman, S. (1951b). The influence of geographic isolation on the skull of the green monkey (*Cercopithecus aethiops sabaeus*). II. The cranial dimensions of the St Kitts and the African green monkey. *Proc. R.*

Soc. (B) 138: 204—213.

Ashton, E. H. & Zuckerman, S. (1951c). The influence of geographic isolation on the skull of the green monkey (Cercopithecus aethiops sabaeus). III. The developmental stability of the skulls and teeth of the St Kitts and African green monkey. Proc. R. Soc. (B) 138: 213—218.

Ashton, E. H. & Zuckerman, S. (1951d). The influence of geographic isolation on the skull of the green monkey (Cercopithecus aethiops sabaeus). IV. The degree and speed of dental differentiation in the St Kitts Green monkey. Proc. R. Soc. (B) 138: 354—374.

Ashton, E. H. & Zuckerman, S. (1952a). Age changes in the position of the occipital condyles in the chimpanzee and gorilla. Am. J. phys. Anthrop. (n.s.) 10: 277—288.

Ashton, E. H. & Zuckerman, S. (1952b). Overall dental dimensions of hominoids. Nature, Lond. 169: 571—572.

Ashton, E. H. & Zuckerman, S. (1954). The anatomy of the articular fossa (fossa mandibularis) in man and apes. Am. J. phys. Anthrop. (n.s.) 12: 29—61.

Ashton, E. H. & Zuckerman, S. (1956a). Age changes in the position of the foramen magnum in hominoids. Proc. zool. Soc. Lond. 126: 315—325.

Ashton, E. H. & Zuckerman, S. (1956b). Cranial crests in the Anthropoidea. Proc. zool. Soc. Lond. 126: 581—634.

Ashton, E. H. & Zuckerman, S. (1958). The infraorbital foramen in the Hominoidea. Proc. zool. Soc. Lond. 131: 471—485.

Blackith, R. E. & Reyment, R. A. (1971). Multivariate morphometrics. London & New York: Academic Press.

Bronowski, J. & Long, W. M. (1951). Statistical methods in anthropology. Nature, Lond. 168: 794.

Bronowski, J. & Long, W. M. (1952). Statistics of discrimination in anthropology. Am. J. phys. anthrop. (n.s.) 10: 385—394.

Bronowski, J. & Long, W. M. (1953). The australopithecine milk canines. Nature, Lond. 172: 251.

Broom, R. & Robinson, J. T. (1947). Further remains of the Sterkfontein ape-man, Plesianthropus. Nature, Lond. 160: 430—431.

Broom, R. & Robinson, J. T. (1952). Swartkrans ape-man: Paranthropus crassidens. Transv. Mus. Mem. 6: 1—123.

Broom, R., Robinson, J. T. & Schepers, G. W. H. (1950). Sterkfontein ape-man: Plesianthropus. Transv. Mus. Mem. 4: 1—117.

Broom, R. & Schepers, G. W. H. (1946). The South African fossil ape-men: the Australopithecinae. Transv. Mus. Mem. 2: 1—272.

Buxton, L. H. D. & Morant, G. M. (1933). The essential craniological technique. J. R. anthrop. Inst. 63: 19—47.

Clark, W. E. Le Gros (1940). Palaeontological evidence bearing on human evolution. Biol. Rev. 15: 202—230.

Clark, W. E. Le Gros (1947a). The importance of the fossil Australopithecinae in the study of human evolution. Sci. Progr., Lond. 139: 377—395.

Clark, W. E. Le Gros (1947b). Observations on the anatomy of the fossil Australopithecinae. J. Anat. 81: 300—333.

Clark, W. E. Le Gros (1949). New palaeontological evidence bearing on the evolution of the Hominoidea. Q. Jl geol. Soc. Lond. 105: 225—264.

Clark, W. E. Le Gros (1955). The os innominatum of the recent Ponginae with special reference to that of the Australopithecinae. Am. J. phys. Anthrop. (n.s.) 13: 19—27.

Clark, W. E. Le Gros (1962). *The antecedents of man.* (2nd edn). Edinburgh: University Press.

Clark, W. E. Le Gros (1967). *Man-apes or ape-men?* New York: Holt, Rinehart & Winston.

Colyer, F. (1948). Variations of the teeth of the green monkey in St Kitts. *Proc. R. Soc. Med.* 41: 845—848.

Corruccini, R. S. (1978). Morphometric analysis: uses and abuses. *Yearb. phys. Anthrop.* 21: 134—150.

Dart, R. A. (1925). *Australopithecus africanus*: the man-ape of South Africa. *Nature, Lond.* 115: 195—199.

Darwin, C. (1871). *The descent of man and selection in relation to sex* (2 vols). London: Murray.

Exner, H. E. (1978). Shape characterisation. *M.O.P. News* No. 6: 1—5.

Fisher, R. A. (1936a). The use of multiple measurements in taxonomic problems. *Ann. Eugen.* 7: 179—188.

Fisher, R. A. (1936b). "The coefficient of racial likeness" and the future of craniometry. *J. R. anthrop. Inst.* 66: 57—63.

Gregory, W. K. (1920). The origin and evolution of the human dentition — a palaeontological review. IV. The dentition of the higher Primates and their relationships with man. *J. dent. Res.* 2: 607—717.

Gregory, W. K. & Hellman, M. (1939). The dentition of the extinct South African man-ape *Australopithecus (Plesianthropus) transvaalensis* Broom. A comparative and phylogenetic study. *Ann. Transv. Mus.* 19: 339—373.

Grossman, J. W. & Zuckerman, S. (1955). An X-ray study of growth changes in the base of the skull. *Am. J. phys. Anthrop.* (n.s.) 13: 515—519.

Hotelling, H. O. (1936). Relations between two sets of variates. *Biometrika* 28: 321—377.

Howells, W. W. (1973). Cranial variation in man: a study by multivariate analysis of patterns of difference among recent human populations. *Pap. Peabody Mus.* 67: 1—259.

Huxley, T. H. (1863). *Evidence as to man's place in nature.* London: Williams & Norgate.

Jones, F. Wood (1919). On the zoological position and affinities of *Tarsius. Proc. zool. Soc. Lond.* 1919: 491—494.

Keith, A., Smith, G. Elliot, Woodward, A. Smith & Duckworth, W. L. H. (1925). The fossil anthropoid ape from Taungs. *Nature, Lond.* 115: 234—236.

Kern, H. M. Jr & Straus, W. L. Jr (1949). The femur of *Plesianthropus transvaalensis. Am. J. phys. anthrop.* (n.s.) 7: 53—78.

Leakey, R. E. F. (1973). Australopithecines and hominines: a summary on the evidence from the early Pleistocene of eastern Africa. *Symp. zool. Soc. Lond.* No. 33: 53—69.

Lisowski, F. P., Albrecht, G. H. & Oxnard, C. E. (1974). The form of the talus in some higher Primates: a multivariate study. *Am. J. phys. Anthrop.* (n.s.) 41: 191—215.

Mahalanobis, P. C. (1936). On the generalised distance in statistics. *Proc. natn. Inst. Sci. India* 2: 49—55.

McHenry, H. M. (1973). Early hominid humerus from East Rudolf, Kenya. *Science, N.Y.* 180: 739—741.

McHenry, H. M. & Corruccini, R. S. (1976). Fossil hominid femora and the evolution of walking. *Nature, Lond.* 259: 657—658.

Medawar, P. B. (1945). Size, shape and age. In *Essays on growth and form presented to D'Arcy Wentworth Thompson*: 157—187. Clark, W. E. Le

Gros & Medawar, P. B. (Eds). Oxford: Clarendon Press.
Medawar, P. B. (1950). Transformation of shape. *Proc. R. Soc.* (B) 137: 474–479.
Mivart, St. G. (1873). Man and apes. *Pop. Sci. Rev.* 12: 113–137 and 243–264.
Moore, W. J., Adams, L. M. & Lavelle, C. L. B. (1973). Head posture in the Hominoidea. *J. Zool., Lond.* 169: 409–416.
Morant, G. M. (1926). Studies of palaeolithic man. I. The Chancelade skull and its relation to the modern Eskimo skull. *Ann. Eugen.* 1: 257–276.
Morant, G. M. (1927). Studies of palaeolithic man. II. A biometric study of neanderthaloid skulls and of their relationships to modern racial types. *Ann. Eugen.* 2: 318–381.
Morant, G. M. (1928). Studies of palaeolithic man. III. The Rhodesian skull and its relations to neanderthaloid and modern types. *Ann. Eugen.* 3: 337–360.
Morant, G. M. (1930). Studies of palaeolithic man. IV. A biometric study of the upper palaeolithic skulls of Europe and of their relationship to earlier and later types. *Ann. Eugen.* 4: 109–214.
Morant, G. M. (1936). A biometric study of the human mandible. *Biometrika* 28: 84–122.
Napier, J. R. (1959). Fossil metacarpals from Swartkrans. *Fossil Mammals Afr.* 17: 1–18.
Napier, J. R. (1963). Brachiation and brachiators. *Symp. zool. Soc. Lond.* No. 10: 183–195.
Napier, J. R. & Napier, P. H. (1967). *A handbook of living primates.* London & New York: Academic Press.
Oxnard, C. E. (1963). Locomotor adaptations in the primate forelimb. *Symp. zool. Soc. Lond.* No. 10: 165–182.
Oxnard, C. E. (1967). The functional morphology of the primate shoulder as revealed by comparative anatomical, osteometric and discriminant function techniques. *Am. J. phys. Anthrop.* (n.s.) 26: 219–240.
Oxnard, C. E. (1968a). A note on the fragmentary Sterkfontein scapula. *Am. J. phys. Anthrop.* (n.s.) 28: 213–218.
Oxnard, C. E. (1968b). A note on the Olduvai clavicular fragment. *Am. J. phys. Anthrop.* (n.s.) 29: 429–432.
Oxnard, C. E. (1973). *Form and pattern in human evolution: some mathematical, physical and engineering approaches.* Chicago & London: University of Chicago Press.
Oxnard, C. E. (1974). Primate locomotor classifications for evaluating fossils: their inutility and an alternative. In *Symp. 5th congr. int. Primate Soc.*: 269–286. Kondo, S., Kawai, M., Ehara, A. & Kawamura, S. (Eds). Tokyo: Japan Science Press.
Oxnard, C. E. (1975). *Uniqueness and diversity in human evolution: morphometric studies of australopithecines.* Chicago & London: University of Chicago Press.
Oxnard, C. E. (1976). Primate quadrupedalism: some subtle structural correlates. *Yearb. phys. Anthrop.* 20: 538–554.
Pearson, K. (1926). On the coefficient of racial likeness. *Biometrika* 18: 105–117.
Pearson, K. & Bell, J. (1919). A study of the long bones of the English skeleton: the femur of the primates. *Drapers Co. Res. Mem. Biom. Ser.* 11: 1–7.
Rightmire, G. P. (1972). Multivariate analysis of an early hominid metacarpal from Swartkrans. *Science, N.Y.* 176: 159–161.

Ripley, S. (1967). The leaping of langurs: a problem in the study of locomotor adaptation. *Am. J. phys. Anthrop.* (n.s.) **26**: 149–170.

Siegel, A. F. & Benson, R. H. (1979). Resistant fitting as a basis for estimating allometric change in animal morphology. *Tech. Rep. Univ. Wisconsin* No. 552: 1–33.

Smith, G. Elliot (1924). *Essays on the evolution of man.* Oxford: University Press.

Sneath, P. H. A. (1967). Trend-surface analysis of transformation grids. *J. Zool., Lond.* **151**: 65–122.

Straus, W. L. Jr (1948). The humerus of *Paranthropus robustus. Am. J. phys. Anthrop.* (n.s.) **6**: 285–312.

Thompson, D'Arcy W. (1917). *On growth and form.* Cambridge: University Press.

Tobias, P. V. (1971). *The brain in hominid evolution.* New York & London: Columbia University Press.

Weidenreich, F. (1937). The dentition of *Sinanthropus pekinensis*: a comparative odontography of the hominids. *Palaeont. sin.* **101**: 1–180.

Weidenreich, F. (1943). The skull of *Sinanthropus pekinensis*: a comparative study on a primitive hominid skull. *Palaeont. sin.* **127**: 1–485.

Weiner, J. S. (1955). *The Piltdown forgery.* London & New York: Oxford University Press.

Weiner, J. S., Oakley, K. P. & Clark, W. E. Le Gros (1953). The solution of the Piltdown problem. *Bull. Br. Mus. nat. Hist.* (Geol.) **2**: 139–146.

Yates, F. & Healy, M. J. R. (1951). Statistical methods in anthropology. *Nature, Lond.* **168**: 1116–1117.

Zuckerman, S. (1926). Growth changes in the skull of the baboon, *Papio porcarius. Proc. zool. Soc. Lond.* **1926**: 843–873.

Zuckerman, S. (1928). Age changes in the chimpanzee, with special reference to growth of brain, eruption of teeth, and estimation of age; with a note on the Taungs ape. *Proc. zool. Soc. Lond.* **1928**: 1–42.

Zuckerman, S. (1933). *Sinanthropus* and other fossil men: their relations to each other and to modern types. *Eugen. Rev.* **24**: 273–284.

Zuckerman, S. (1950). Taxonomy and human evolution. *Biol. Rev.* **25**: 435–485.

Zuckerman, S. (1951a). An ape or the ape. *J. R. anthrop. Inst.* **81**: 57–68.

Zuckerman, S. (1951b). Some features of the anthropoid skull of apes. *Proc. Linn. Soc. Lond.* **164**: 9–10.

Zuckerman, S. (1954). Correlation of change in the evolution of higher primates. In *Evolution as a process*: 300–352. Huxley, J. S., Hardy, A. C. & Ford, E. B. (Eds). London: Allen & Unwin.

Zuckerman, S. (1955). Age changes in the basicranial axis of the human skull. *Am. J. phys. Anthrop.* (n.s.). **13**: 521–539.

Zuckerman, S. (1966). Myths and methods in anatomy. *J. R. Coll. Surg. Edinb.* **11**: 87–114.

Zuckerman, S., Ashton, E. H., Flinn, R. M., Oxnard, C. E. & Spence, T. F. (1973). Some locomotor features of the pelvic girdle in primates. *Symp. zool. Soc. Lond.* No. 33: 71–165.

Zuckerman, S., Ashton, E. H., Oxnard, C. E. & Spence, T. F. (1967). The functional significance of certain features of the innominate bone in living and fossil primates. *J. Anat.* **101**: 608.

Symp. zool. Soc. Lond. (1981) No. 46, 127—167

Beyond Biometrics: Studies of Complex Biological Patterns

CHARLES E. OXNARD and HAROLD C. L. YANG

Departments of Biology and Anatomy, The Graduate School, University of Southern California, Los Angeles, California, USA, and Department of Anatomy, Division of Biological Sciences. Pritzker School of Medicine, University of Chicago, Illinois, USA

SYNOPSIS

Consideration of the use of multivariate statistical methods for the study of the external form of bones has led to many other attempts to quantify form and pattern in bones. In particular the complex patterns that exist within the cancellous portions of bones are not easily studied using more conventional biometrics and attempts have therefore been made to investigate them using Fourier transformations. Such transforms, made in two-dimensional form, and using optical rather than computational equipment, are easily able to supply succinct descriptions of cancellous networks as revealed in sections, laminographs and radiographs of bone. Some vertebrae, e.g. the fourth lumbar vertebra in man, show generally orthogonal arrangements of trabeculae that are conventionally associated with the orthogonal network of stresses that exist in such vertebrae during function. But the optical analyses are sensitive enough to demonstrate that in some other vertebrae, e.g. the second lumbar vertebra in man and many of the lower vertebrae in the gorilla, chimpanzee and bonobo, off-orthogonal elements are superimposed upon a basically orthogonal arrangement. Such a pattern is still consistent with an association with the orthogonal stress pattern if it is assumed that the off-orthogonal variations exist for purely local stress-bearing reasons.

But the finding of architectural patterns in the orang-utan that are not related in any way to an orthogonal network suggests that a new relationship with stress bearing must be sought. It must be one that does not depend upon any particular orthogonal stress pattern, whether stemming from any particular function or from the resultant pattern of all functions, but rather one that renders the architecture capable of bearing each of the different orthogonal stress patterns in turn. Such an idea is entirely compatible with the findings in the other hominoids but has been hidden because in the other hominoids the basic posture is such that many of the individual stress patterns during function are likely to be very close to the resultant orthogonal stress pattern.

In the orang-utan, in contrast, the many very different activities of which the vertebral column is capable (suspension or support by one, many or all limbs, in vertical, horizontal or inclined positions) must result in a markedly different series of orthogonal stress patterns. Though such a series of orthogonal networks will have a resultant orthogonal pattern that can be calculated, it is to the temporal succession of stress patterns that the architecture must be related.

In each case the findings using the optical Fourier transforms, not at all apparent from visual observation of the sectional and radiographic materials, have been corroborated by sample inspections of actual trabecular networks in dissected specimens.

The new theory thus elaborates on the relationship between architecture and stress bearing in cancellous bone. It has implications, therefore, for the relationship between stress-bearing and architecture in cortical bone and, indeed, in the entire external form of bones. It provides further information for studies of the mechanism of adaptation of bone form both from the viewpoint of the changes that can be wrought in bone by experiment and from the point of view of the electromechanical basis of such changes. Finally it renders more sound the basis upon which it is possible to assess stress from architecture in specimens whose stress patterns are unknown, i.e. fossils; and it demonstrates that the assessment of function from such fossil specimens cannot go beyond a determination of the basic nature of the stress-bearing patterns (however these may have been produced during the life processes of the individual).

THE BIOMETRIC BACKGROUND TO THE PRESENT STUDIES

Biometrics has been mostly concerned with the characterization of biological shapes through measurements and their statistical handling. In earlier investigations the constraints of physical measuring apparatus such as calipers, and of computational equipment such as calculating machines and earlier generations of computers, severely limited the scope of the studies that could be undertaken. More recently, however, the development of semi- and fully-automated techniques of obtaining measurements using tools like photogrammetry, and the evolution of the new generations of computers providing enormous statistical capability, make biometry most powerful, indeed, for the study of shape. These possibilities were recognized by Lord Zuckerman many years ago and it is to the initiation, continuing stimulus and interest provided by him that our studies owe so very much.

These relatively new abilities for handling the problems inherent in characterizing and comparing biological shapes, especially, for instance, for dealing with the external shapes of bones as described in Professor Ashton's contribution (this volume, p. 67), invite us to consider whether the even greater complexity of some biological patterns, for example the complex internal arrangements that are presented by the cancellous architectures of those same bones, may yield to quantitative handling. Recently, methods from many different fields have been applied to the complex problems presented by such patterns (for review see Oxnard, 1978).

Many of these methods involve attempting to isolate individual features in shapes and patterns such as edges, corners, nodes, tangents, vectors, inflection points, central axes, curve-fitted images, skeletal transforms and so on. The use of tangents and inflection points is described by Attneave & Arnoult (1966) for characterizing nonsense patterns for use in psychological tests; their possible application for describing the shape of a bone is illustrated by Oxnard (1973a). Likewise, central axis functions are presented by Blum (1962, 1967) and Philbrick (1968) using simplified biological shapes as some of their examples; a demonstration of how they might work if applied to the form of bones is supplied by Oxnard (1973b); their possibilities are emphasized by Waddington (1977). Curve-fitted images have been applied, if not to a real biological object, at least to a representation of one (a "Barbie" doll, Agin & Binford, 1976). Many of these methods are based upon measurements between individual points upon objects, however those points may have been defined.

Other techniques have been devised to attempt to assess directly the form or pattern of a picture or an object. Some of these stem from studying the contours of objects and these can be obtained in a variety of ways, many of which involve applications of moiré fringe analysis. Takasaki (1970) has shown that contour moiré fringe pictures of full sized living bodies can be readily obtained using incoherent light, cameras and grating shadows. His results are an enormous improvement over the rather crude characterizations obtained for the pelves of living and fossil apes and men in a demonstration by Oxnard (1973a). It is also possible to produce such contours using coherent light (Greguss, 1976). These methods have mainly been applied to those anatomical structures whose complicated three-dimensional surfaces have proven especially difficult to quantify using regular biometric methods, e.g. the surfaces of teeth (Chmielewski & Varner, 1969; Zelenka & Varner, 1968; Eick, Johnson & McGivern, 1971; Elliott & Morris, 1978). Methods such as these produce field views of overall shape that do not depend upon defining precise points upon an object although they sometimes utilise reference orientations.

Yet other techniques assess shapes by comparing the difference between two field views of shapes. One of the most well-known of these is the method of transformations as described long ago by D'Arcy Thompson (1917) and recently developed in an automated computer-visual mode by Appleby & Jones (1976) and Gabrielson (1977). Other ways of comparing two forms or patterns using related ideas include trend surface analysis borrowed from geology and applied to hominid skulls by Sneath (1967) and bio-orthogonal grids also developed in other disciplines (e.g. geography) and applied to biological shapes, also primate crania, by Bookstein (1978). Though these methods as currently explicated are confined to comparisons of pairs of objects and two-dimensional views only, it is not difficult to see how they might be further evolved to study populations of objects and in three dimensions. Such evolution, though not requiring new theoretical knowledge, is not by any means an easy task.

Finally, however, the information inherent in a shape or pattern can be studied by a whole range of methods that involve Fourier and related transformations. Most of such studies have been aimed at characterizing external shape by treating the curve of the shape as a one-dimensional wave to be transformed into its simple harmonic functions. Usually this is done computationally using fast Fourier transforms. Examples here include studies of the outline of the face by Lu (1965) followed in subsequent years, using similar methods for describing the outlines of the shapes of bones such as the skull and the ends of long bones, by studies by Lestrel (e.g. Lestrel, 1976; Lestrel & Brown, 1976; Lestrel, Kimbel, Prior & Fleischmann, 1977). Although the forms analysed here are two-dimensional (the projected outline of some object) the Fourier tranforms are one-dimensional because the outline is treated as a single wave. Similar methods have been used in other areas of biology to characterize the shape of zooecial chambers in fossil bryozoans (Anstey & Delmet, 1972), of ostracode margins (Kaesler & Waters, 1972) and of blastoid form (Waters, 1977).

Of course, it is also possible to perform such analyses in two-dimensional form for the study of two-dimensional patterns. Although this can be done computationally after the pattern has been rendered digitally using a method such as a television scanning system, it can also be done optically using the two-dimensional transforming properties of a simple lens system. Most of the examples of this usage are in sciences other than biology; for instance, Cutrona (1965) provides an early description showing how the method can be generally used for the detection of patterned arrangements within apparently randomly arranged designs. Several authors have shown how the method can be used for the detection of patterns within seismographic data (e.g. Dobrin, Ingalls & Long, 1965; Jackson, 1965). Many geologists have also used such methods for analysing a variety of patterns such as those seen in thin sections of rocks (Pincus, 1969; Dobrin, 1968; Davis & Preston, 1972; Davis, 1973) and in aerial photographs and maps of the Earth's surface (Pincus, Power & Woodzick, 1973; Srivastava, 1977). Holeman (1968) illustrates the use of two-dimensional Fourier analyses for the automatic reading of printed characters and words. A variety of related transformations (Walsh transforms, Haar transforms) are used to analyse satellite photographs in improving picture quality (Andrews & Pratt, 1969; Kennedy,

1971; Rao, Narasimhan & Revuluri, 1975). Rather more recently these methods have been taken up by biological microscopists to enhance images and to analyse patterns hidden within complex visual representations (e.g. Taylor & Ranniko, 1974; Hillenkamp, Hutzler & Kinder, 1976; Kopp, Lisa, Mendelsohn, Pernick, Stone & Wohlers, 1976).

Although, therefore, a wide variety of methods are available for quantifying the information contained within the outer envelope of a bone, it is to Fourier transforms as such descriptors that we may turn for characterizing the more complex internal structure of bone. An early investigation (Oxnard, 1973a, b) uses laser diffraction to study macroscopic trabecular patterns in cancellous bone. A more recent work (Bacon, Bacon & Griffiths, 1979) applies neutron diffraction to study microscopic alignments of hydroxy appatite crystals in bone cortex.

THE PROBLEM AND THE METHOD

The problem is represented by the difficulty of handling, quantifying and analysing patterns like that presented by Fig. 1. In this transverse

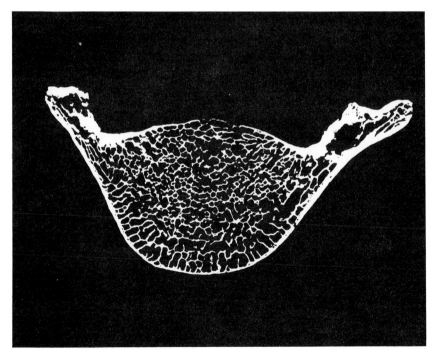

FIG. 1. Face of a transverse cut through the fifth lumbar vertebra in man.

section of a human lumbar vertebra, knowledge of the precise ways in which the various cancellous plates are arranged is of considerable biological importance given the idea, long since proven (reviewed in Murray, 1936; Evans, 1957), that there is an association between such cancellous architectures and the stresses that they bear during function. With the more usual methodology of biometrics, characterizing such a complex structure could be achieved by measuring the lengths, widths, and angles of orientation of each individual trabecular plate in the section and drawing some overall summary from the resulting large data set by techniques such as those of multivariate statistical analysis. Using manual methods such as rulers or calipers for measurement makes such a study time-consuming in the extreme; even with image-analytical aids such as the Quantimet, obtaining so many data is no light task. And however the data are obtained, the problem of analysis of those data, even if only for summary purposes, becomes difficult, especially when it is remembered that analysis of many such sections would be necessary to characterize even only a single vertebra, examination of many such bones to describe a single vertebral column, study of several columns to define the parameters of a single population of animals, and investigation of many such populations in order to tackle problems of functional and evolutionary relationships. But the problem yields rather more easily to methods such as optical data analysis that can view a structure like that in Fig. 1 as a field problem.

For instance, a single transect across the picture in Fig. 1 can be represented as a somewhat complicated square wave; upland plateaus are equivalent to the bright areas (trabecular plates), lowland planes model the dark areas (intertrabecular spaces). Once we visualize the analogy of waves, we are reminded that Fourier functions may be an appropriate way to analyse them. And though Fourier analysis usually means fast Fourier transforms performed by a computer, it can also involve the optical analogy shown in Fig. 2. Here a single, complicated wave is split into its simple components; in optical terms, a pattern of a complex white light is divided into its individual, simpler colour components; in data-analytical terms, complex data are separated into their individual, simpler elements. Of course, in order to characterize a two-dimensional picture like that in Fig. 1 we would still have to analyse an infinite number of such transects across the object (or at the very least, a reasonably large number of them) if sensitivity were not to be lost.

However, those who are involved in microscopy will immediately recall that the one-dimensional Fourier transformation that is produced by a prism has a two-dimensional counterpart when a lens is

THE NATURE OF THE POWER SPECTRUM

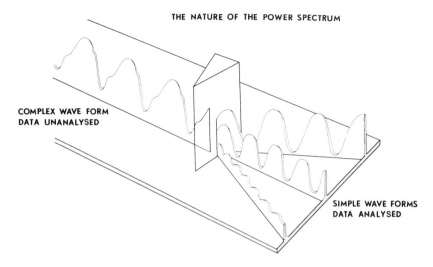

COMPLEX WAVE FORM
DATA UNANALYSED

SIMPLE WAVE FORMS
DATA ANALYSED

FIG. 2. The optical analogue of one-dimensional Fourier analysis.

used that is equivalent to transforming simultaneously all possible transects across Fig. 1. The effect of the first lens system in a microscope is to produce exactly such an analysis though it is only in very recent years that biologists have started to use this property of a lens for the analysis of microscopic patterns. In order to perform such two-dimensional Fourier analyses optically for macroscopic patterns such as those inherent in a photograph of a section or compressed into the shadows of a laminogram or a radiograph, coherent light must be used together with a simple system of lenses.

A pattern such as that in Fig. 1 can be studied by passing through it the beam from a small laser, suitably modified using a collimator. After further passage of the light through a lens, a Fourier transform results which can itself be photographed for analysis and study (Fig. 3). If the transform is not interrupted a second lens will, of course, reproduce an image of the original, a third lens a repeat of the transform, and so on. The existence of multiple images and transforms in such a system means that the transforms can be filtered at any particular stage, and the effect of filtering upon the subsequent reconstruction of the image can be examined. In addition, therefore, to analysing the information present in the original pattern, it is also possible to "dissect" that pattern to help reveal more clearly the different elements of which it is composed (Oxnard, 1973b, 1977).

In order to have some intuitive notion of how the images and the transforms are related to one another, let us look at some simple patterns (Pincus, 1969). Figure 4 demonstrates the Fourier transform

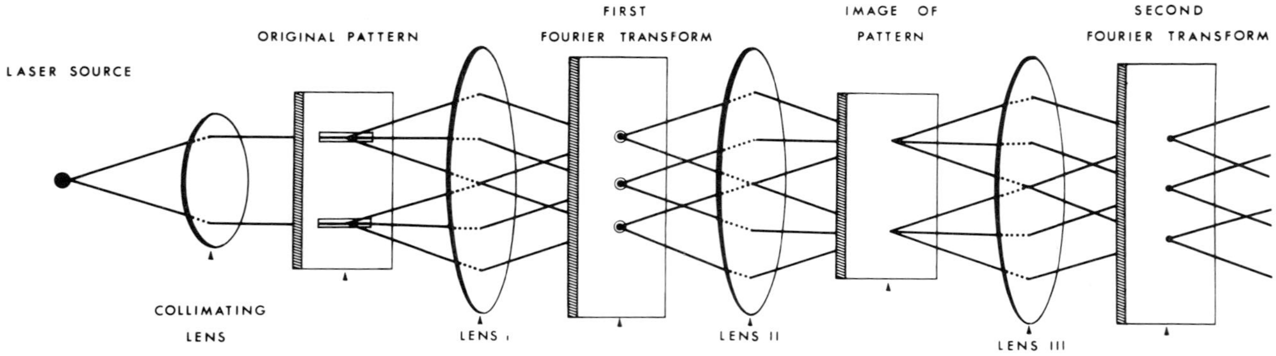

FIG. 3. Optical two-dimensional Fourier analysis.

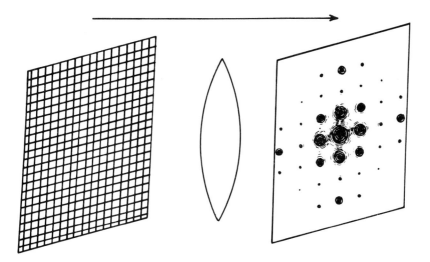

FIG. 4. A grid and its optical Fourier transform (for diagrammatic purposes the intensities of light in the transform are shown as dark spots).

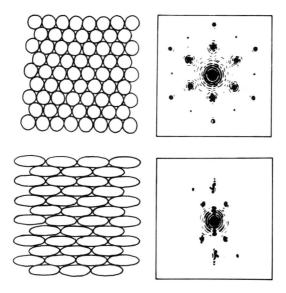

FIG. 5. The optical transforms of a set of touching circles and a set of touching ellipses. The effect of squashing the circles to ellipses in the vertical direction produces a squashing of the transform in the horizontal direction.

of a square grid (in the diagram the bright spots of the transform
have been rendered dark). The sizes and angles of the different
elements of the grid can be easily obtained from the transform. Figure 5 demonstrates the Fourier transforms of a pattern of circles and
a pattern of ellipses obtained by distorting the circles. This demonstrates further that sizes (transformed inversely) and orientations
(transformed by 90°) can be obtained from the transforms. These are
precisely the data that we might require about a structure as complex
as the internal architecture of bone.

Of course, if we really wished to know about such simple architectural features as in Figs 4 and 5 we would never actually use Fourier
analysis; we would make the measurements directly upon the original.
But when our interest is directed to more complex patterns it turns
out that the necessary summarizing parameters may be obtained
enormously more easily through measurement of the transformation
rather than directly from the pattern.

Thus the Fourier transform associated with a complex pattern like
that in Fig. 1 is not a discrete pattern of spots as in our simple
example. The transform rather appears as a bright cloud (or flux) of
light at the transform plane as in Fig. 6. However, even on visual
inspection it is clear that the cloud of light has a complex internal
structure of its own. And it is that internal structure that provides
the information about the pattern. One way of obtaining the necessary information is to contour the brightness or intensity of the flux
across the transform plane using a densitometer. For a pattern like
that in Fig. 1 and shown in the transform of Fig. 6 the contoured
flux of light may be thought of as shown in Figs 7 and 8. Determination of the "volume" of light contained within a band-shaped
element of the spectrum provides information about particular sizes
within the pattern; determination of the "volume" of light contained
within some pie-shaped sector of the spectrum provides information
about particular orientations within the pattern.

What can be achieved by such a method is best understood by
examining two particular patterns. Figure 9 shows sagittal sections
through the second and fourth lumbar vertebrae in man. Although it
is readily apparent that there are differences between these two
sections (the trabecular elements in the second lumbar vertebra are,
for instance, rather larger than those in the fourth), our general
assessment of these patterns is that they are similar overall. Each
comprises a large number of elements with the main orientation
craniocaudal, and a secondary orientation dorsoventral. Less material
is oriented at angles intermediate between these two orthogonal
directions.

FIG. 6. The transform of Fig. 1: the cut surface of the fifth lumbar vertebra.

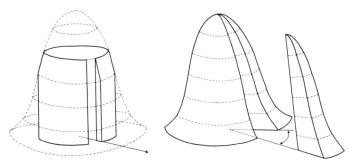

FIGS 7 and 8. (7) Left: a diagrammatic representation of the flux of light in Fig. 6 with the intensities marked as contours. The total volume of the flux is the volume of the cone; the volume erected over the band on the base is a transformed measure relating to information about particles in the original between the sizes associated with the radii of the band.

(8) Right: a diagrammatic representation of the flux of light in Fig. 6. In this case the volume of the pie-shaped piece is a transformed measure relating to information about particles in the original between the angles associated with the radii of the sector.

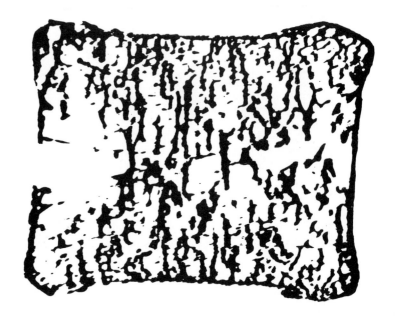

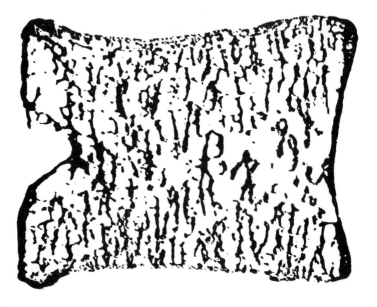

FIG. 9. Surfaces of sagittally cut human lumbar vertebrae: (A) second; (B) fourth.

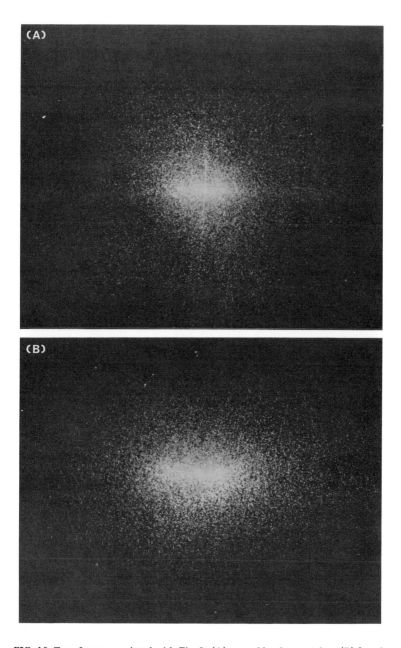

FIG. 10. Transforms associated with Fig. 9: (A) second lumbar vertebra; (B) fourth.

The Fourier transforms of these two sections are shown in Fig. 10 and it is immediately obvious that they are different. One difference is that the flux of light extends further away from the central reference axis in the spectrum of the fourth vertebra as compared with that for the second. This means that the elements in the fourth lumbar vertebra are smaller than those in the second; this information, however, is readily available to us on visual inspection of the original sections. The quantitative measure of this difference can now be determined by a small number of measurements on the transform as compared with a large number of measurements on the original; already, therefore, a useful summarization may be obtained from the transform.

There is a second difference. It is that the internal structures of the two transforms are different. The transform of the fourth lumbar vertebra is an oval or elliptical form that has its short axis horizontal and its long axis vertical. This implies that the main orientations in the original picture are of elements with their long axes vertical and their short axes horizontal. We have learnt nothing new here: this was apparent from visual inspection of the original. But the transform of the second lumbar vertebra, though generally a horizontal oval like the fourth and therefore generally orthogonally arranged like the fourth, has a different internal structure. The eye can readily appreciate it in the transform, but it requires the contoured power spectrum to provide the detailed information.

Figure 11 shows the contoured flux of light from each of these transforms. It confirms the simple vertical–horizontal arrangement already noted about the transform of the fourth lumbar vertebra.

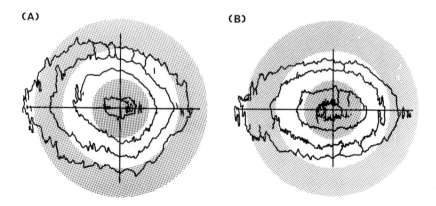

FIG. 11. Contoured power spectra (Fourier transforms) associated with (A) the second lumbar vertebra, and (B) the fourth.

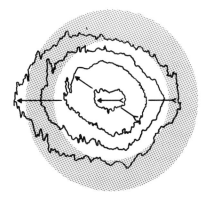

FIG. 12. Contoured power spectrum of the second lumbar vertebra with arrows emphasizing the orthogonally directed inner and outer contours and the special distortion of an intermediate contour at approximately 30°.

But it demonstrates that additional off-orthogonal information is available for the second lumbar vertebra that is lacking for the fourth. This is reflected in a different orientation of some of the intermediate contours and attention is especially drawn to this feature using arrows in Fig. 12. The largest and smallest contours, providing information about the smallest and largest elements in the section, are oriented in a way consistent with the craniocaudal dorsoventral arrangement. But an intermediate contour is oriented at a different angle; and this implies the existence of elements in the second lumbar vertebra, not present in the fourth and lying at about 30° to the vertical.

FURTHER TESTS

Those many scientists interested in stereology (the prediction of three-dimensional structures from information about two-dimensional sections through them) will not be at all surprised at seeing oblique elements arise from a section of a three-dimensional orthogonal network. All that is required is that the section be not exactly perpendicular to the framework of pattern. Such sections often suggest spurious architectural features resulting from artefacts due to the angle of the plane of sectioning (reviewed in Underwood, 1970). In an attempt, therefore, to check if the specific plane of the section is responsible for the result just presented, and bearing in mind that with museum materials sections normally cannot be made, we may compare the patterns that are obtained in passing from vertebral

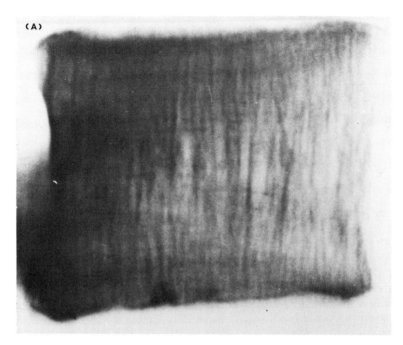

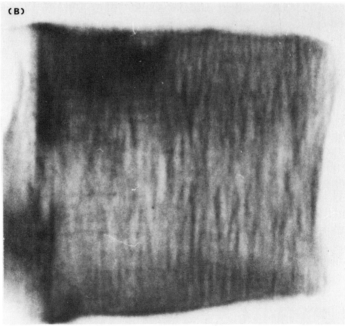

FIG. 13. Tomograms (laminographs) of sagittal views of the same second (A) and fourth (B) human lumbar vertebrae.

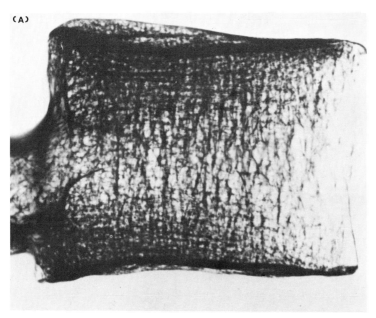

(A)

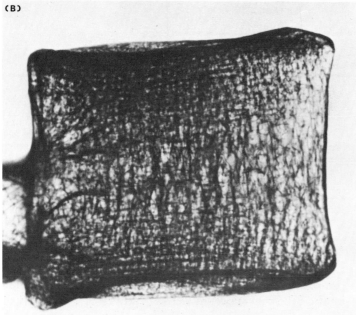

(B)

FIG. 14. Radiographs of these same human lumbar vertebrae: (A) second and (B) fourth.

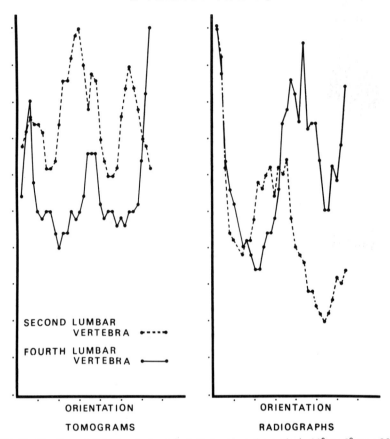

SECOND LUMBAR
        VERTEBRA •----•

FOURTH LUMBAR
        VERTEBRA •——•

ORIENTATION   ORIENTATION

TOMOGRAMS   RADIOGRAPHS

FIG. 15. The flux of light (vertical axis) plotted against the angle (−90° to 0° to +90° in 30° steps) at which the flux is measured (horizontal axis) for the tomograms and radiographs in Figs 13 and 14. The shapes of the curves for tomograms differ completely from those for radiographs because the information in the two is pictorially different. But both sets show peaks that lie at the same angles: zero and 90°; and for each pair, the curve for the second lumbar vertebra shows additional high points on the curves about 30° out of phase with the peaks for the respective fourth lumbar vertebra.

sections, in theory, at any rate, of infinitesimal thickness, through laminograms which are radiographic shadows of some small finite thickness, usually a few millimetres, to radiographs which are shadows of the full thickness of the vertebra. Both of these latter can, of course, be made without destroying the specimens and are therefore useful in the study of museum representatives of specimens of rare living species, or priceless fossils. Figure 13 shows laminograms for these same two vertebrae; Fig. 14 shows the respective radiographs.

Visual inspection of Figs 13 and 14 instantly confirms the idea that results from visual inspection of the sections alone. Vertically

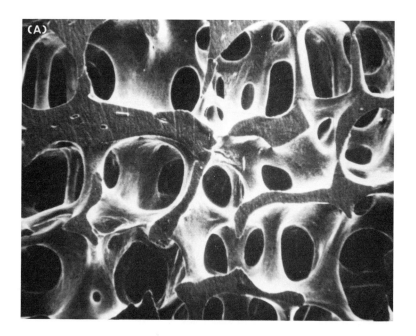

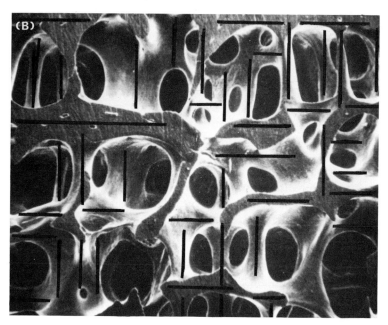

FIG. 16. A view of a small portion of the cancellous network of the second lumbar vertebra viewed from the median plane. The lower section indicates diagrammatically those primary elements of this picture that are dorsoventral and craniocaudal in orientation — a major portion of the structure.

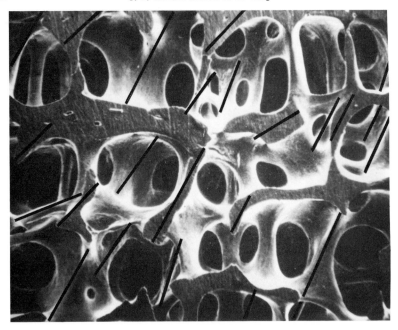

FIG. 17. The same view as Fig. 16 for the second lumbar vertebra. In this case those elements oriented at a specific angle in one direction have been emphasized. Although, of course, some materials are oriented at all possible directions, those oriented in this special way seem to be considerably more numerous.

placed (craniocaudal) elements predominate not only in the sections, but also in the laminograms and the radiographs. But plots of the flux of light (vertical axis) against the angle of orientation (horizontal axis) for the Fourier transforms of the laminogram and the radiograph, show that, in addition to the peaks corresponding to the preferred zero and 90° directions in the original pictures, there are extra peaks for the second lumbar vertebra that are aligned at 30°. Though the individual forms of these curves differ totally (the laminogram is a different picture from the radiograph) the extra 30° peak is a constant element in the second lumbar vertebra in each (Fig. 15). The 30° elements are real.

And study of a small part of the trabecular network using scanning electron microscopy demonstrates where they lie. The main elements in the trabecular networks are, indeed, vertical and horizontal (Fig. 16) as indicated by the heavy lines. But there are a significant number of elements lying at approximately 30°. These too are emphasized with heavy lines in Fig. 17.

FUNCTIONAL IMPLICATIONS OF THESE FINDINGS

These findings have implications for theories of stress-bearing by bone. Ever since the early part of the nineteenth century there has been a great deal of interest in the mechanical significance of bone architecture. At first, most of the attention was centred on cancellous bone which was interpreted in terms of the "trajectorial theory of bone architecture". This associated the alignment of the individual trabecular elements of cancellous bone with the pattern of principal stress trajectories thought to represent the stress situation in the bone during function. The cancellous elements were believed to be aligned in the same way as the stresses that they resisted (Meyer, 1867; Wolff, 1870, 1892). The theory has been criticized upon a number of different grounds since the end of the last century.

Some of the criticisms relate to a better understanding of cancellous architecture. Thus, though the alignment of trabeculae seems to be an orthogonal network like that of the principal stresses in some regions of bones, in other regions it seems to be somewhat different from such a pattern. A closer view of small parts of the cancellous network seems especially to deny the theory for it is well known that the local details of cancellous bones involve trabecular intersections that are anything but the right-angled corners that would be demanded by strict adherence to the trajectorial theory (e.g. Whitehouse & Dyson, 1974).

Other criticisms relate to a better understanding of mechanisms of stress-bearing. For though one particular orthogonal stress network stands revealed if assumptions are made that cancellous bone bears stress like a beam acting under infinite beam theory, a quite different orthogonal network results from treating the bone as though it were acting under finite element theory (Rybicki, Simonen & Weiss, 1972). And though the stress network within a solid, homogeneous structure provides orthogonal stress networks like those posited by most workers (and it is this network to which the cancellous framework is supposed to be a stress resisting model), the stress network within a real solid containing many holes is actually completely different (Fig. 18).

There are many other objections and it is thus almost certain that the trajectorial theory of bone architecture in its original form is incorrect. Yet it remains most attractive. For instance, relatively recent studies have shown how, when corrections are made for special features of loading, the form of epiphyseal plates seems closely related to the network of principal stresses (Smith, 1962). And the modifications of cancellous structures that have long been

FIG. 18. A comparison between the degrees of stress in a solid beam (A) a beam with a single large hole (B) and a beam with many small holes (C). The effect of a single hole is to change the stresses greatly in regions near that hole but to lesser degrees away from the hole; the effect of many small holes is to produce a totally new stress pattern.

known to occur in natural or artificial experiments that modify loading upon bone imply most powerfully that loading is, indeed, closely
related to architecture (e.g. Murray, 1936; Evans, 1957). Modern
discoveries of mechanoelectric phenomena in bone and other biological materials begin to suggest something of the possible causal
mechanisms of the adaptation of bone (e.g. Bassett, 1971).

It is therefore rather likely that the trajectorial theory of bone
architecture as originally expressed by Culmann and Meyer (Meyer,
1867) and later modified by Wolff (1870, 1892; for excellent reviews
see Murray, 1936 and Evans, 1957) requires modification. Some
other theory linking stress-bearing and architecture must be closer to
reality. Yet it is at least possible that there is some general truth in a
notion that links the orthogonal pattern of stress trajectories that
result from function with the architecture of the cancellous elements
bearing those stresses; it may be that it is only the details of the
association that require modification. Certainly the generality of the
trajectorial theory seems so interesting that it continues to draw the
attention of investigators (e.g. Pauwels, 1968; Kummer, 1959; Smith,
1962; Oxnard, 1971) and to be included in our general teaching of
bone biology (Young, 1957; Alexander, 1968; Wainright, Biggs,
Currey & Gosline, 1976).

The new information provided here about the cancellous network
in a human vertebral body bears upon this problem. Studies have
long suggested that in the body of the vertebrae the cancellous networks consist chiefly of larger craniocaudal bars bearing the main
stresses associated with an upright vertebral column under compression, together with smaller dorsoventral bars (as seen in lateral
view) crossing this system approximately orthogonally. This architecture seems clearly associated with a similar set of orthogonal stresses
existing in the body of the vertebra during function and demonstrated
by a number of workers but especially Pauwels (1968, 1965) and
Kummer (1959). That the trabeculae individually do not meet at
sharp corners but at rounded junctions does not necessarily negate
the general theory; rather it suggests that specific local stresses are
merely more complex and associated with more complex specific
local architecture. The general theory thus need not be abandoned if
it is supposed that the association between stress and architecture is
one between an averaged orthogonal architecture that is associated
with an averaged orthogonal network of stresses. This certainly seems
quite compatible with the view of vertebral architecture seen in
laminographs and radiographs which by definition are an average
view of the shadows of many individual trabeculae in inter-trabecular
spaces. And it would also seem compatible with the notion of some

pattern of orthogonal principal stresses that was the average of the many different orthogonal networks that can be calculated from the many different functions of which the vertebral column is capable.

Such a more general notion is supported by the findings here for the fourth lumbar vertebra. Visually the architectural pattern seems an orthogonal network; optical data analysis confirms that, though there is much material laid down at every different angle within the bone, the preferential arrangement is, indeed, orthogonal.

But what are we to make of the findings here for the second lumbar vertebra? There may indeed be a general orthogonal architectural network in that vertebra; but there is also the subsidiary pattern of intermediately sized elements at $30°$. One possibility is that these $30°$ elements are the special local elements that do not have a major impact upon the general orthogonal pattern that exists. This bends the data to fit the modified theory.

But another is that the $30°$ elements are associated with some complexity in the nature of the association between stress and architecture that we have so far not considered. For instance, their existence may be related to the possibility that, though on average the fourth lumbar vertebra bears mostly simple compressive forces (it lies almost on the line of gravity), the second lumbar vertebra mainly bears stresses associated with bending because it lies away from the line of gravity owing to the normal curvature of the lumbar column in man. This possibility is supported by the finding of similar off-orthogonal elements at slightly different angles in several of the lower vertebrae.

However, more tangible support for a further modification of these ideas would best come from a study of cancellous architecture in a bone which had to bear, during function, a very wide variety of different stresses, the existence of which, even if they could not be measured, could be in no doubt. This chance presented itself as our examination of vertebral architecture was extended into the various great apes.

STUDIES OF VERTEBRAL TRABECULAE IN OTHER HOMINOIDS

Evidence speaking to these problems has resulted from more extensive investigations of human and ape lumbar columns that have been carried out by the present authors, C. E. Oxnard and Harold C. Yang, originally a doctoral student at the University of Chicago and now, during his collaboration, a medical student in the Pritzker School of Medicine. These new studies have not taken for granted anything suggested by the prior investigation; indeed, in these recent studies an extensive suite of tests has been carried out in order to make quite certain that the results are free of technical artefacts. The process utilizes initial radiography with a Hewlett Packard Faxitron X-Ray System. Optical Fourier transforms are produced using a Rank Precision Industries Image Analyser 3000. Quantification of the transforms is carried out with an Optronics International Densito-meter.

At each stage, tests were conducted not only to discover the degree to which the parameters of the particular instruments might affect the results, but also to monitor the effects of the various optical transfers required by the different technical specifications of the equipment. Tests were conducted to determine the degree to which the methods are sensitive enough to detect differences between different vertebrae within a single column, between homologous vertebrae of the same species with especial note of the differences between sexes, and among vertebrae from the different species that form the primary problem in the study. These various tests are out-lined in Yang (1978) and Yang & Oxnard (in preparation a, b, c), and they demonstrate that the instruments and methods have the sensitivity to recognize the large differences that exist between speci-mens and individuals as compared with the very small differences that are found in replicate studies of the same specimens.

Lateral radiographs of vertebral bodies are examined at each position along the lumbar vertebral column for gorilla, man, orang-

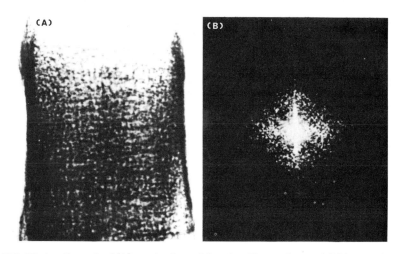

FIG. 19. A radiograph of (A) a sagittally positioned gorilla vertebra, and (B) its transform.

utan, chimpanzee and bonobo. Figure 19 shows a typical radiograph and its resulting Fourier transform. Figure 20 provides a picture of an intermediate stage in the analysis of the transform in which its density is rendered quantitatively and Fig. 21 presents a computer-drawn contoured power spectrum obtained from the densitometric output of the prior figure. Such an output can be further quantified. One simple method is shown in Fig. 22 where the ratio of the dorso-ventral and craniocaudal dimensions of the transform provides

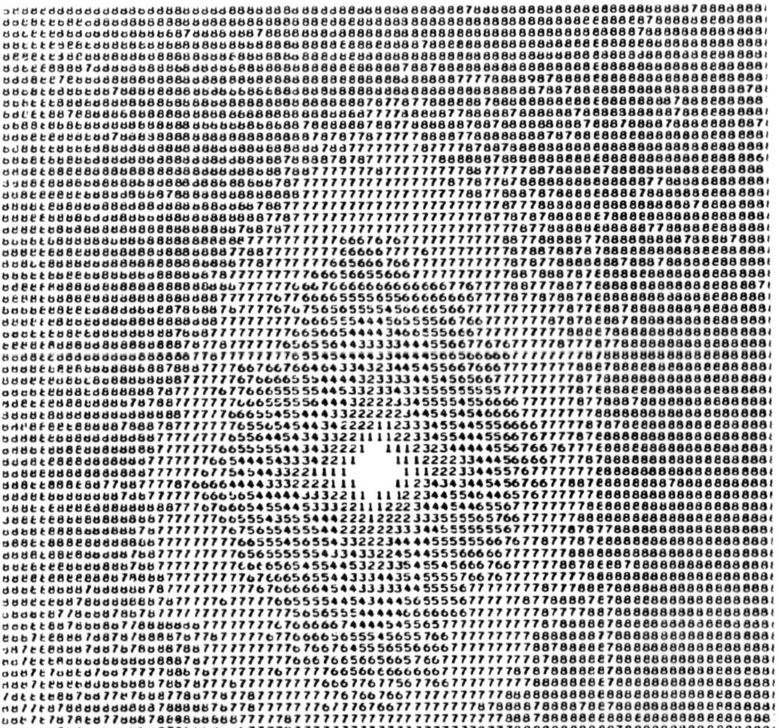

FIG. 20. A densitometric map of the transform of Fig. 19.

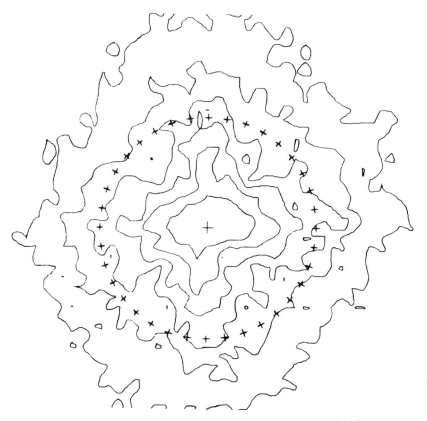

FIG. 21. The contoured map of the densitometric map of Fig. 20.

information about the proportions of the orthogonal craniocaudal
and dorsoventral elements of the radiographic pattern. Ratios of
1:10 and 10:1 are displayed. Of course, much additional information
about off-orthogonal elements is also present in such a contoured
diagram; they can be quantified, for example, using angular measurements (as plotted in Fig. 15).

The most obvious finding of the study relates to the extreme
sensitivity of the technique. Thus Fig. 23 demonstrates lateral radiographs for sample specimens of chimpanzee and orang-utan. Visual
inspection of these radiographs does not provide any special features
by which the two may be distinguished; nor does it inform us about
the nature of the patterns of radiographic shadows that result from
the shadows of the criss-crossing patterns of trabeculae and inter-
trabecular spaces other than that they seem to present the same
craniocaudal and dorsoventral elements that we know also exist in

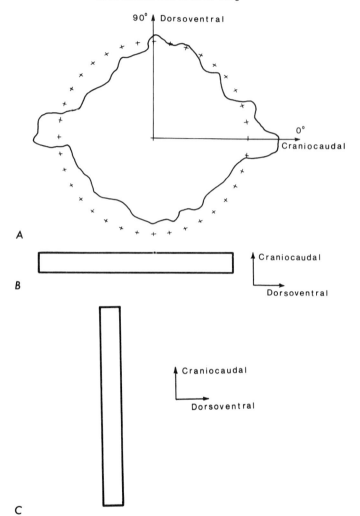

FIG. 22. Showing how ratios of measures taken from a particular contour of a power spectrum may give information about the shape of the pattern in an original (note the 90° transformation and the inversion in size).

man (compare Fig. 15). In order to remove such distinguishing clues as are present in the external forms of the bones (and also for technical reasons related to the production of transforms) the patterns are covered by circular masks. Figure 24 demonstrates that, when the masks are in place, the problem of distinguishing chimpanzees from orang-utans is even more difficult and the existence of the orthogonal pattern of radiographic shadows seems equally confirmed.

(A)   (B)

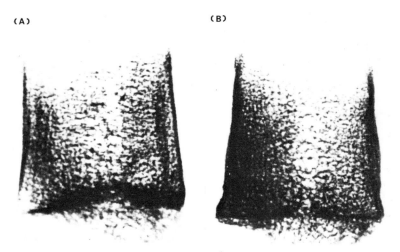

FIG. 23. Lateral radiographs of vertebra of (A) chimpanzee and (B) orang-utan.

(A)   (B)

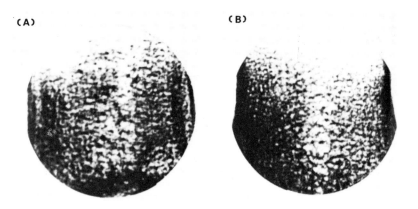

FIG. 24. The same radiographs as in Fig. 23 with the external features masked.

Optical data analysis provides the Fourier transforms illustrated in Fig. 25. Even without further analysis, it is immediately obvious how different the chimpanzee is from the orang-utan. An experienced anatomist can scarcely differentiate the primary radiographs upon which the transforms are based; a child can compare the transforms and distinguish them with ease. The sensitivity of the method is very great indeed; and we can readily understand why it has proven so useful in the comparison of complex patterns such as those of finger-prints (Weinberger & Almi, 1971) or those of nuclear substructure (Adolph, Caspar, Hollinshead, Lattman, Philips & Murakami, 1979).

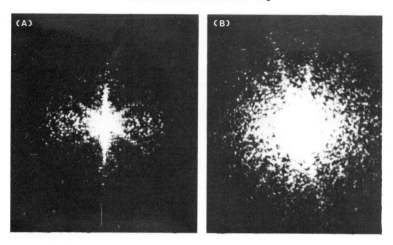

FIG. 25. The power spectra of the same radiographs as in Fig. 23: (A) chimpanzee; (B) orang-utan.

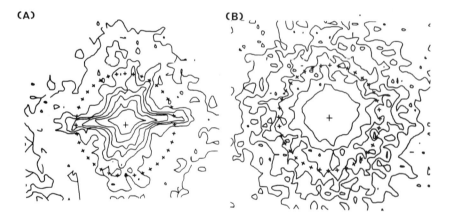

FIG. 26. The contoured diagrams for the chimpanzee compared with the orang-utan.

The major differences that are so obvious in the raw transforms themselves are equally obvious from the contoured power spectra (Fig. 26) which also provide quantitative information as required.

SIGNIFICANCE OF THESE TRANSFORMS FOR TRABECULAR ARRANGEMENT

The transform for the chimpanzee is generally similar to that for man, gorilla, and bonobo (Figs 27, 28). A generally cruciate transform

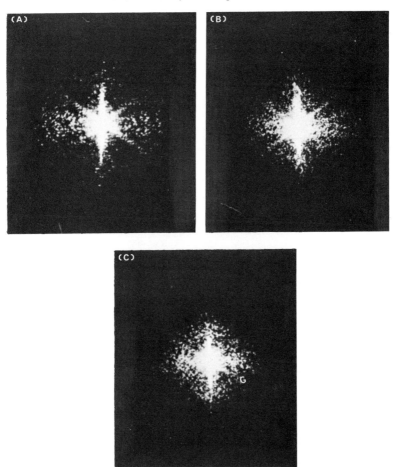

FIG. 27. The power spectra for examples from the sequence (A) gorilla, (B) chimpanzee and (C) bonobo.

indicates that a generally orthogonal arrangement of trabecular shadows exists within the vertebrae of all these species. In this sense these results confirm those previously presented for the section of the lumbar vertebrae in man. However, it is also apparent that for each of these vertebrae in each of these species, considerable off-orthogonal elements exist, thus further confirming the findings for the section of the human second lumbar vertebra: that the pattern may not be entirely or simply orthogonal. The deviations from orthogonality in all these cases are of similar degree, even though humans, gorillas, chimpanzees and bonobos can be clearly differentiated from one another. In other words, this new information is

(A)   (B)

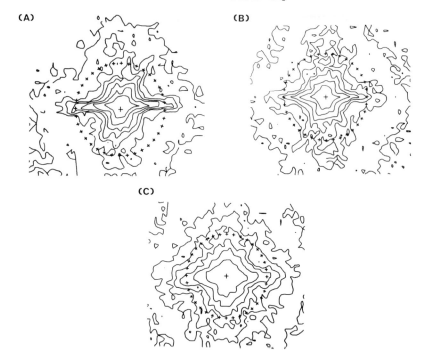

(C)

FIG. 28. The contoured diagrams from the same sequence as Fig. 26; gorilla, chimpanzee and bonobo.

compatible with both of the modified hypotheses about the load-bearing significance of bone architecture that were posited earlier: either (a) that a generally orthogonal arrangement is the major reality with small local variations from it, or (b) that the arrangement is not orthogonal but of some other form. The departure from orthogonality in each of the above cases is not large enough for us to make certain determination between these ideas.

However, the picture is completely different when we come to examine the transform and contoured power spectrum for the orang-utan (Figs 25 and 26). Here, there is no semblance of ortho-gonality whatsoever. Indeed, the star-shaped nature of the transform and the irregularly circular form of the contoured power spectrum for the orang-utan imply some arrangement of trabeculae involving materials laid down at a wide variety of different angles and with little or no special preference for material laid down orthogonally (although, of course, there will still be much material so arranged). Such an arrangement might include, among others, a star-shaped design, a honeycomb arrangement, or a spongiform pattern.

This new information is so startling that it seems most important, despite the rarity of the various bony materials, to check upon the transforms by examining actual sections of representative vertebrae. Figure 29 shows the trabecular pattern in *Pan* associated with the most obviously orthogonal of the various vertebral transforms. These confirm absolutely that, though bony spicules are, indeed, laid down at all sorts of angles, the general arrangement is one in which craniocaudal and dorsoventral elements predominate.

In complete contrast is the picture in *Pongo*. This is shown in Fig. 30 and demonstrates unequivocally not only that there is little evidence of a predominating craniocaudal dorsoventral orthogonal pattern such as that seen in all the other species, but also that, of the various non-orthogonal possibilities suggested above, the spongiform model fits most closely. It would thus appear that at least one example exists where arrangements of trabeculae that depart very far from orthogonal are the norm.

SIGNIFICANCE OF THESE TRABECULAR ARRANGEMENTS FOR THEORIES OF STRESS BEARING

The trajectorial theory in its originality suggested that: an orthogonal network of stresses resulting from function is associated with an orthogonal network of architectural elements bearing those stresses. This idea can be represented by the equation:

orthogonal stress is associated with orthogonal architecture

Previous discussion has indicated, however, that the trajectorial theory in its originality is unlikely to be correct.

One modification that is offered, and covertly acknowledged without discussion in many studies, is the following: (a) a resultant orthogonal network of stresses stems from the many different functions of the bone during life; (b) the architectural structures that exist, although non-orthogonal in their immediate local arrangements, nevertheless exhibit an overall orthogonal arrangement when averaged by some such mechanism as visual inspection of an entire bony section, or of radiographic or laminographic patterns of whole bones; (c) the modified theory represents the association between these two, new, somewhat more abstract, averaged patterns. These ideas might be expressed in the following equations:

stress 1 + stress 2 + stress 3 + . . .

= resultant orthogonal stress pattern

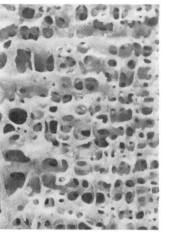

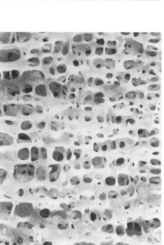

FIG. 29. Views of two small portions of sagittally sectioned lumbar vertebrae in the chimpanzee.

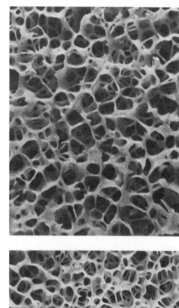

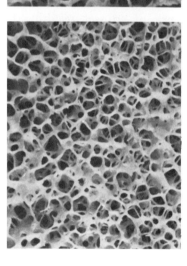

FIG. 30. Views of two small portions of sagittally sectioned lumbar vertebrae in the orang-utan.

and

actual architecture at site a

+ actual architecture at site b

+ actual architecture at site c

+ ...

= overall orthogonal architectural pattern

The final equation representing the modified theory would then be one in which:

$$\frac{\text{orthogonal}}{\text{resultant stress}} \quad \text{is associated with} \quad \frac{\text{orthogonal}}{\text{overall architecture}}$$

The findings presented here for the second lumbar vertebra in man are compatible with this modified theory. For instance, the forces impinging upon the vertebral column possess a general resultant. For though the human vertebral column of man is capable of a very wide variety of functions, it is the case that the great majority of these functions are carried out with the trunk vertical and supported only by the lower limbs. Therefore, whatever smaller differences in stress pattern may exist as a result of all the different functions, they will be lesser modifications upon a single major theme resulting from the upright position of the trunk. This could mean a resultant stress that is not badly modelled by a columnar system under simple compression; and this stress system could well be associated with an architectural system that was mainly designed to bear such primary orthogonal stresses, whatever other minor elements might exist.

In a similar way, the vertebral columns of the gorilla, chimpanzee and bonobo also exhibit a primary mode of loading whatever the various activities that the animals may undertake. All of these creatures are, indeed, capable of moving bipedally at times, and all of them also are capable of a variety of acrobatic activities in the trees, the bonobo perhaps most, the gorilla perhaps least. Yet there can be little doubt that the majority of the activities that they practise are carried out with a vertebral column lying in an approximately horizontal position and being supported by both upper and lower limbs. Whatever other activities we may discern, this is the basic rhythm that contains most of the biomechanical forces of the day for these animals (sleeping and sitting postures scarcely count as the forces involved in them, though acting for considerable periods of time, must be many times smaller than those involved in locomotion). Again, therefore, whatever differences in stress patterns may result from all the different functions, they will in general be modifications of a single major orthogonal pattern of stresses dependent upon a usually quadrupedal vertebral column in these creatures. This could be associated with an overall orthogonal architecture, whatever other minor architectural features we may have demonstrated exist. These findings also, therefore, do not totally deny the modified theory.

It is the case of the orang-utan that settles the matter. For by the accidents of evolution, the orang-utan does not adopt any one particular mode of vertebral load bearing especially more than any other. The vertebral column is equally likely to be supported or suspended by any or all combinations of four different limbs. Thus, though we have not so far studied the actual stresses existing in the orang-utan vertebral column and such a study would be very difficult (e.g. Lanyon, 1971, 1972, 1973), the possible range of patterns must be very large and quite different from one another.

Yet, if the modified theory were correct, there would still be some resultant orthogonal network that could be calculated and there should still be some approximate orthogonal architecture associated with it. There is not. The spongiform arrangement denies orthogonality.

On the other hand, such a structure might well be capable of bearing a series of quite different loading patterns in turn over time. The equation, therefore, may read:

$$\text{stress } 1 + \text{stress } 2 + \text{stress } 3 + \ldots$$

is associated with

a single architectural network capable of

bearing each stress network in turn and

not, therefore, itself necessarily orthogonal

This new hypothesis is not at variance with the picture provided for man, gorilla, chimpanzee and bonobo. It is just that stress networks 1, 2, 3, . . . for each of the functions in those creatures may be similar enough that an approximate orthogonal architectural network is the one most mechanically efficient for supporting each in turn.

The new theory, thus, is not incompatible with any data resulting from this study. Neither is it denied by any data from prior studies of which the present authors are aware. It does not deny, further, that there is, indeed, a most important association between stress-bearing and architecture, and it is, therefore, compatible (a) with all those investigations that show how cancellous architectures change when new biomechanical demands are placed upon them, and (b) with all those examinations that, acknowledging the property of the adaptation of bone form, seek to elucidate the nature of the causal relationship between stress-bearing and architectural modification. In each case, however, the new theory provides a new stepping off point for these experimental studies of morphogenetic causality.

And if, in the process of carrying out such studies, it ever proves possible to examine radiographs of vertebrae from various fossils alleged to have been bipedal in general habit (e.g. *Homo habilis, Australopithecus africanus, A. robustus, A. afarensis*, or any other of the fossils that are almost daily being discovered in Africa) then given the major differences between (a) bipedal man, (b) quadrupedal African apes, and (c) arboreal, quadrumanal orang-utans, we may be in a position to make better assessments of what these fossils may have been doing.

ACKNOWLEDGEMENTS

The work described in this paper was carried out in the Department of Anatomy, The University of Chicago, during the years 1968 to 1977, and subsequently in the Departments of Biological Sciences and Anatomy, and The Graduate School, The University of Southern California. It has been supported by NSF grant GS 30508 and NIH grant HD 02852; it is currently supported by NSF grant DEB 81939 and by funds from the Associates of The University of Southern California.

We are obliged to several museums and University Departments for permitting access to their collections of primate material: first Dr Karel Liem, then Dr Luis de la Torre and now Dr Patricia Freeman and the Field Museum of Natural History, Chicago; Drs Prue Napier and Theya Mollison, and the British Museum of Natural History, London; Mr L. R. Barton and the Powell-Cotton Museum, Birchington, U.K.; and Dr Thys van den Audenaerde and the Musée Royale de l'Afrique Central, Tervuren, Belgique. Materials from the Department of Anatomy, University of Chicago, and from the private collections of the authors were also utilized.

I (C.E.O.) remain indebted to John C. Davis who first introduced me to optical data analysis and who performed the first of the analyses on his equipment at the University of Kansas. The continuing stimulus to my studies in quantitative biology continues to spring from Professor Lord Zuckerman who first introduced me to biometrics during my period as a student in his department at the University of Birmingham, U.K., and to Professor Eric H. Ashton with whom I have collaborated for many years in multivariate morphometric investigations.

REFERENCES

Adolph, K. W., Caspar, D. L. D., Hollinshead, C. J., Lattman, E. E., Philips, W. C. & Murakami, W. T. (1979). Polyoma virion and capsid crystal structures. *Science, Wash.* 203: 1117–1120.

Agin, G. J. & Binford, T. O. (1976). Computer description of curved objects. *IEEE Trans. Comput.* C-25: 439–49.

Alexander, R. McN. (1968). *Animal mechanics.* Seattle: University of Washington Press.

Andrews, H. C. & Pratt, W. K. (1969). Transform image coding. In *Symposium on computer processing in communications*: 63–84. Andrews, H. C. (Ed.). Brooklyn: Polytechnic Institute.

Anstey, R. L. & Delmet, D. A. (1972). Genetic meaning of zooecial chamber shapes in fossil bryozoans: Fourier analysis. *Science, Wash.* 177: 1000–1002.

Appleby, R. M. & Jones, G. L. (1976). The analogue video reshaper – a new tool for palaeontologists. *Palaeontology* 19: 565–586.

Attneave, F. & Arnoult, M. D. (1966). The quantitative study of shape and pattern perception. In *Pattern recognition*: 123–141. Uhr, L. (Ed.). New York: J. Wiley.

Bacon, G. E., Bacon, P. J. & Griffiths, R. K. (1979). The orientation of apatite crystals in bone. *J. appl. Cryst.* 12: 99–103.

Bassett, C. A. (1971). Biophysical principles affecting bone structure. In *The biochemistry and physiology of bone.* 3. *Development and growth*: 1–76. Bourne, G. H. (Ed.). London: Academic Press.

Blum, H. (1962). An associative machine for dealing with the visual field and some of its biological implication. In *Biological prototypes and synthetic systems* I: 244–260. Bernard, E. E. & Kare, M. R. (Eds). New York: Plenum Press.

Blum, H. (1967). A transformation for extracting new descriptors of shape. In *Models for the perception of speech and visual form*: 362–380. Walthen-Dunn, W. (Ed.). Boston, Mass.: Data Sciences Lab., and Air Force Cambridge Research Labs.

Bookstein, F. L. (1978). *The measurement of biological shape and shape change.* New York: Springer-Verlag.

Chmielewski, N. & Varner, J. R. (1969). An application of holographic contouring in dentistry. *Biomed. Sci. Instrum.* 6: 72–79.

Cutrona, L. J. (1965). Recent developments in coherent optical technology. In *Optical and electro-optical information processing*: 83–123. Tippett, J. T., Berkowitz, D. A., Clapp, L. C., Koester, C. J. & Vanderburgh, A., Jr (Eds). Cambridge, Mass.: MIT Press.

Davis, J. C. (1973). *Statistics and data analysis in geology.* New York: J. Wiley.

Davis, J. C. & Preston, F. W. (1972). Optical processing: an alternative to digital computing. *Spec. Pap. geol. Soc. Am.* No. 146: 49–68.

Dobrin, M. B. (1968). Optical processing in the earth sciences, *IEEE Spectrum* 5: 59–66.

Dobrin, M. B., Ingalls, A. L. & Long, J. A. (1965). Velocity and frequency filtering of seismic data using laser light. *Geophysics* 30: 1144–1178.

Eick, J. D., Johnson, L. N. & McGivern, R. F. (1971). *Stereoscopic measurements on the micro-scale by combining scanning electron microscopy and photogrammetry.* Tech. Papers ASP – Univ. Ill. Symp. Close-Range Photogrammetry, Urbana.

Elliott, S. B. & Morris, W. J. (1978). *Holographic contouring of fossil teeth.* Mimeographed. Los Angeles: Occidental College.

Evans, F. G. (1957). *Stress and strain in bones.* Springfield, Ill.: Thomas.

Gabrielson, V. K. (1977). Mesh generation for two-dimensional regions using a DVST (Direct View Storage Tube) graphics terminal. *Comput. Graphics* 2: 59–66.

Greguss, P. (1976). Holographic interferometry in biomedical sciences. *Optics & Laser Techn.* 8: 153–159.

Hillenkamp, F., Hutzler, P. & Kinder, J. (1976). Coherent and quasi-coherent optical methods in biology and medicine. *Ann. N.Y. Acad. Sci.* 267: 216–229.

Holeman, J. M. (1968). Holographic character reader. In *Pattern recognition*: 63–78. Kanal, L. N. (Ed.). Washington, D. C.: Thompson Book Company.

Jackson, P. L. (1965). Analysis of variable density seismograms by means of optical diffraction. *Geophysics* 30: 5–23.

Kaesler, R. L. & Waters, J. A. (1972). Fourier analysis of the ostracode margin. *Bull. geol. Soc. Am.* 83: 1169–1178.

Kennedy, J. D. (1971). Experimental Walsh-transform image processing. In *Proceedings fourth Hawaii international conference on system sciences*: 233–235. Lin, S. (Ed.). Honolulu: Univ. Hawaii.

Kopp, R. E., Lisa, J., Mendelsohn, J., Pernick, B., Stone, H. & Wohlers, R. (1976). Coherent optical processing of cervical cytologic samples. *J. Histochem. Cytochem.* 24: 122–137.

Kummer, B. (1959). *Bauprinzipien des Saugerskeletes.* Stuttgart: Springer-Verlag.

Lanyon, L. E. (1971). Strain in sheep lumbar vertebrae recorded during life. *Acta orthop. Scandinav.* 42: 102–112.

Lanyon, L. E. (1972). *In vivo* bone strain recorded from thoracic vertebrae of sheep. *J. Biomech.* 5: 277–281.

Lanyon, L. E. (1973). Analysis of surface bone strain in the calcaneus of sheep during normal locomotion. Strain analysis of the calcaneus. *J. Biomech.* 6: 41–49.

Lestrel, P. E. (1976). Some problems in the assessment of morphological size and shape differences. *Yb. phys. Anthropol.* 18: 140–62.

Lestrel, P. E. & Brown, H. D. (1976). Fourier analysis of adolescent growth of the cranial vault: a longitudinal study. *Hum. Biol.* 48: 517–528.

Lestrel, P. E., Kimbel, W. H., Prior, F. W. & Fleischmann, M. L. (1977). Size and shape of the hominoid distal femur: Fourier analysis. *Am. J. phys. Anthropol.* 46: 281–290.

Lu, K. H. (1965). Harmonic analysis of the human face. *Biometrics* 21: 491–505.

Meyer, G. H. (1867). Die Architectur der Spongiosa. *Arch. Anat. Physiol. Wiss. Med.* 1867: 615–628.

Murray, P. D. F. (1936). *Bones: a study of the development and structures of the vertebrate skeleton.* Cambridge: University Press.

Oxnard, C. E. (1971). Tensile forces in skeletal structures. *J. Morph.* 135: 425–435.

Oxnard, C. E. (1973a). Some problems in the comparative assessment of skeletal forms. In *Human evolution*: 103–125. Day, M. (Ed.). London: Taylor & Francis.

Oxnard, C. E. (1973b). *Form and pattern in human evolution: some mathematical, physical, and engineering approaches.* Chicago: Univ. Chicago Press.

Oxnard, C. E. (1977). Human fossils: the new revolution. In *The great ideas today 1977*: 92—153. Hutchins, R. M. & Adler, M. J. (Eds). Chicago: Encyclopaedia Britannica, Inc.

Oxnard, C. E. (1978). One biologist's view of morphometrics. *A. Rev. Syst. Ecol.* 9: 219—241.

Pauwels, F. (1965). *Gesammelte Abhandlungen zur funktionellen Anatomie des Bewegungsapparates*. Berlin: Springer-Verlag.

Pauwels, F. (1968). Beitrag zur funktionellen Anpassung der corticalis der Rohrenknocken. *Z. Anat. Entwickl.-Gesch.* 127: 121—137.

Philbrick, O. (1968). Shape description with the medial axis transformation. In *Pictorial pattern recognition*: 395—407. Cheng, G. C., Ledley, R. S., Pollock, D. K. & Rosenfeld, A. (Eds). Washington, D.C.: Thompson Book Company.

Pincus, H. J. (1969). Sensitivity of optical data processing to changes in rock fabric. Parts I, II, III. *Int. J. Rock Mech. Min. Sci.* 6: 259—268; 269—272; 273—276.

Pincus, H. J., Power, P. C., Jr. & Woodzick, T. (1973). Analysis of contour maps by optical diffraction. *Geoform* 14: 39—52.

Rao, K. R., Narasimhan, M. A. & Revuluri, K. (1975). Image data processing by Hadamard-Haar transform. *IEEE Trans. Comput.* C-24: 888—896.

Rybicki, E. F., Simonen, F. A. & Weis, E. B., Jr. (1972). On the mathematical analysis of stress in the human femur. *J. Biomech.* 5: 203—215.

Smith, J. W. (1962). The structure and stress relations of fibrous epiphyseal plates. *J. Anat.* 96: 209—225.

Sneath, P. H. A. (1967). Trend-surface analysis of transformation grids. *J. Zool., Lond.* 151: 65—122.

Srivastava, G. S. (1977). Optical processing of structural contour maps. *Math. Geol.* 9: 3—38.

Takasaki, H. (1970). Moiré topography. *Appl. Optics* 9: 1457—1472.

Taylor, C. A. & Ranniko, J. K. (1974). Problems in the use of selective optical spatial filtering to obtain enhanced information from electron micrographs. *J. Microscopy* 100: 307—314.

Thompson, D. A. W. (1917). *On growth and form*. Cambridge: University Press.

Underwood, E. E. (1970). *Quantitative stereology*. Reading, Mass.: Addison-Wesley Publishing Co.

Waddington, C. H. (1977). *Tools for thought*. New York: Basic Books.

Wainright, S. A., Biggs, W. D., Currey, J. D. & Gosline, J. M. (1976). *Mechanical design in organisms*. New York: Wiley.

Waters, J. A. (1977). Quantification of shape by use of Fourier analysis: the Mississippian blastoid genus *Pentremites*. *Paleobiology* 3: 288—299.

Weinberger, H. & Almi, U. (1971). Interference method for pattern comparison. *Appl. Optics* 10: 2482—2487.

Whitehouse, W. J. & Dyson, E. D. (1974). Scanning electron microscope studies of trabecular bone in the proximal end of the human femur. *J. Anat.* 118: 417—444.

Wolff, J. (1870). Ueber die Innere Architektur der Knochen und ihre Bedeutung für die Frage vom Knochenwachstum. *Virchows Arch. path. Anat. Physiol.* 50: 389—453.

Wolff, J. (1892). *Das Gesetz der Transformation der Knochen*. Berlin.

Yang, H. C. L. (1978). *The development of optical data analysis for the investigation of radiographic patterns: its application to the study of the lumbar vertebral column in man and the great apes*. Doctoral dissertation: University of Chicago.

Yang, H. C. L. & Oxnard, C. E. (In preparation a). *An approach to the study of trabecular patterns in radiographs: Optical data analysis.*

Yang, H. C. L. & Oxnard, C. E. (In preparation b). *Optical data analysis of lumbar vertebrae in some hominoids: implications for the study of primate locomotion.*

Yang, H. C. L. & Oxnard, C. E. (In preparation c). *The structure of cancellous bone as revealed by optical data analysis of radiographs: associations with stress bearing.*

Young, J. Z. (1957). *The life of mammals.* Oxford: University Press.

Zelenka, J. S. & Varner, J. R. (1968). A new method for generating depth contours holographically. *App. Optics* 7: 2107–2110.

Symp. zool. Soc. Lond. (1981) No. 46, 169—188

Functional Morphology of the Joints of the Evolving Foot

O. J. LEWIS

Department of Anatomy, St Bartholomew's Hospital Medical College, London, UK

SYNOPSIS

This chapter is a synoptic account of certain aspects of joint structure and function in primate feet, with the emphasis upon the subtalar and transverse tarsal joint complexes. The unique actions of these joints in bipedal man are contrasted with their arboreally adapted functions in other primates. An attempt is made to define the key human apomorphic features.

The compromise axis of the human subtalar joint has become re-orientated into a very elevated position and more nearly in line with the functional foot axis than it is in subhuman primates. This has been achieved partly by changes in the conarticular surfaces and partly by remodelling of the foot about the axis. The forefoot has been effectively adducted by refashioning of the cuneiforms, cuboid, bases of the metatarsals and the intervening joints; in contrast the heel has been abducted.

The subtalar joint complex in Early Miocene apes retained a primitive, putatively arboreal morphology. In the foot of Olduvai Hominid 8 the subtalar axis retained an essentially primitive disposition, with little evidence of the uniquely human remodelling.

The variety of forms presented by the cuboid articular surfaces of the calcanei of African Miocene apes provide reasonable precursors for the characteristic morphologies exhibited by *Pan*, *Gorilla* and even *Pongo*. The calcaneocuboid joint of man has become remodelled into a specialized sellar form which provides for an arcuate swing moving the heel laterally during supination of the lamina pedis, and thus bringing the weight-bearing foot into close-packed position. This contrasts strikingly with the action in subhuman primates. The complexly interlocking conarticular surfaces of this joint in OH8 could perhaps be considered as a plausible precursor for the human condition but lacked the unique functional attributes found in man.

INTRODUCTION

For some time the author has been engaged in a detailed study of the structure and function of the feet of extant primates, paying particular attention to the possible evolutionary precursors of such structure and the modifications which would be entailed in converting the grasping foot of an arboreal primate into that of erect bipedal man. These studies have provided the basis for a reassessment of the quite abundant fossil evidence. This work will be published in greater anatomical detail and with a more extensive survey of the relevant literature, in a series of papers elsewhere.

This present chapter is a preliminary report of certain aspects of that work and is restricted in its scope largely to a consideration of the hominoid foot, fossil and extant. Particular attention will be paid to the identification of derived, specialized characters which might have critically influenced the evolutionary fortunes of the great apes and emergent man. Following the current vogue such characters could be described as apomorphic and the value of the identification of such derived character states in the modern construction of phylogenies is being quite impressively demonstrated (Delson, 1977; Tattersall & Eldredge, 1977). This methodology is in marked contrast to that of multivariate morphometrics which, by its very nature, tends to emphasize overall resemblance. Indeed, important apomorphic characters may be overlooked entirely in such studies either because the complex shapes involved may be difficult to express metrically, or more commonly because the basic functional morphology has simply not been adequately analysed.

The present chapter is concerned, almost exclusively, with identifying derived character states, and their probable functional import, in the subtalar and transverse tarsal joint complexes.

THE PRIMATE SUBTALAR JOINT COMPLEX

Terminology

Because of varying usage and changes in terminology the above term is here used for those articulations underlying the talus which have commonly been referred to by comparative anatomists as the lower ankle joint. In eutherian mammals there are here two anatomically distinct articulations — the posterior talocalcaneal joint (the subtalar joint of the human Nomina Anatomica) and anteriorly the talocalcaneonavicular joint. Both are involved simultaneously in movement of the remainder of the foot — aptly termed the lamina pedis (footplate) by MacConaill & Basmajian (1969) — upon the talus, which then effectively becomes part of the leg.

The movements at this joint complex are best defined as inversion and eversion; the alternative terms supination and pronation will be restricted to movements of the forefoot at the transverse tarsal articulation.

New World Monkeys

Arboreal platyrrhine monkeys such as *Cebus* or *Pithecia* show clearly what seem to be characteristic primate heritage features of the subtalar joint complex, with its typical eutherian posterior and anterior anatomical subdivisions (Fig. 1). Posteriorly, the calcaneus articulates

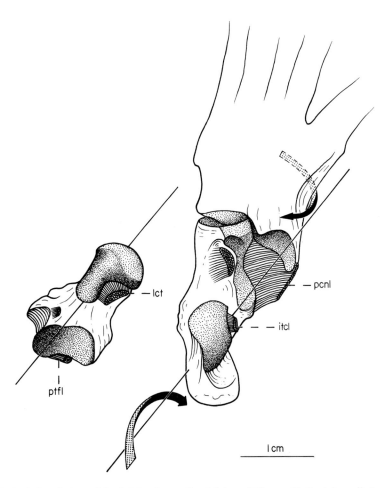

FIG. 1. A dorsal view of the left lamina pedis of *Cebus albifrons* with the talus rolled away laterally from its subtalar articulations. The conarticular surfaces of the subtalar joint complex are thus displayed and in each case the compromise subtalar axis is shown. The calcaneocuboid joint has been opened dorsally to expose the articular surfaces. The lower arrow indicates the screwing motion of the lamina pedis about the subtalar axis during inversion; the upper arrow indicates the accompanying movement of the forefoot during supination at the calcaneocuboid joint. lct, ligamentum cervicis tali; ptfl, posterior talofibular ligament; pcnl, plantar calcaneonavicular (spring) ligament; itcl, interosseous talocalcaneal ligament. In this and subsequent figures the bar represents 1 cm.

with the talus by a convex facet which is markedly prolonged back-wards and medially. The head and neck of the talus, entering into the anterior joint, is essentially cylindrical, capped at its extremity by a convexity for the navicular. On the surface of this cylinder is an L-shaped area for articulation with facets, variably confluent, on the body and sustentaculum tali of the calcaneus; the ligamentum cervicis tali attaches into the angle of the L-shaped area. The calcaneus is quite markedly prolonged forwards beyond the talus to achieve a considerable articulation with the navicular.

Movements of inversion—eversion of the lamina pedis occur about a compromise axis which traverses the head and neck of the talus and then enters the calcaneus below its posterior talar facet. This axis occupies quite a flat plane, rising little when traced anteriorly but deviating far medially from the long axis of the foot.

It is immediately obvious that the posterior talocalcaneal articu-lation has a helical orientation in relationship to the subtalar axis; in a left foot it is a segment of a left-handed screw. As the lamina pedis moves into inversion the calcaneus is screwed forward beneath the talus, its exposed extremity then contacting the navicular. As will be seen later this movement is subtly correlated with that occur-ring in the calcaneocuboid joint in the overall grasping function of the foot.

Old World Monkeys

Morphology and function are in general comparable to the condition in platyrrhine monkeys but the screwing action at the joint complex is reduced, and in correlation with this, articular contact of the calcaneus with the navicular is relatively restricted. Although the screwing action is reduced, particularly in the more terrestrial cerco-pithecines, the contention that this particular type of movement has been lost in most cercopithecoids (Szalay, 1975), seems to be in error.

Anthropoid Apes

Hylobates

Gibbons retain the essentials of the structure and function exhibited by arboreal platyrrhine monkeys and the subtalar axis is similarly very oblique and lies in a flat plane. There is quite marked screwing action at the subtalar joint complex but contact between the cal-caneus and navicular is limited to a narrow articular strip. The cal-caneal surfaces on the head and neck of the talus have the expected L-shaped configuration.

Pongo

The conservative primate structural and functional attributes of the joint complex found in gibbons are also retained in the orang-utan.

Pan and Gorilla

The chimpanzee and gorilla (Figs 2, 18B) share certain derived features which are clearly grafted onto a morphology essentially similar to that in the primates described above. The subtalar axis is markedly oblique but its anterior end is now more elevated. The talus, perhaps in association with this feature, is more squat in form with a short, broad neck, this being particularly evident in *Gorilla*. The component limbs of its L-shaped articular surface are partially merged into a somewhat confluent single area.

A typical screwing action accompanies inversion and the calcaneus then attains a small articular contact with the navicular.

Homo sapiens

The human subtalar complex (Fig. 3) has been drastically remodelled but nevertheless retains the hallmarks of the basic arboreal primate heritage and the specializations seem to be exaggerations of trends already incipient in the largely terrestrial great apes *Pan* and *Gorilla*.

The joint complex has the usual anterior and posterior subdivisions. Traversing the foot between these is the compromise subtalar axis which is strikingly reorientated towards the long axis of the foot (Fig. 18c) and elevated into a much more vertical position; as described by Shephard (1951) it runs upwards, forwards and medially through calcaneus and talus from the lateral side of the heel.

The posterior articulation, as in other primates, is inclined to this axis so as to form a segment of a left-handed screw in the left foot, as was elegantly demonstrated by Manter (1941), and this was apparently the first suggestion of a screw-like action at this joint in any primate. It is thus particularly surprising that Szalay (1975) should have maintained that this type of action has been lost in the hominid joint.

The calcaneal articulation in the anterior joint is refashioned from an L-shape into a smoothly confluent arc forming also the segment of a screw, in this case with the pitch right-handed in a left foot. This helical calcaneal facet, which may be varyingly subdivided (Bunning & Barnett, 1965) contributes to the articular cup — aptly termed "acetabulum pedis" by MacConaill (1945) — which receives the head of the talus.

The combination of these counter-rotating screwing movements on the talus at the two joints produces a rotatory torque on the bone as a whole about a vertical axis which is reflected in the very upright inclination of the human compromise axis.

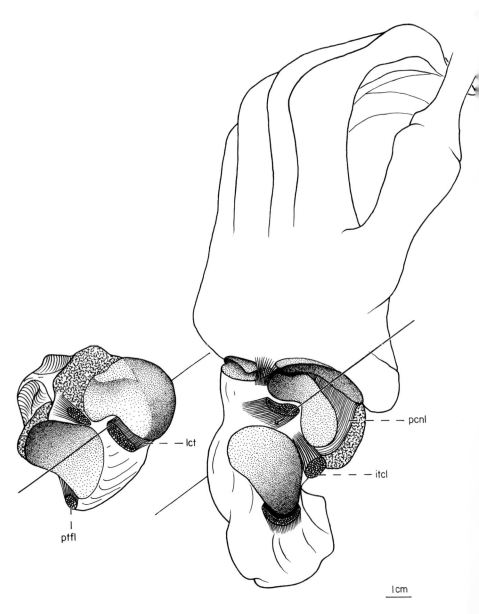

FIG. 2. A dorsal view of the left lamina pedis of *Gorilla gorilla* ("Guy") with the talus rolled away, revealing the conarticular surfaces of the subtalar joint complex; in each case the compromise subtalar axis is shown. Labels as in Fig. 1.

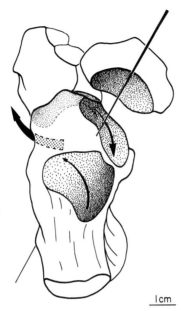

FIG. 3. A dioptograph tracing of the hinder part of the articulated foot of *Homo sapiens* shown in Fig. 18 after removal of the talus to expose the articular surfaces of the subtalar joint complex whose compromise axis is indicated; the direction of the helical movements at the anterior and posterior joint surfaces are indicated by arrows. The large upper arrow indicates the direction of movement of the calcaneus at the calcaneocuboid joint during supination of the lamina pedis.

THE PRIMATE TRANSVERSE TARSAL JOINT

Terminology

This joint complex again is not a single anatomical entity but includes the talonavicular part of the talocalcaneonavicular joint (itself part of the subtalar complex) and the calcaneocuboid joint; it is the form of the latter component which effectively determines the movements.

At this joint complex occur the movements of supination and pronation of the forefoot upon the hindfoot with the former accentuating the inturning of the sole which accompanies inversion at the subtalar joint complex.

Monkeys

In both platyrrhine and catarrhine monkeys the cuboid presents a kidney-shaped articular surface bearing a slightly protuberant convexity adjacent to the plantar "hilum" where the fibrous capsule of the joint is reinforced by a massive ligament (the homologue of the short and long plantar ligaments of human anatomy) radiating forwards from the anterior plantar tubercle of the calcaneus.

Serial radiographs of ligamentous preparations demonstrate that as the forefoot supinates at this joint it is also angulated medially.

Anthropoid Apes

Hylobates

Form, and apparently function, are quite comparable to that occurring in monkeys.

Pan

The great apes and man all possess derived morphologies in the form of the conarticular surfaces of calcaneus and cuboid and that shown by the chimpanzee appears to be central to the understanding of these differing shapes.

Whereas monkeys and gibbons show a relatively unobtrusive ventral convexity on the cuboid, *Pan* (Figs 4 & 5) exhibits a prominently exaggerated articular beak-like projection, fitting into a corresponding deep depression in the calcaneus. The articular beak is usually rather eccentrically situated, being displaced to a varying degree towards the medial side, and it blends insensibly into the remainder of the cuboid surface which is flat or slightly concave.

Despite this altered form, function appears very similar to that in monkeys with the joint providing the rotatory motion involved in supination of the forefoot which again includes a component of angular displacement — slight adduction of the forefoot.

Gorilla

In the great majority of gorillas the form of the articulating surfaces is in striking contrast to that shown by the chimpanzee. The joint surfaces are almost flat with only the slightest suggestion of an elevation on the cuboid received into a correspondingly shallow depression on the calcaneus (Fig. 17).

Occasionally, however, a very different morphology is found. The calcaneus is excavated to accommodate the cuboid to a greater extent (Fig. 16) than is seen in chimpanzees and this deep depression is surrounded by a relatively flattened articular rim. As will be seen, it is likely that this represents the ancestral *Gorilla* pattern and that the common flattened form is derived from it.

However, in both variants a pivotal action occurs which appears to be essentially like that in *Pan*.

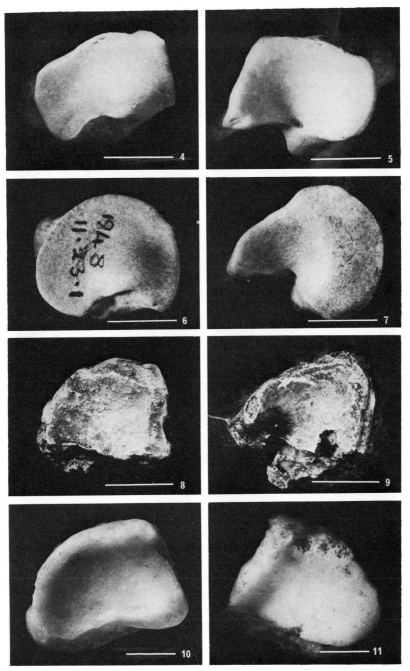

FIGS 4–11. (4) The proximal articular surface of the left cuboid of *Pan troglodytes* (B.M. 76.437). The scale bar is 1 cm. (5) The distal articular surface of the left calcaneus of *Pan troglodytes* (B.M. 76.437). (6) The proximal articular surface of the left cuboid of *Pongo pygmaeus* (B.M. 1948.11.23.1). (7) The distal articular surface of the left calcaneus of *Pongo pygmaeus* (B.M. 1948.11.23.1). (8) The proximal articular surface of the left cuboid (cast) of OH8. (9) The distal articular surface of the left calcaneus (cast) of OH8. (10) The proximal articular surface of the left cuboid of *Homo sapiens*. (11) The distal articular surface of the left calcaneus of *Homo sapiens*.

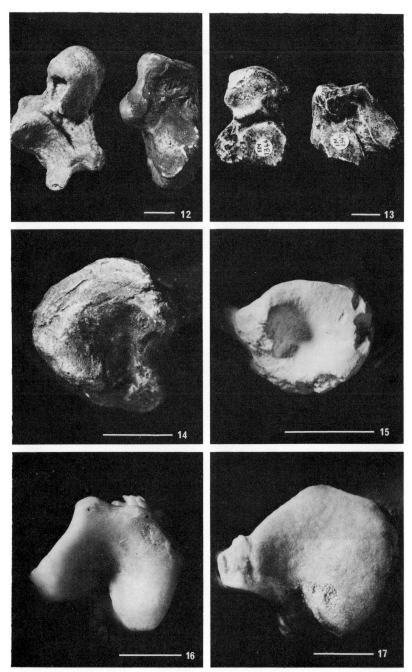

FIGS 12–17. (12) The right talus (KNM-SO-389) and calcaneus (KNM-SO-390) of *Proconsul major*; the talus is rolled away medially so as to expose the conarticular surfaces of the subtalar joint complex. The scale bar is 1 cm. (13) Casts of the left talus and calcaneus of OH8 with the talus rolled away laterally to expose the conarticular surfaces of the subtalar joint complex. (14) The distal aspect of the right calcaneus (KNM-SO-390) of *Proconsul major*. (15) The distal aspect of the cast of a left calcaneus (KNM-SO-427) from Songhor. (16) The distal aspect of the left calcaneus of *Gorilla gorilla* ("Guy"). (17) The distal aspect of the left calcaneus of *Gorilla gorilla* (B.M. 1948.2.27.1).

Pongo

The orang-utan joint surfaces present a characteristically unique form of their own. The calcaneus bears a deep hemispherical depression which lodges a peg-like protuberance on the cuboid and is sharply demarcated from a broad flat articular surface located to its lateral side (Figs 6 & 7). This morphology is well suited to the pivotal action involved in supination of the forefoot.

Homo sapiens

The human joint surfaces are uniquely remodelled (Figs 10 & 11) yet retain clear residual evidence of affinities with the pongids. They are noteworthy for their asymmetrical character: the convexity on the cuboid and its receptive concavity on the calcaneus now form the medial boundary of the articulation, whilst laterally an enlarged flaring convexity on the calcaneus lodges into a concavity on the cuboid. The joint has effectively been remodelled into a sellar one.

Function is significantly altered and under the ordinary conditions of walking consists of an arcuate swing of the calcaneus upon the cuboid in the weight-bearing foot (Fig. 3). Pivotal movement (conjunct rotation) still occurs, but the associated angular movement (swing) between forefoot and hindfoot accompanying it has been reversed in direction when compared with subhuman primates. Thus as the heel swings laterally the calcaneus rotates in such a way that the torsion, which is an integral part of the architecture of the human lamina pedis, is untwisted; by analogy with the forearm this position has been called supination by MacConaill & Basmajian (1969). Indeed, if conditions had been reversed, with the calcaneus stationary and the forefoot moving, the latter would then be in a supinated attitude.

FOOT FUNCTION AS A WHOLE

Subhuman Primates

The actions of the subtalar and transverse tarsal joint complexes are neatly integrated in the grasping feet of arboreal primates. Movement of the lamina pedis at the subtalar joint complex inverts it into an appropriate grasping attitude, but at the same time it is screwed forward; this provides scope for the full expression of movement at the transverse tarsal joint. Here the forefoot is supinated and the accompanying medial folding or adduction of the forefoot is accommodated by the helical movement about the subtalar axis. When this grasping attitude is attained both sets of joints have been brought into a stable close-packed condition.

Homo sapiens

In bipedal man the subtalar joint complex is concerned, not so much with movements of foot upon leg, as with movement of the talus (carrying with it the leg and so the whole body) upon the lamina pedis during the stance or support phase of gait; during this phase compensatory movements at the transverse tarsal joint effectively tension the footplate bringing it into close-packed position for its weight-bearing role.

As the body is propelled forwards over the weight-bearing foot the centre of gravity veers laterally and then medially (Carlsöö, 1972) and the head of the talus is effectively screwed downwards and medially into the "acetabulum pedis" (MacConaill, 1945; MacConaill & Basmajian, 1969) rotating around the comparatively upright subtalar axis. Concurrently the lamina pedis becomes untwisted (supinated), the major movement occurring at the calcaneocuboid joint; the accompanying lateral movement of the heel tensions the obliquely disposed, and uniquely human, short plantar ligament in the sole and also the plantar calcaneonavicular (spring) ligament supporting the head of the talus. The whole lamina pedis is then in close-packed position.

EARLY MIOCENE FOSSILS

A number of fossil tali and calcanei, few of them formally described, are known from two early Miocene inter-rift localities of East Africa — Songhor and Rusinga Island. No satisfactorily documented accounts of monkey fossils are known from either site and the tarsal bones therefore belong, presumptively, to apes. Cranial and dental remains suggest that the common species present at Songhor were *Proconsul major, Rangwapithecus gordoni* and *Limnopithecus legetet*; at Rusinga the common species were *Proconsul africanus, Proconsul nyanzae* and *Dendropithecus macinnesi* (Andrews & Van Couvering, 1975). Good casts of four specimens previously described and illustrated by Clark & Leakey (1951) were available from the British Museum (Natural History). These were: a large articulating calcaneus (KNM-SO-390; C.M.H.146) and talus (KNM-SO-389; C.M.H.145) from Songhor which have been reasonably attributed to *Proconsul major*; a rather small talus from Rusinga (KNM-RU-1743; C.M.H.147) attributed to *Proconsul nyanzae*; a small talus (KNM-RU-1745; B2) which has been suggested as belonging to *Proconsul africanus*. In addition casts of nine additional tali and six calcanei from these two sites were made available by Dr Peter Andrews of the British Museum (Natural History). These fossil specimens will not be described in any detail here (although certain are illustrated in this paper) nor will any comprehensive attempt be made to assign them to a particular taxon. Certain generalities of their structure are, however, relevant to a consideration of hominoid, and hominid, origins.

It is abundantly clear that all of these fossils present the predictable morphology to be expected in emergent unspecialized arboreal apes, lacking even derived features such as the foreshortening of the neck of the talus seen in the largely terrestrial extant African great apes.

All the tali conserve an L-shaped calcaneal surface on the markedly deviated neck, which itself is indicative of a primitively oblique subtalar axis, and the posterior calcaneal surface has a clear-cut helical disposition in relation to this axis. The form and disposition of the comparable articular surfaces on the calcanei confirm these conclusions (Fig. 12).

The form of the cuboid articular surface on the calcanei is particularly informative. With one conspicuous exception (KNM-SO-390) the surface is deeply excavated, somewhat towards the medial side, and clearly lodged a prominent beak-like process on the cuboid (Fig. 15). With virtually no modification such structure could easily be ancestral to the morphology seen in the extant chimpanzee. It seems quite plausible, also, that with minor evolutionary change it could have provided the stem form for those unusual variants among *Gorilla* calcanei which are also deeply excavated (Fig. 16); the much more common flattened form of *Gorilla* articular surface (Fig. 17) is likely to be a relatively late further specialization.

The notable exception in calcaneal structure is the large specimen (KNM-SO-390; C.M.H.146) which was recovered together with an articulating talus (KNM-SO-389; C.M.H.145). The surface for the cuboid in this specimen (Fig. 14) strikingly resembles that described above for the orang-utan (Fig. 7). It is tempting to suggest, therefore, that here is the ape ancestor of the Asiatic orang-utan. This derivation would accord also with other morphological evidence (Lewis, 1972a). The timing would appear to be right, for a little later, at the beginning of the middle Miocene, a land bridge to Eurasia was completed (Andrews & Van Couvering, 1975). This very tentative hypothesis demands a caveat. The uncommon excavated form of calcaneus in

Gorilla (Fig. 16) also has at least some resemblance to the form in *Pongo* and to that of KNM-SO-390. Pilbeam (1969) basing his conclusions on a multivariate study of the talus (KNM-SO-389) of this pair of fossils suggested that it (and by implication the associated calcaneus) should be ascribed to *Proconsul major*. This seems reasonable but he further suggested that *P. major* was ancestral to *Gorilla*; this may be an alternative possibility but on balance seems less likely.

Le Gros Clark & Leakey (1951) decided from their study of the Rusinga and Songhor fossils available to them that the nearest modern counterparts are to be found among the more terrestrial cercopithecoid monkeys. This view cannot be sustained by the present work.

Several of the fossil tali from these sites have been subjected to multivariate analysis, and the use of this technique culminated in two sophisticated studies by Lisowski, Albrecht & Oxnard (1974, 1976). These authors allocated the *Proconsul* tali to the envelope of arboreal species with the smallest closest to *Hylobates* and the others to *Pongo*. This result appears to be in broad agreement with the conclusions reached in the present chapter, using a very different approach. Caution is needed, however, before putting too much emphasis on such agreement. The results of a multivariate study are only as reliable and meaningful as the functional and evolutionary importance of the underlying morphological data. In the studies cited the eight dimensions utilized seem to contain such information only in a limited and rather indeterminate way.

CONVERSION OF APE FOOT TO HUMAN FOOT

The usual but simplistic view of this evolutionary progression is that the primitive ape-like foot coped with its new terrestrial and bipedal role by becoming everted and adducting the hallux. Such a change, however, would have no effect on the disposition of the subtalar axis, yet this is markedly altered in the human foot. Moreover it seems likely that the emergent terrestrial foot would be remodelled about a substantially abducted hallux, for then, with the great toe in grasping posture the first tarsometatarsal joint is screwed into close-packed position (Lewis, 1972b) and some semblance of an arched structure would be created. A rather mechanistic, but largely testable hypothesis can be constructed from this first premise.

There is little doubt that the human hallux has been realigned somewhat from the ape condition but this seems to be largely the resultant of a changed disposition of its joint surface on the medial cuneiform (Schultz, 1930). More significantly, it seems that the remainder of the forefoot has been realigned towards the stabilized hallux and so towards the subtalar axis. One of the factors involved in this has been reconstruction of the cuboid and its articulation with the calcaneus so that together they are arched plantarwards and medially. Similarly the lateral cuneiform has become bent medially towards the subtalar axis. The remodelling has been completed by angulation of the bases of the metatarsals, especially the second and third (Figs 18B, C) and the shafts of the reorientated metatarsals have apparently undergone torsion so that the plantar surfaces of their heads remain in contact with the substrate. These changes have clearly involved major, and characteristically human, remodelling of the articulations between the distal tarsals and the metatarsal bases which will be described in detail elsewhere. The hinder part of the foot has been similarly remodelled about the subtalar axis. Effectively, the heel has been laterally deviated altering the orientation of the subtalar articular surfaces in relationship to the calcaneus as a whole. In compensation for these changes which have realigned the functional anteroposterior axis of the lamina pedis, the trochlea of the talus has been rotated medially effectively diminishing the neck angle of the bone and altering the way in which the long flexor tendons enter the sole. Thus the functional anteroposterior axis of the foot as a whole has been realigned to more nearly coincide with the subtalar axis.

OLDUVAI HOMINID 8

These foot bones, from site FLK NN, dated at about 1.7 million years have, following the initial study by Day & Napier (1964), been almost uncritically accepted as belonging to a hominid with a perfected bipedal gait and they have usually then been assigned to the taxon *Homo habilis*. When analysed, however, even in a rather elementary fashion in the light of apomorphic features distinguishing the subtalar complex and the calcaneocuboid joint of the human foot, this conclusion seems to be quite indefensible. The excellent casts at the British Museum (Natural History) have been available for study and some points, but not all, have been verified on the original fossils in Nairobi.

The subtalar axis (Fig. 18A) clearly retains the relatively flat and oblique disposition, in relation to the foot as a whole, which is found in *Pan* and the subtalar joint surfaces (Fig. 13) are not remodelled into characteristic human form. The talus itself (Fig. 13) shows the

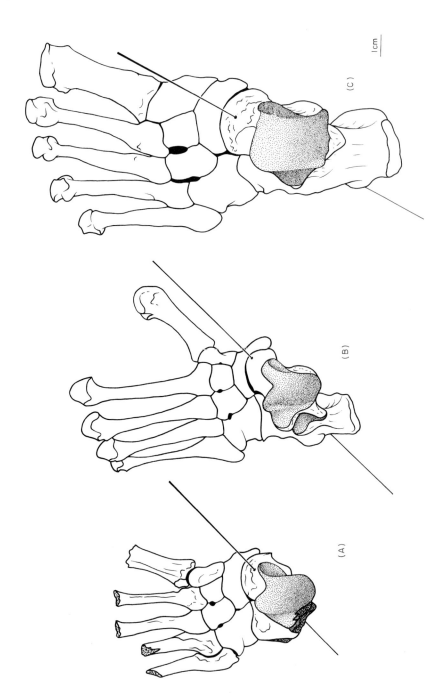

FIG. 18. Dioptograph tracings of the articulated tarsus and metatarsus of the casts of (A) OH8; (B) *Pan troglodytes*, B.M.76.437; (C) *Homo sapiens*, viewed from dorsally and with the subtalar axis shown in each case.

1cm

(A)

(B)

(C)

squat foreshortened appearance seen in the extant African apes (Fig. 2) and the calcaneal articular surfaces on its head and neck are similarly merged and no longer present a clearcut L-shaped pattern. Multivariate studies (Lisowski *et al.*, 1974) have suggested that the talus is most similar to that of the orang-utan and this probable false result may be attributed to the lack of incorporation of such morphological features as those above in the analysis.

The hallux is certainly approximated to the rest of the metatarsus to an appreciable degree but this is clearly the result of remodelling of the medial cuneiform. However, when the joint at the base of the first metatarsal is in close-packed position the great toe would seem to be quite appreciably divergent and possibly retained some grasping capability; this supposition is not excluded by the evidence for the presence of an adventitious articulation between the bases of the first and second metatarsals. The remainder of the anterior tarsus and the metatarsal bases show little evidence of the remodelling characteristic of the human condition and in contrast, a detailed consideration of the involved articulations indicates that an essentially ape-like morphology was retained. Although the heel is missing, its disposition is clearly indicated by the fractured surface, and there is little doubt that it was orientated as in *Pan*, and not man, and that the mode of entry of the long flexor tendons to the sole was similarly chimpanzee-like. Taken overall, it is clear that the subtalar complex functioned as in *Pan* and *Gorilla* and lacked the important human functional specializations which are important in the transfer of weight during walking.

Although the heads of the metatarsals are missing the degree of torsion of at least the medial three bones can be quite reliably estimated. The direction and degree of torsion of the first metatarsal in OH8 seems to have been comparable to that of *Pan* and *Gorilla* and to be less accentuated than in man; this is further suggestive evidence that some grasping function was retained. In contrast the second and third metatarsals appear to have achieved at least an approximation to the human condition.

The conarticular surfaces of the calcaneocuboid joint are probably the most remarkably individual feature of the OH8 foot yet have excited virtually no comment. In general, hominoid affinities in form are unmistakable (Figs 8 & 9) but the initial impression is of greatest similarity to man. More detailed analysis, however, leads to a rather different conclusion. The cuboid is prolonged into an exceedingly protuberant beak, larger than that of man, which forms the ventromedial corner of the bone and is lodged into a deep excavation of the calcaneus which undercuts the sustentaculum tali. Dorsolaterally this prong is continuous with a flattened area of limited extent but it is not expanded as far laterally into a concave cup as in man. Function seems to contrast markedly with that of man. Manipulation of the casts indicates that movement is restricted and that a small amount of supinatory action quickly impacts the cuboid prong under the sustentaculum tali welding the whole into a very stable close-packed structure. There is no hint of the specialized arcuate swing occurring in the human foot which carries the heel laterally in supination and so effects close-packing of the lamina pedis. A joint of this form could readily be evolved from a structure such as that seen in *Pan* which, as noted above, seems to possess a calcaneocuboid morphology central to the divergent patterns found in other hominoids.

One dimension which has helped fuel controversy over the affinities of the OH8 foot is the talar neck torsion angle which has been cited as 24.5°, within subhuman primate limits (Lisowski, 1967), and as 40.0°, similar to man (Day & Wood, 1968). This wide discrepancy illustrates the problems involved in biological measurement and that the impression of scientific precision may sometimes be quite spurious. This torsion angle is traditionally defined as the angle between the median axis of the head and the plane of the trochlea; this plane is obviously an important parameter determining function in the ankle joint. Thus, the traditional angle fails to express clearcut information about either the ankle joint or the subtalar joint complex. In fact, when the talus is removed from the articulated foot bones of OH8, *Pan* and *Homo sapiens* shown in Fig. 18 there appears to be little difference in the orientation of the navicular receptive surface for the talar head. It is not surprising then that if the tali are orientated upright upon a horizontal surface (on the standard basal talar plane) and viewed from distally there is little apparent difference in the angulation of the greatest diameter of the head. This is exactly the result one would expect in view of the type of remodelling described above for the evolution of the human foot.

CONCLUSIONS

The early Miocene apes of Africa, represented in part by the populations at Songhor and Rusinga Island, form a perfectly reasonable stock, at least as regards foot structure, for the evolutionary derivation of the great apes, even including the orang-utan. Further, the hominid line must surely have its roots within this stock. From the evidence of foot structure, as from other points of view, the closest affinities of man seem to be with the evolving African great apes, *Pan* and *Gorilla*.

It is clear from the evidence that the OH8 foot could not have participated in anything like the highly perfected bipedal gait of modern man. The overall impression is of the essential ape-like primitiveness of its structure, with affinities perhaps closest to *Pan*. The relatively adducted hallux is, of course, one distinguishing feature but there is suggestive evidence that it had not completely forsaken its grasping function. Another progressive feature is the stabilized form of the calcaneocuboid joint. It is of interest that Elftman & Manter (1935) postulated that in the evolution of the human foot the transverse tarsal joint had become relatively fixed in a position of plantar flexion. This has been partly vindicated by the fossil evidence: the joint of the OH8 foot is certainly relatively fixed, but not in a position of plantar flexion. If the OH8 foot is indeed on the human line this must have been followed by remodelling of the cuboid as described above to create a lateral longitudinal arch and the impression of fixation in plantar flexion.

The OH8 foot taken overall could quite credibly represent a transitional stage, still perhaps with some arboreal capability, in the evolution of a foot perfected for terrestrial bipedal locomotion. Alternatively, it may merely be an abortive hominoid evolutionary experiment, albeit an instructive one, in terrestrial locomotion.

ACKNOWLEDGEMENTS

The author wishes to express his thanks to the staff of the British Museum (Natural History), and particularly to Dr Peter Andrews, for allowing access to the casts of fossil material in their care, and also to Mr Richard Leakey, Director of the National Museum of Kenya, for permitting examination of the original fossil foot bones of Olduvai Hominid 8. The photographs are the result of the skilful work of Mr A. J. Aldrich.

REFERENCES

Andrews, P. & Van Couvering, J. A. H. (1975). Palaeoenvironments in the East African Miocene. In *Approaches to primate paleobiology*: 62–103. Szalay, F. S. (Ed.). Basel: Karger.

Bunning, P. S. C. & Barnett, C. H. (1965). A comparison of adult and foetal talocalcaneal articulations. *J. Anat.* 99: 71–76.

Carlsöö, S. (1972). *How man moves. Kinesiological studies and methods.* London: Heinemann.

Clark, W. E. Le Gros & Leakey, L. S. B. (1951). *The Miocene Hominoidea of Africa. Fossil mammals of Africa* No. 1: 1–117. London: British Museum (Natural History).

Day, M. H. & Napier, J. R. (1964). Fossil foot bones. *Nature, Lond.* 201: 969–970.

Day, M. H. & Wood, B. A. (1968). Functional affinities of the Olduvai Hominid 8 talus. *Man* 3: 440–445.

Delson, E. (1977). Catarrhine phylogeny and classification: principles, methods and comments. *J. hum. Evol.* 6: 433–459.

Elftman, H. & Manter, J. (1935). The evolution of the human foot with especial reference to the joints. *J. Anat.* 70: 56–67.

Lewis, O. J. (1972a). Osteological features characterizing the wrists of monkeys and apes, with a reconsideration of this region in *Dryopithecus (Proconsul) africanus*. *Am. J. phys. Anthrop.* 36: 45–58.

Lewis, O. J. (1972b). The evolution of the hallucial tarsometatarsal joint in the Anthropoidea. *Am. J. phys. Anthrop.* 37: 13–34.

Lisowski, F. P. (1967). Angular growth changes and comparisons in the primate talus. *Folia primatol.* 7: 81–97.

Lisowski, F. P., Albrecht, G. H. & Oxnard, C. E. (1974). The form of the talus in some higher Primates: a multivariate study. *Am J. phys. Anthrop.* 41: 191–216.

Lisowski, F. P., Albrecht, G. H. & Oxnard, C. E. (1976). African fossil tali: further multivariate morphometric studies. *Am. J. phys. Anthrop.* 41: 5–18.

MacConaill, M. A. (1945). The postural mechanism of the human foot. *Proc. R. Ir. Acad.* 50: 265–278.

MacConaill, M. A. & Basmajian, J. V. (1969). *Muscles and movements. A basis for kinesiology*. Baltimore: Williams and Wilkins.

Manter, J. T. (1941). Movements of the subtalar and transverse tarsal joints. *Anat. Rec.* 80: 397–410.

Pilbeam, D. (1969). Possible identity of Miocene tali from Kenya. *Nature, Lond.* 223: 648.

Schultz, A. H. (1930). The skeleton of the trunk and limbs of higher primates. *Hum. Biol.* 2: 303–438.

Shephard, E. (1951). Tarsal movements. *J. Bone Jt Surg.* 33: 258–263.

Szalay, F. S. (1975). Haplorhine phylogeny and the status of the Anthropoidea. In *Primate functional morphology and evolution*. 1: 3–22. Tuttle, R. H. (Ed.). The Hague: Mouton.

Tattersall, I. & Eldredge, N. (1977). Fact, theory, and fantasy in human paleontology. *Am. Scient.* 65: 204–211.

Symp. zool. Soc. Lond. (1981) No. 46, 189—218

Molecular Evolution and Vertebrate Phylogeny in Perspective

K.A. JOYSEY

University Museum of Zoology, Cambridge, UK

SYNOPSIS

Cladists and pheneticists differ about "evolutionary relationship" because morphological distance is not proportional to time of divergence and because parallel evolution is commonplace.

When amino acid sequencing of proteins became available it was hoped that this new evidence would provide a golden key to unlock phylogenetic problems. Furthermore, it has been claimed that the rate of evolution for each protein is constant enough to provide a "molecular clock" suitable for dating divergences.

Investigation of the evolution of the muscle protein myoglobin among mammals (especially primates) has shown that parallel evolution is commonplace also at the molecular level. The maximum parsimony procedure, which has been equated with the cladistic method, often provides several different most economical solutions, some, or all, of which may be unacceptable on the agreed weight of other evidence. Those species which are most unstable in their placing are often those which were included in the hope of resolving a phylogenetic dispute. This inability of morphological and molecular evidence to solve the same problems may have common causes. When pooling evidence from several molecules, the quest for objectivity may result in useful information being swamped; it is suggested that progress may lie in choosing particular substitutions in different molecules as being phylogenetically significant characteristics.

Investigation of absolute rates of molecular evolution requires estimates of dates of divergence based on the combined evidence of comparative anatomy and the fossil record. The basis of such dating is discussed in detail because there has been some misunderstanding of the criteria used by different groups. For myoglobin, the rate of fixation of mutations differs between lineages to such an extent that the so-called "molecular clock" is completely unreliable.

The myoglobin data have also been subjected to phenetic analysis by several different procedures; each produces a different clustering because each is based on different assumptions.

It has become evident that the functional morphology of the molecule itself places constraints upon possible change, and circumstantial evidence suggests that some of the observed parallel evolution may be of adaptive significance in physiological terms. It is concluded that the analysis of molecular data is useful for highlighting possible parallel evolution but that until the functional significance of amino acid substitutions is better understood, at both the molecular and physiological levels, such analysis is unlikely to contribute much to our knowledge of phylogeny.

INTRODUCTION

Bearing in mind the overall title of this symposium my original intention was to prepare a review article on "Perspectives in molecular evolution and primate phylogeny". I soon realized that this would be pointless because, relatively recently, we have been treated to a whole volume on *Molecular anthropology* (Goodman & Tashian, 1976), based on papers given at one of the Burg Wartenstein Symposia. Accordingly, rather than attempting to summarize the progress which has been made in recent years, and avoiding the temptation of trying to forecast further advances which may lie just over the horizon, I have decided to look back and take stock. Furthermore, because I believe that studies on primate phylogeny need to be judged in a wider framework, the title of my contribution has been revised to become "Molecular evolution and vertebrate phylogeny in perspective". Rather than singing praises I wish to seek re-appraisal.

If I were to set out to criticize the work of others then this essay would soon seem carping and somewhat ill mannered, and so I have chosen to draw largely upon team work with which I have been involved. I have been privileged to work in Cambridge jointly with Professor H. Lehmann, Dr A. E. Romero-Herrera and Dr A. E. Friday on the evolution of myoglobin among mammals, and especially among primates.

In our first paper (Romero-Herrera, Lehmann, Joysey & Friday, 1973), based on only 18 mammalian myoglobins, we expressed misgivings about the reality of the so-called "constant" rate of molecular evolution, but we were encouraged by the apparent potential of molecular data as a tool for elucidating evolutionary relationships. For the species involved in this early study, the pattern of relationship suggested by this single molecule was concordant with the generally accepted pattern of evolutionary relationship based largely on the evidence of comparative anatomy. We had high hopes that myoglobin might help to solve some outstanding problems involving species of disputed evolutionary relationships.

Our further work has been fascinating for its own sake in so far as it has contributed to our understanding of molecular evolution, but it has not been very effective in resolving evolutionary relationships. I feel that some of the problems which we have encountered in the course of our work on myoglobin are of general application, although this has not yet been generally admitted by those involved with other molecules. You must reach your own judgement on other work in the light of this self-criticism.

EVOLUTIONARY RELATIONSHIPS

First it is necessary to emphasize that the term "evolutionary relationship" can mean different things to different people. Comparative anatomy was pursued and found interesting for its own sake well
before the advent of evolutionary theory in the 19th century.
Comparative anatomists have long sought similarities and differences
in the way in which animals are constructed (and function). When
evolutionary theory emerged these similarities and differences took
on a new meaning and became measures of evolutionary relationship between species. Intermediate forms became "connecting links"
and the degree of man's relationships to other primates was judged in
terms of similarities and differences.

During the past few months there has been a heated correspondence in the pages of *Nature* on the evolutionary relationships of the
salmon, the lungfish and the cow (Halstead, 1978; Gardiner *et al.*,
1979; Halstead, White & MacIntyre, 1979) which has high-lighted
different usages of the term "evolutionary relationship".

If Fig. 1 correctly shows the pattern of evolutionary connection
between the salmon, the lungfish and the cow, then proponents of
phylogenetic systematics (cladistics) would advocate that the lungfish and the cow are more closely related because they share more
recent common ancestry, whereas those who attach more importance
to the degree of similarity (phenetics) would advocate that the
lungfish and the salmon are the more closely related because the

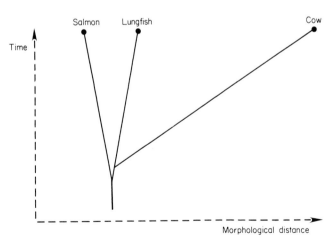

FIG. 1. Possible pattern of relationship between the salmon, the lungfish and the cow,
indicating a period of common ancestry between the lungfish and the cow.

morphological distance between them is least. It is evident that the two approaches reach different conclusions because morphological distance is not directly related to the time of divergence. If it were, then morphological distance could be used to provide an estimate of the time of divergence and there would be no disagreement between cladists and pheneticists. In the real world, it is evident from fossils that there have been widely different rates of morphological change; for example, lungfish have undergone relatively little morphological change since the time when mammals first appeared. Hence, differences in the rate of evolution along different lines of descent are fundamental to the difference of opinion between phylogenetic and phenetic systematists. Given the situation that morphological distance does not reflect time of divergence, the pheneticist attaches most importance to morphological distance, in the hope that this will best reflect genetical distance, whereas the cladist attaches most importance to the order (succession) of divergence, because this can be equated with a genealogical pattern of relationship.

The previous paragraph perhaps oversimplifies the issue because, in practice, the pheneticist tries to take into account as many morphological characters as possible, whereas the cladist rejects shared primitive characters and uses only shared derived characters in reaching his judgements on the genealogical affinity of "sister groups".

Apart from these differences in principle, the practice of both groups is confounded if they should fail to recognize, or fail to take into account, the phenomena of convergent and parallel evolution. There can be no doubt that parallel evolution is commonplace at the morphological level. Different groups of animals may have similar characteristics, not as a result of common ancestry, but as a result of parallel adaptation. Quite often it has been found that the use of different sets of characters as a basis for classification has resulted in "crossing classifications". We may not know which of a pair of characters has been the subject of parallel evolution, even though we can be sure that at least one of the characteristics has evolved more than once; this has often resulted in disputes as to which are the "best" characters to use. For the pheneticist, parallel evolution can contribute to similarity which may not reflect genetic similarity and for the cladist parallel evolution may not be "recognized", so that a shared derived state may not, after all, be unique (i.e. a synapomorphy). For this reason, compatibility analysis has been used to detect and eliminate those parts of the evidence which do not fit in with the majority view.

Returning now to molecular evolution, one of the special interests of our work on myoglobin is that the data have been treated both from a genealogical (cladistic) and a phenetic point of view, and that special attention has been given to the problems of rate of evolution and of parallel evolution.

MYOGLOBIN AND MAMMALIAN PHYLOGENY

During the early 1970s Professor H. Lehmann and his research student, Dr A. E. Romero-Herrera, were engaged in determining the primary amino acid sequence of several mammalian myoglobins, especially those of primates.

Myoglobin is a protein which is found as an intracellular component of red muscle. It combines reversibly with oxygen and it is believed to act as an oxygen carrier and reservoir within the cells. It consists of a single chain of about 153 amino acids, combined with a single haem group. Some sections of the peptide chain have a helical form (secondary structure), and the whole chain is looped in a very definite pattern. This conformation of the molecule (or tertiary structure) is maintained by a combination of Van der Waal's forces between non-polar side chains, by seven saltbridges and by 22 specific hydrogen bonds, some intra-chain and others inter-chain in position. We are here concerned with changes (or substitutions) in the primary structure of the molecule (i.e. the sequence of amino acids), which represent fixed mutations in the DNA coding for this protein.

Cladistic Analysis

In a sample of 18 mammal species representing six orders, it was found, for example, that at position 118, all of the species represented had the residue Lys, except for three cetacean species all of which had Arg at that position (Table I). Similarly, at position 31, all had Arg except for three species of New World monkeys which had Ser at that position. Obviously, one would conclude in each case that the exceptions were likely to be sharing a derived character state, probably the result of a single fixed mutation.

Again at position 110, all species had Ala, except for the two species of Old World monkeys which both had Ser, and the two apes and man which had Cys at that position. It so happens that in terms of the genetic code Ala (GCU/C) could have been changed to Ser (UCU/C) by a single point mutation and Ser could have been changed to Cys (UGU/C) by a further single point mutation. Hence, Ala can be interpreted as the primitive condition found in the majority of

TABLE I

Variation in the amino acid residue at nine chosen positions among 18 mammalian myoglobins.

	19	23	27	31	51	66	110	116	118
Man	Ala	Gly	Glu	Arg	Ser	Ala	Cys	Gln	Lys
Chimpanzee	Ala	Gly	Glu	Arg	Ser	Ala	Cys	His	Lys
Gibbon	Ala	Ser	Glu	Arg	Ser	Ala	Cys	Gln	Lys
Baboon	Ala	Ser	Glu	Arg	Ser	Ala	Ser	Gln	Lys
Macaque	Ala	Ser	Glu	Arg	Ser	Val	Ser	Gln	Lys
Woolly monkey	Ala	Ser	Glu	Ser	Ser	Thr	Ala	Gln	Lys
Squirrel monkey	Ala	Ser	Glu	Ser	Ser	Val	Ala	Gln	Lys
Marmoset	Ala	Ser	Glu	Ser	Ser	Val	Ala	Gln	Lys
Galago	Ala	Gly	Asp	Arg	Thr	Val	Ala	Gln	Lys
Sportive lemur	Ala	Gly	Glu	Arg	Thr	Thr	Ala	His	Lys
Harbour seal	Thr	Gly	Glu	Arg	Ser	Asn	Ala	His	Lys
Porpoise	Ala	Gly	Asp	Arg	Thr	Asn	Ala	His	Arg
Dolphin	Ala	Gly	Asp	Arg	Thr	Asp	Ala	His	Arg
Sperm whale	Ala	Gly	Asp	Arg	Thr	Val	Ala	His	Arg
Horse	Ala	Gly	Glu	Arg	Thr	Thr	Ala	His	Lys
Ox	Ala	Gly	Glu	Arg	Thr	Asn	Ala	His	Lys
Sheep	Ala	Gly	Glu	Arg	Thr	Asn	Ala	His	Lys
Kangaroo	Thr	Gly	Asp	Arg	Ser	Ile	Ala	Gln	Lys

mammals and Ser, retained in the Old World monkeys, can be interpreted as an intermediate step towards the Cys found in apes and man.

Using such methods as illustrated in the previous two paragraphs, and taking the genetic code into account, it is possible to reconstruct the most economical ancestral residue at each position and to pinpoint on a pattern of relationships the stems along which substitutions are most likely to have taken place. It is conventional to seek the most economical solution because it would appear to be capricious to do otherwise, and contrary to Occam's razor. In such reconstructions, preference has always been given to those solutions which require single step changes rather than double hits at the codon level. Bearing in mind that some amino acids may be coded in alternative ways, all interpretations of the pathway of change are undertaken by minimizing the amount of change at the codon level. It is to be expected that the introduction of new species into the cladogram may cause some changes in the interpretation of the nature of the ancestral residues (or their codons).

Even with a small number of species such as the original 18 myoglobin sequences, it is possible to observe probable examples of parallel substitution. For example, at position 19, all have Ala except for the kangaroo and the harbour seal, both of which have Thr. In such a case it would be conventional to take the majority view and regard Ala as ancestral and Thr as having been derived twice by parallel evolution, but it is equally economical to regard Thr as being ancestral, giving rise to Ala in the stem leading to the eutherians, but reverting to Thr in the habour seal as the result of a back mutation. In the absence of any other information, there would be nothing to choose between these two interpretations. On the other hand, if one were to adopt a less conventional pattern of relationships and place the Carnivora as the sister group of all other eutherians, then a more economical reconstruction would be available, with Thr being the ancestral condition retained in kangaroo and harbour seal, and Ala being the derived state present in all other eutherians as the result of a single fixed mutation. Such adjustment of the phylogenetic tree, which may be referred to as "branch-swapping", will often allow a more economical (parsimonious) reconstruction of the pathway of substitutions at a single position but, for example, it might prove that such a placement of the Carnivora required a more expensive interpretation of the events at other positions in the sequence. In such a case it is conventional (and reasonable) that the most parsimonious interpretation for the single residue would be rejected in favour of the most parsimonious solution overall.

Another likely example of parallel evolution was presented by position 27, where all species had Glu except the kangaroo, the three cetaceans and the bush baby, all of which had Asp at that position. Either Glu or Asp can be regarded as ancestral and in each case the observed distribution can be accounted for by parallel mutations to the alternative residue.

At position 23, all 18 species had Gly except the Old World monkeys, the New World monkeys and the gibbon, all of which had Ser. Accepting Gly as the ancestral residue for mammals, and accepting the usual pattern of relationship within the primates, the presence of Gly in man and chimpanzee can best be accounted for by a single back mutation.

Similarly, at position 51, where the anthropoid primates, the harbour seal and the kangaroo had Ser, and the prosimian primates, the cetaceans and the ungulates had Thr, it would seem that a minimum of two fixed mutations is necessary to account for the observed distribution on any generally accepted pattern of mammalian phylogeny and that these substitutions must involve either parallel evolution or a back mutation.

Similarly, at position 116, the kangaroo and all the primates had Gln, except for the chimpanzee and the sportive lemur which shared His with the rest of the eutherian mammals. Once again, on any generally accepted pattern of mammalian relationships, the most economical solution requires a minimum of three substitutions involving either parallel mutations, or back mutations, or both.

At all the positions mentioned so far only two or three alternative residues were found among the 18 species represented in the initial study and it has become evident that, even considering these few species, the most economical solutions have required numerous parallel and/or back mutations to account for the observed distribution of residues. At some other positions several residues were represented, the extreme being the hypervariable position 66 at which the following situation was found; Ala (man, chimpanzee, gibbon and baboon), Val (macaque, woolly monkey, marmoset, bush baby and sperm whale), Thr (squirrel monkey, sportive lemur and horse), Asn (ox, sheep, porpoise and harbour seal), Asp (dolphin) and Ile (kangaroo). Casual inspection reveals that some apparently useful phylogenetic information has been overlaid by parallel change resulting in a difference between baboon and macaque, between squirrel monkey and the other two ceboid species, between bush baby and sportive lemur, and between each of the three cetaceans.

When dealing with relatively few species it is feasible to examine the variable positions one at a time, in the manner illustrated in the preceding paragraphs. Taking the generally accepted pattern of relationships of the species as a starting point it is possible to fit the molecular evidence to it and, in so doing, to seek molecular evidence which is not concordant with it. Experiments in branch swapping can be performed and the new pattern can be applied to other variable positions. It soon becomes apparent whether or not the proposed economy for one position demands more expensive interpretations elsewhere.

In our initial study of the myoglobin of 18 mammal species (Romero-Herrera, Lehmann, Joysey et al., 1973) we set up a putative phylogenetic tree based on conventional systematic grouping, and we sought evidence for the branching point from the fossil record. Where fossils failed to resolve the sequence of two successive dichotomies we left their order unresolved as a trichotomy. After the molecular data had been fitted to this phylogenetic pattern it was reassuring to find that the trichotomies were resolved in a manner which was compatible with the evidence of comparative anatomy. For example, the apes and the cercopithecoids were found to share several substitutions which were not present in the ceboids. Feeling satisfied with the reconstructed pathways of mutational change in the myoglobin molecule we went on to consider rates of molecular evolution, which will be discussed in a later section of this chapter.

It will be appreciated that this somewhat laborious method of reconstructing hypothetical ancestral protein sequences is amenable to solution by computers. Programs have been developed for this purpose, some of which can be constrained to provide a minimum reconstruction of those substitutions which are required to account for the distribution of residues on a fixed phylogenetic pattern, whereas others can indulge in branch swapping procedures to determine that pattern which provides the most parsimonious solution. (Moore, Barnabas & Goodman, 1973; Moore, Goodman & Barnabas, 1973). Through the courtesy of Professor Goodman, and with the help of Dr R. Hanka, we have experimented with such programs and have found that the most parsimonious solutions resulting from branch swapping procedures produce phylogenetic patterns which include features that are totally unacceptable on zoological grounds (for example, the sportive lemur is often separated from the rest of the primates and is placed as sister group to the horse). Goodman and others have stressed that the most parsimonious solution provides a best estimate of gene phylogeny rather than animal phylogeny and have suggested that some bizarre "relationships" can be explained in terms of gene duplication.

I can see no biological basis for the assumption that a most parsimonious solution will reflect the actual course of evolution. It is evident both from the literature and from our own experience with branch swapping programs that a most parsimonious solution is often equalled by several other solutions representing different phylogenetic patterns. There is only one true phylogeny. In seeking a best estimate of that pattern, advocates of parsimony procedures have often found it necessary to make a choice between equally economical reconstructions. It is evident from the literature that preference has usually been shown for those solutions which best fit the pattern of relationships based on other evidence. While it would be perverse to do otherwise, this convention gives the impression that the morphological and molecular approaches have reinforced one another, and one tends to lose sight of the fact that an equally economical molecular solution has been submerged (on morphological grounds).

It must also be mentioned that for large numbers of species one can never be sure that the computer has found the most parsimonious solution, because it is impracticable to explore all theoretical possibilities. It is conventional to start the program with a pattern, often derived from a clustering procedure, which is believed to approximate to the actual phylogeny, and then allow branch swapping to seek a more parsimonious solution.

During the early 1970s more myoglobin sequences became available as a result of work in many different laboratories. In Cambridge we deliberately included some animals of disputed evolutionary relationship, such as the tree shrew and the hedgehog. By 1975 the complete myoglobin sequences of 29 mammals and two birds were available and in our contribution to the Burg Wartenstein Symposium we deliberately included a pentachotomy in our "starting pattern" to emphasize the degree of uncertainty regarding the relationships of several orders of mammals (Romero-Herrera, Lehmann, Joysey & Friday, 1976). We hoped that this pentachotomy would be resolved by the evidence of myoglobin. Using manual methods (rather than the computer) several possible resolutions of the pentachotomy were explored with a view to finding that pattern which allowed the most economical reconstruction of the evolution of myoglobin. In effect we were trying to reach a compromise with the principle of parsimony. The logic of the compromise seemed hopeful but we found that two different phylogenetic patterns provided equally parsimonious solutions. In one of these patterns the tree shrew and hedgehog were linked as sister groups on a common stem whereas in the other pattern the hedgehog was separated from the tree shrew by the stem leading to the Carnivora. Although our analysis of one of these two cladograms provided a detailed account of the evolution of myoglobin among the primates the most important finding of that paper is contained in the last paragraph where we stated:

> The analysis of the cladogram presented here has revealed that parallel evolution is commonplace. Indeed, of the 278 mutational events reconstructed, 139 changes occur in parallel (i.e. 50%). Parallelism occurs at 38 out of 83 positions which have accepted change (and at four of these positions the parallel event was a back mutation). The extent to which this high incidence of parallelism reflects functional constraint on the possibilities for change clearly deserves further investigation, but meanwhile it is evident that the unguided use of similarity as a basis for phylogeny is likely to lead to some false conclusions.

With the help of Dr R. Hanka we also tried computer analysis of the data using Moore, Barnabas & Goodman's (1973) maximum parsimony method. At each stage of branch swapping the computer presented several different steps, each of which was more economical than the previous pattern. In order to save computer time, rather than following all possibilities and rather than allowing the computer to choose each step at random, we edited the output and eliminated those steps which appeared to be leading towards unacceptable patterns. We were then disconcerted to find that the starting pattern appeared to influence the final result. For example, when tree shrew and hedgehog were started as sister groups on a common stem, they were never split apart regardless of where that stem was moved during branch swapping. On the other hand, when tree shrew and hedgehog were started separately they were never brought together on a common stem regardless of where each was moved during branch swapping. In retrospect, we have appreciated that the editorial constraint which we imposed at each step might have influenced the outcome in so far as, by not allowing the pattern to pass through an "unacceptable" stage, we may have prevented the computer from reaching the most economical solution. (As a topographical analogy, by not allowing the facility to climb out of a local hollow we may have prevented the program from finding the lowest point on the landscape.) However, each of the editorial decisions which we made might equally well have been chosen at random by the computer, so producing the same endpoint, and we have subsequently learned that Goodman himself indulges in a comparable "editing" procedure (Romero-Herrera, Lieska, Goodman & Simons, 1979).

Romero-Herrera, Lieska *et al.* (1979) have claimed recently that, having generated the ancestral protein chain in such a manner as to obtain the most economical solution for that particular tree, "it is possible, on the basis of the sequence alone, to position those species whose tree locations, as determined by other disciplines, are ambiguous". I believe that this is an exaggerated claim and that it is damaging to the reputation of studies in this field to try to preserve such a myth. Lehmann, Romero-Herrera, Joysey & Friday (1974) used myoglobin to try to elucidate the phylogenetic position of the tree shrew (*Tupaia*); we found that it could be accommodated equally well as a primate or as a non-primate because the alternative interpretations led to an identical set of changes from the postulated ancestral chain.

Elsewhere in their paper, Romero-Herrera, Lieska *et al.* (1979) correctly state: "A major problem introduced by the use of the computer in reconstructing phylogenetic trees is that an overwhelming number of equally parsimonious solutions may be produced". It has been our experience in Cambridge, when comparing such equally parsimonious solutions, that those species which are unstable in their placing are those which are of uncertain phylogenetic position on other criteria. We have elsewhere suggested that the inability of different criteria (i.e. anatomical and molecular) to solve the same problem may result from a similar cause or causes (Joysey, Friday, Romero-Herrera & Lehmann, 1977; Joysey, 1978), and I shall return to this point later in this section.

Our suspicion of the maximum parsimony technique was further deepened when the myoglobin sequence of the orang-utan became available (Romero-Herrera, Lehmann, Castillo, Joysey & Friday, 1976). As man, chimpanzee, gorilla and gibbon were all known to possess Cys at position 110 it was surprising that orang-utan, like the two Old World monkeys, had Ser at this position. The most parsimonious solution required that the lineage leading to the orang-utan should branch off before that leading to the gibbon (Fig. 2). Such a phylogenetic pattern, suggesting that the gibbon shared more recent common ancestry with man than did the orang-utan, receives very little support from other sources of evidence. We explored other phylogenetic patterns and found that either a back mutation to 110 Ser in the orang-utan or a parallel mutation to 110 Cys in the gibbon, each of which would increase the "cost" by only one fixed mutation, would allow alternative phylogenetic patterns which were more in keeping with morphological, behavioural and immunological evidence. In retrospect, it was ridiculous to expect that the most parsimonious solution based on a single molecule would provide the best estimate of the true phylogeny. In just the same way that the most parsimonious solution for the phylogeny at a single position within myoglobin may be subjugated by the weight of evidence from other positions, so it is reasonable to expect that the evidence of a single molecule may be subjugated by the weight of evidence from several other molecules.

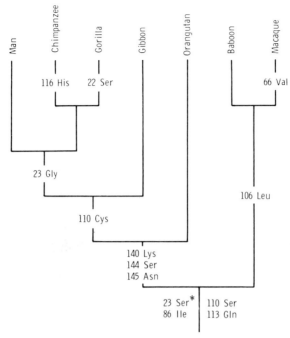

FIG. 2. A pattern which allows the most parsimonious solution of the amino acid substitutions leading to the known myoglobin sequences among catarrhines. (The asterisk indicates an ancestral condition brought forward from an earlier time.) It is now known (Bruce *et al.*, 1977) that the myoglobin sequences of gibbon and siamang are identical. (After Joysey, 1978.)

In a continued attempt to compromise with parsimony procedures (still dealing with the myoglobin sequences of 29 mammals and two birds) we sought the minimum solution for each of eight widely different possible phylogenetic patterns, all of which have been suggested in the literature (Romero-Herrera, Lehmann, Joysey & Friday, 1978). To summarize our results, there was little to choose between them; two alternatives were equal at 281 fixed mutations, one solution required 282 fixed mutations and five were equal at 283 fixed mutations. In other words, none of the possible phylogenies had a convincing advantage over any other; it again became apparent that the evidence of myoglobin had failed to resolve those same problems which had been intractable to other lines of evidence.

Apart from the obvious difficulty of separating those shared derived characters which truly indicate monophyletic groups from the parallel acquisition of such characters (especially where parallel evolution is commonplace), it seemed possible that the sequence of branching which we were attempting to resolve might have occurred over quite a short period of time relative to the known average rate of fixation of substitutions in myoglobin. For example, if the four branching events which were required to resolve the pentachotomy had all occurred within a period of ten million years (as is possible) then myoglobin, with an average rate of fixation of one mutation in about four million years (in any given lineage), is unlikely to resolve the pattern. Such a problem will arise either when several branching events occur in a short period of time relative to the average rate of fixation of mutations for that particular protein, or when a particular lineage passes through a period when the rate of fixation of mutations is well below the average rate for that molecule (i.e. when the sequence of a modern species is not very different from the reconstructed ancestral chain, as is the case for tree shrew myoglobin).

It would seem that the view expressed above is contrary to that of Romero-Herrera, Lieska et al. (1979). In the caption to their Fig. 3 (which is a cladogram based on the amino acid sequences of the alpha haemoglobin chain as determined by the maximum parsimony technique) they state: "Some uncertainty in phylogenetic placement surrounds the mammalian non-primate species and orders. A plausible explanation for this is the high rate of fixation of mutations which occurred during the adaptive radiation of mammals". It appears to me that a high rate during that period would have helped rather than hindered the situation provided that the high rate was not subsequently maintained.

In recent years attempts have been made to synthesize the evidence from several different molecules by pooling their sequences (Fitch & Langley, 1976a, b). It is well known that different proteins have evolved at different average rates: for example, fibrinopeptides are relatively fast, haemoglobin and myoglobin are intermediate and cytochrome C is relatively slow. Hence, fibrinopeptides would be unsuitable for resolving dichotomies which took place a hundred million years ago because the evidence is likely to have been overwritten by many subsequent events (including back mutations), and cytochrome C is unsuitable for resolving relatively recent speciation because few differences are found. Clearly, different parts of the phylogenetic tree can best be resolved by different lines of evidence and the pooling of fibrinopeptide data with myoglobin data seems likely to swamp the usefulness of the latter. In making this assertion it is, of course, recognized that within any one molecule some positions are highly conservative (slow changing) and others are hypervariable (fast changing). Thus, little importance can be attached to the evidence of hypervariable sites in the resolution of older dichotomies; it is of particular interest that for myoglobin the only residue reconstructed as being shared on the putative common stem of the anthropoid and prosimian primates is at position 66, which is the most variable myoglobin residue (Romero-Herrera, Lehmann, Castillo et al., 1976). In short, the myoglobin molecule provides no hard evidence that anthropoids and prosimians are sister groups, and when branch swapping is allowed to proceed to the most parsimonious solution the prosimians sometimes become the sister group of those mammalian orders of condylarthran derivation, rather than of the anthropoids.

It seems possible that problems of this nature may be solved in the future by combining the evidence from different molecules rather than by pooling it. Evidence might be chosen from different molecules bearing in mind the observed variation at each position. With such motives in mind we have sought a better understanding of the problems of assessing rates of molecular evolution.

Rates of Fixation of Mutations

It is possible to investigate relative rates of evolution, by comparing the amount of change incorporated into two lineages arising from a single branching point, without knowing the absolute date of that dichotomy. In our earliest study (Romero-Herrera, Lehmann, Joysey et al., 1973) we found that there were differences of this nature in the rate of fixation of mutations, and that even when considered over quite long periods of time (about 80 million years) there were still considerable differences in the overall rate in different lineages. Having some misgivings regarding the accuracy of this particular "molecular clock" we suggested that "average" rate was a more appropriate term than "constant" rate.

Estimates of the rate of molecular evolution are of interest because its supposed constancy has been one argument in favour of the hypothesis that a high proportion of fixed mutations are neutral, or nearly so, as far as selection is concerned. In order to investigate absolute rates of evolution, and possible fluctuations in rate along a single lineage, it is necessary to seek best estimates of the absolute date of each branching point on the phylogeny.

When we first sought to discover from the literature the time of origin of various groups of mammals we were dismayed to find that the dates provided were often based on a total absence of direct evidence. In principle, it is valid to suppose that direct fossil evidence of the presence of a particular group indicates that divergence from its sister group has already occurred at some earlier time, but it is a matter of opinion how much earlier. For example, McKenna (1969) stated that the earliest known bats were Eocene and listed the origin of the order as "probably Cretaceous", and he stated that the earliest known rodents were late Palaeocene and listed their origin as "late Cretaceous or Paleocene". Similarly, the earliest known Carnivora are Palaeocene but McKenna listed their origin as "late Cretaceous", the earliest recorded cetaceans are Eocene but he listed their origin as "Paleocene" and the earliest known Australian marsupials are probably Oligocene but he listed their origin as "late Cretaceous or early Tertiary". It is possible that, in due course, McKenna may be proved correct in all his "guesstimates" but, in trying to establish the most probable date for each of the branching events on our clado-grams we have, as a point of policy, rigorously demanded actual fossil evidence and we have rejected all speculation about fossils not yet found. In consequence, our dates should be regarded as minimum dates, because some are likely to be later than the actual event. Even this statement needs to be qualified because the choice of a particular fossil as indicative of a particular event is often a matter of opinion. For example, in dating the origin of hominids we took *Ramapithecus* as the relevant fossil in our first paper (Romero-Herrera, Lehmann, Joysey *et al.*, 1973), and we used *Australopithecus* in a more recent paper (Joysey *et al.*, 1977) not because of any particular commit-ment to either view but rather to explore the consequence of such differences of opinion. A full justification of the dates we have adopted in the Cambridge myoglobin studies is given in Romero-Herrera, Lehmann, Joysey *et al.* (1978). Before that paper was fully written Romero-Herrera left Cambridge to work with new colleagues in Detroit; this accounts for the somewhat different dating criteria adopted by Romero-Herrera, Lieska *et al.* (1979), which deserve some comment.

As indicated above, even when an actual fossil has been found it is necessary to weigh different opinions regarding its affinities and to sort out different opinions regarding the age of the deposit in which it was found. In our first paper (Romero-Herrera, Lehmann, Joysey *et al.*, 1973) we deliberately drew attention to these problems by giving several alternative opinions for the earliest known members of some groups and we also indicated the degree of stratigraphical uncertainty on our phylogenetic chart (reproduced here as Fig. 3). When the direct fossil evidence failed to resolve two successive dichot-omies we illustrated a trichotomy. We stated: "Trichotomies do not represent our opinion of the phylogenetic pattern but reflect a failure in the time resolution of the available fossil evidence and result from the superimposition of two adjacent dichotomies; their order of occurrence can be suggested on the basis of comparative anatomy".

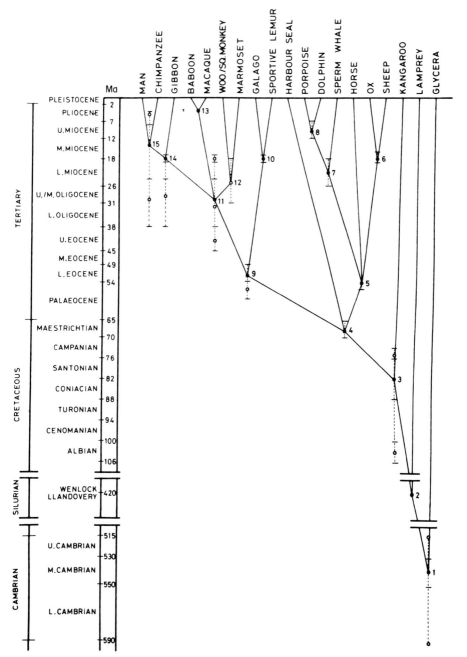

FIG. 3. Phylogenetic chart of 18 mammalian species, the points of divergence being based on the fossil record and numbered for reference purposes. The most probable point is indicated by a solid circle, possible alternatives by open circles. Horizontal lines indicate the time range within which a fossil cannot be definitely fixed; where the centre point of this range was used, fractions were rounded to the nearest million years older. Some radio-metric dates have been placed on the absolute time scale regardless of stratigraphical corre-lations. (After Romero-Herrera, Lehmann, Joysey et al., 1973.)

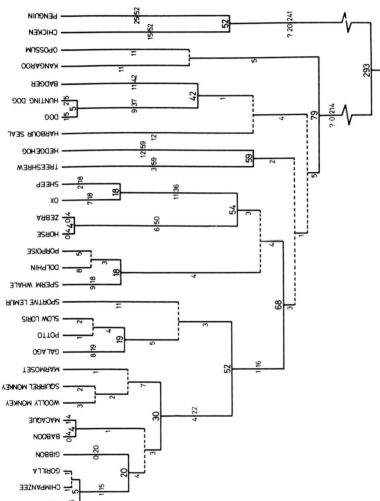

FIG. 4. A cladogram based on one of the two patterns which provided a most economical reconstruction of the pathway of amino acid substitutions in myoglobin. The date of divergence (where available) is shown above each dichotomy, and to the right of each branch the time span between successive dichotomies. The number of mutations reconstructed along each lineage between successive dichotomies is shown on the left of each branch. Horizontal dotted lines represent dichotomies for which no reliable date was available. (After Romero-Herrera, Lehmann, Joysey et al., 1978. The actual reconstruction of the mutations is shown in Fig. 3 of that paper; see footnote on p. 68 regarding sequence errors.)

It later became apparent that such a phylogenetic chart was likely to be misunderstood by those who do not read the text and so in papers subsequent to that by Lehmann *et al.* (1974) we have abandoned this form of presentation.

The trichotomies on our 1973 phylogenetic chart were resolved by the biochemical evidence in a manner concordant with an acceptable pattern of mammalian relationships, and they were illustrated as successive dichotomies in our cladogram showing the number of fixed mutations along each branch (see Fig. 5 in Romero-Herrera, Lehmann, Joysey *et al.*, 1973). In retrospect, it is unfortunate that the length of each branch was not drawn proportionately to the time elapsed, but we have adopted this improvement in our later cladogram (reproduced here as Fig. 4) which shows differences in the rate of fixation of mutations in a larger number of lineages (Joysey *et al.*, 1977; Joysey, 1978; Romero-Herrera, Lehmann, Joysey *et al.*, 1978). When we plotted a cumulative graph of the number of reconstructed base substitutions against time (see Fig. 31 of Romero-Herrera, Lehmann, Joysey, *et al.*, 1978) we bypassed those dichotomies for which no satisfactory date existed and joined the adjacent dated points in order to obtain the maximum information about rates. This created some polychotomies and we pointed out that these had no phylogenetic significance.

This preamble on our methods of presentation has been necessary because the Detroit group have presented a diagram (Fig. 2 of Romero-Herrera, Lieska *et al.*, 1979) purporting to be: "Phylogenetic reconstructions based on the fossil record and comparative anatomy (A) and the fossil record alone (B)". So far as the primates are concerned the rates of evolution shown in their Fig. 2B are derived from the Cambridge group's cladogram (reproduced here as Fig. 4) with the addition of information on the orang-utan (Romero-Herrera, Lehmann, Castillo *et al.*, 1976) and the siamang (Bruce, Castillo & Lehmann, 1977).

Despite a superficial similarity, it must be stressed that the Detroit group's Fig. 2B (Romero-Herrera, Lieska *et al.*, 1979) is constructed on quite different principles from our earlier phylogenetic chart (Romero-Herrera, Lehmann, Joysey *et al.*, 1973). In the accompanying text they state that: "When insufficient evidence was found to date a particular dichotomy the divergence was taken from an earlier more reliable point. Hence, several lineages are represented as having diverged from the same point." In contrast, when the Cambridge group (Romero-Herrera, Lehmann, Joysey *et al.*, 1978) used an earlier, more reliable point to calculate the rate of change between successive dated points they did not imply divergence from that earlier point. It is quite wrong to describe the Detroit Fig. 2B as a phylogenetic reconstruction; it is a corrupted version of the Cambridge cladogram from which any phylogenetic significance has been removed.

In principle the Cambridge cladogram is directly comparable with the Detroit Fig. 2A, both being based on the fossil record and comparative anatomy. The only difference lies in the matter of opinion regarding the dates themselves. Romero-Herrera, Lieska *et al.* (1979) state:

> ... it is obvious that the identification of a fossil with anatomical affinities to a given species indicates that divergence from an ancestor in common with another species had already occurred. Figure 2A takes this fact into consideration by designating the date of divergence at an inferred date earlier than the fossil. Where there was no fossil record the pattern of branching was derived from comparative anatomy and the dates of divergence were deduced from other sources such as continental drift and biogeographical information (Simons, 1976).

Hence, we differ only in the level of reliability of the evidence which we are prepared to accept. When I have given my opinion on the most probable date of a particular branching point I have usually discussed alternative possible dates and given reasons for not accepting them (Romero-Herrera, Lehmann, Joysey *et al.*, 1973, 1978). It is likely that opinions on some of these dates will change as more evidence is collected, but in Cambridge we have always felt that there should be available in the literature at least one set of dates based on actual material rather than circumstantial evidence.

Romero-Herrera, Lieska *et al.* (1979) stressed that they had no intention whatsoever of creating a confrontation between the two sets of dates. I have no objection to the Cambridge dates being regarded as a conservative set of possible dates, nor am I concerned by the expression of different opinions (provided that the basis of those opinions is clearly defined), but I cannot accept the implication that the branching pattern and comparative anatomy were not taken into account in arriving at the Cambridge dates. Reference to Romero-Herrera, Lehmann, Joysey *et al.* (1978) will immediately make clear that most of the dates which we regarded as being unreliable were rejected because the earliest undoubted fossil evidence that a particular dichotomy had occurred was younger than the evidence for a subsequent dichotomy.

(In parenthesis it must be noted that several confusing errors have crept into Fig. 2 of Romero-Herrera, Lieska *et al.* (1979). In Fig. 2A the numerals on the lines leading to *Papio* and *Erythrocebus* should be 0 rather than 6, and in Fig. 2B the numeral on the line leading to *Galago* should be 2.37 rather than 2, and that on the stem of the three Old World monkeys should be 6.5 rather than 26.)

Returning to Fig. 4 of this chapter, the combination of our dates with the reconstructed fixed mutations derived from one of the two most parsimonious cladograms (Romero-Herrera, Lehmann, Joysey *et al.*, 1978) indicated that there were considerable differences in the amount of change in different lineages. For example, whereas one lineage (to gibbon) appears to have accepted no mutations during the past 20 million years, another lineage (to ox) seems to have fixed seven mutations during the last 18 million years. Goodman, Barnabas, Matsuda & Moore (1971) drew attention to the apparently low rate of molecular evolution among higher primates; a similar observation applies to the myoglobin of the few Old World monkeys so far studied, but the myoglobins of the available New World monkeys and prosimians do not share this feature.

Sampling error will, of course, produce some fluctuations in rate but even over the relatively long period of about 80 million years the fastest rate of fixation of mutations is about three times the slowest rate (based on 34 and 11 fixed mutations, respectively). Hence, we discarded the "molecular clock" as being unreliable for dating divergences, at least within this span of time (Romero-Herrera, Lehmann, Joysey *et al.*, 1978). The degree of inaccuracy of the "clock" is best illustrated in Fig. 5, where the number of base substitutions has been plotted against time. It may be seen that 11 base substitutions can be equated with a period anywhere between 20 and 80 million years.

Phenetic Analysis

It is possible to prepare a difference matrix which summarizes the total number of amino acid differences between the myoglobins of pairs of species (Table II). The difference matrix retains no explicit information about the nature of the molecular changes which have led to the present situation. It is relevant to investigate how much useful information can be extracted from such a matrix because the data is comparable with that which is obtained through serological and immunological studies. Using the myoglobin data Friday (1977) has explored the performance of various procedures in constructing phylogenetic trees. It is sufficient to say here that there are various advantages and disadvantages of the several clustering procedures available and that each procedure provided a slightly different answer, which is to be expected as each is based on different assumptions.

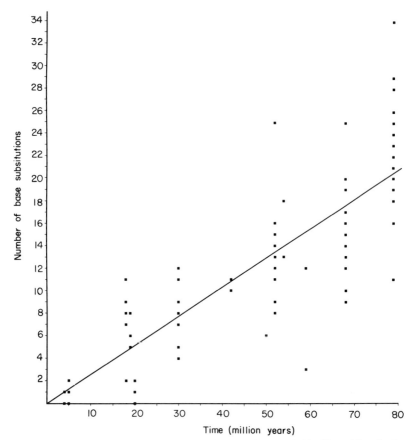

FIG. 5. Number of reconstructed base substitutions in myoglobin (from Fig. 4) plotted against time, the values for each section of a lineage being added to produce composite larger values. The fitted line (y = 0.26x) indicates an average rate of fixation of 1 mutation in approximately 4 million years, but there is considerable variation in the overall rate for different lineages, even over the longer periods of time. (After Romero-Herrera, Lehmann, Joysey *et al.*, 1978.)

The original Wagner method has been elaborated and presented in algorithm form and it is particularly attractive because it is free from assumptions concerning homogeneity of the rate of evolution (Farris, 1972). In the course of development of the network it soon became apparent that, on accepted external criteria, the opossum was discrepant, being assigned near the New World monkeys. The opossum was left out and the procedure tried again but still the placings of several animals were unexpected; for example, the harbour seal clustered with the cetaceans and the horse with the sportive lemur. These results are salutary. Because we know that the opossum is a

TABLE II

Matrix of amino acid differences between the myoglobins of 30 species of mammals and birds.

	Man	Chimpanzee	Gorilla	Gibbon	Baboon	Macaque	Woolly monkey	Squirrel monkey	Marmoset	Galago	Potto	Slow loris	Sportive lemur	Sperm whale	Dolphin	Porpoise	Horse & Zebra	Ox	Sheep	Harbour seal	Dog	Hunting dog	Badger	Hedgehog	Tree shrew	Kangaroo	Opossum	Chicken	Penguin	Lamprey	Glycera
Man	0																														
Chimpanzee	1	0																													
Gorilla	1	2	0																												
Gibbon	1	2	2	0																											
Baboon	6	7	7	5	0																										
Macaque	7	8	8	6	1	0																									
Woolly monkey	16	17	17	15	13	12	0																								
Squirrel monkey	17	18	18	16	12	12	4	0																							
Marmoset	14	15	15	13	11	10	4	4	0																						
Galago	23	24	23	24	22	21	21	23	23	0																					
Potto	16	17	16	17	15	14	20	22	18	12	0																				
Slow loris	19	20	19	20	18	17	19	19	17	11	3	0																			
Sportive lemur	22	21	22	23	21	21	19	17	17	21	18	17	0																		
Sperm whale	24	23	24	25	23	24	28	30	26	26	20	23	24	0																	
Dolphin	26	25	26	27	27	27	31	32	29	30	24	27	23	9	0																
Porpoise	20	19	20	21	21	22	26	26	24	25	20	23	18	15	12	0															
Horse & Zebra	18	17	18	19	17	17	18	18	16	21	14	15	19	21	16	12	0														
Ox	29	28	29	30	29	29	29	29	28	30	26	26	23	29	23	19	16	0													
Sheep	25	24	25	26	25	25	25	26	24	27	24	24	20	28	28	23	19	6	0												
Harbour seal	24	23	24	25	23	24	25	26	26	28	24	24	17	26	24	18	28	27	27	0											
Dog	22	23	22	23	21	21	21	21	21	20	21	20	24	31	29	26	28	28	26	27	0										
Hunting dog	23	24	23	24	22	22	21	21	21	20	21	22	22	30	33	26	21	30	28	22	3	0									
Badger	20	21	21	21	19	19	21	21	18	27	22	22	19	27	27	22	24	24	20	15	18	18	0								
Hedgehog	15	16	16	17	15	15	18	18	18	22	19	20	19	27	30	24	18	30	27	23	21	22	21	0							
Tree shrew	13	14	13	14	12	12	17	17	17	16	13	12	14	22	23	19	24	21	21	19	15	16	16	21	0						
Kangaroo	22	23	23	23	22	22	22	23	22	30	28	29	24	32	32	29	35	31	31	18	23	25	21	22	14	0					
Opossum	15	16	16	16	14	15	18	19	18	26	23	24	23	30	32	28	22	33	29	26	23	24	24	19	16	16	0				
Chicken	34	34	34	35	36	36	37	38	37	37	35	38	37	41	40	38	45	43	37	32	32	35	35	40	36	34	36	0			
Penguin	44	44	44	45	45	44	45	46	43	42	40	42	44	47	48	45	47	45	40	38	40	41	45	43	40	45	44	28	0		
Lamprey	105	105	105	105	106	107	107	107	102	103	102	103	102	104	105	102	104	105	105	104	107	103	103	104	104	107	104	104	107	0	
Glycera	102	102	102	103	103	104	105	106	105	105	102	103	107	102	105	103	105	104	104	105	105	105	106	105	103	103	105	107	104	97	0
Aplysia	107	107	107	107	106	106	106	106	106	109	108	108	106	103	102	105	105	109	106	107	110	110	106	105	108	104	105	108	110	92	106

Inset matrix (number of comparable sites):

	Therians	Chicken	Penguin	Lamprey	Glycera
Chicken	153				
Penguin	152	152			
Lamprey	136	136	135		
Glycera	136	136	135	124	
Aplysia	142	142	141	131	130

The myoglobin of Aplysia and monomeric haemoglobins of Glycera and lamprey are included for comparison. The small inset matrix shows the number of comparable sites. (After Romero-Herrera, Lehmann, Joysey *et al.*, 1978, but see footnote on p. 68 of that paper.)

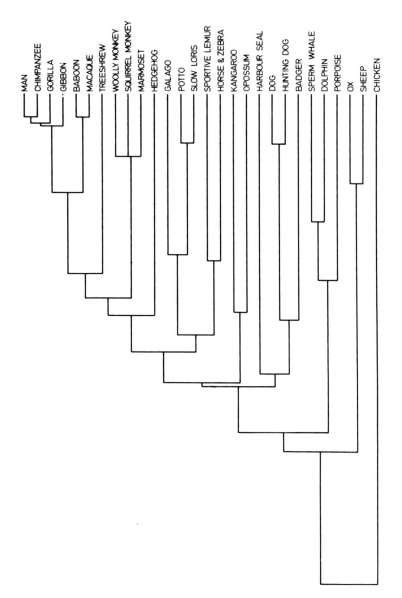

FIG. 6. An unweighted pair group method analysis based on the matrix of amino acid distances for 29 myoglobins. The branches are drawn to scale. (After Romero-Herrera, Lehmann, Joysey *et al.*, 1978).

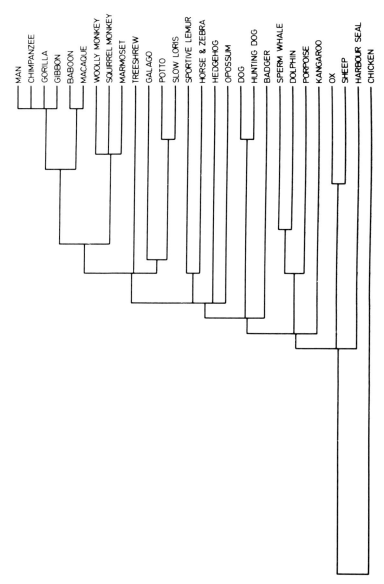

FIG. 7. A single linkage cluster analysis based on the matrix of amino acid distances for 29 myoglobins. The branches are drawn to scale. (After Romero-Herrera, Lehmann, Joysey et al., 1978.)

marsupial we do not read any phylogenetic inference into this similarity with the New World monkeys, but we should bear this in mind when seeking hints on the affinities of such animals as the hedgehog and the tree shrew, of which we do not know the phylogenetic relationships.

Of the other procedures, the unweighted pair group method embedded the two marsupials among the placentals (Fig. 6), and the single linkage clustering method widely separated the opossum and the kangaroo (Fig. 7). The most effective clustering method (i.e. the one which gave the "best" zoological picture!) was a procedure due to McQuitty which was originally devised for use in the social sciences; nevertheless it placed the horse among the primates.

Certain groupings, such as the cetaceans, the apes and the artiodactyls do remain stable through all the various clustering methods mentioned above and in this sense might be regarded as "reliable". On the other hand, the repeated clustering of the horse and sportive lemur remains a mystery and suggests pause for consideration.

MOLECULAR EVOLUTION

It is not difficult to envisage that some of the bonds and linkages which determine the secondary and tertiary structure of the molecule are of considerable functional importance. Indeed, the residues present at about 40% of the positions have been found to be invariant in all the myoglobins studied so far and it seems likely that many of these are of importance either in the functional morphology of the myoglobin molecule itself or in its interaction with other molecules (Fig. 8). At some positions where mutations have been fixed the changes are of a conservative nature; similarity in the properties of the residues involved suggests that the substitutions have been admitted under some constraint either at the molecular or the physiological level. Other positions seem to be able to accept amino acid substitutions quite readily and while one may be tempted to suggest that such positions may be of less functional importance, it must be remembered that changes at any position are likely to have consequences upon adjacent parts of the molecule. In this context, adjacent may be taken to include parts of the molecule which are adjacent in the secondary and tertiary structure, although not necessarily so in the primary structure, and any physiological or immunological consequences of a substitution are likely to be relevant in terms of natural selection.

The haem group lies in an internal cavity formed by hydrophobic

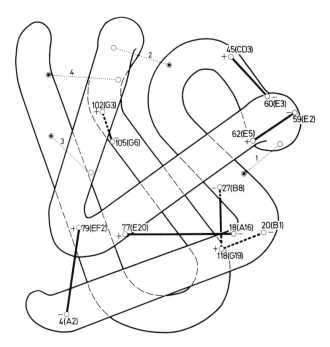

FIG. 8. An outline drawing of a myoglobin molecule, based on that of sperm whale, show-ing the positions of four inter-segmental hydrogen bonds (numbered 1 to 4), five inter-segmental and two intra-segmental salt bridges. (After Romero-Herrera, Lehmann, Joysey *et al.*, 1978.)

residues (the haem pocket), whereas the outside of the molecule in contact with the aqueous environment is composed mainly of hydro-philic residues. Using sperm whale myoglobin as a model (Watson, 1969), 33 residues can be described as internal and 120 as external, and some of the internal residues are of particular importance as haem contacts. It is found that internal residues are less variable than external residues, that those external residues participating in salt bridges are less variable than the rest and that such positions usually admit only conservative changes. It is possible to rationalize the likely chemical consequences of particular substitutions among the other external residues, perhaps affecting the surface charge on the molecule, but direct experimental work on the physiological proper-ties of the myoglobins of different species is mostly lacking.

The main feature which emerged from cladistic analysis of the fixed mutations in myoglobin was that parallel evolution is common-place (see Figs 3, 6 & 7 of Romero-Herrera, Lehmann, Joysey *et al.*, 1978) and the main feature that emerged from the phenetic analysis was that unrelated forms sometimes clustered together because of similarities in their myoglobin. On this basis, if it were found that the myoglobins of diving mammals were similar to one another regardless of their systematic position, it would be natural to seek an explanation in functional terms, and to regard such similarities as the product of parallel adaptation. In fact, there are several parallel substitutions shared by cetaceans and pinnipeds, e.g.: 54 Asp (harbour seal and dolphin); 83 Asp (sea lion and dolphin); 121 Ala (harbour seal, dolphin and porpoise); 122 Glu (harbour seal and porpoise); 152 His (harbour seal, dolphin and porpoise). Furthermore, the myoglobins of diving mammals (cetaceans and pinnipeds) and the penguin contain three to five residues of arginine, whereas those of land mammals contain only one or two residues of this amino acid, and there is one unusual substitution (56 Arg) found only in the harbour seal and the penguin (Romero-Herrera, Lehmann, Joysey et al., 1978). Whether or not these substitutions are related to adaptation to the aquatic environment awaits investigations on the physiological properties of different myoglobins.

It is evident that there are limited possibilities for change, there being a limited number of ways of remaining a functional myoglobin molecule. Given the constraints demanded by the functional morphology of the molecule itself and the constraints of the genetic code, it is to be expected that both will contribute to parallel change. One of the reasons that we have found it difficult to reach any useful phylogenetic conclusions is that parallel evolution is commonplace at the molecular level. Regardless of the cause or causes of parallel evolution, it has contributed to unexpected similarities between the myoglobins of species which we do not believe to be related (e.g. horse and sportive lemur).

Various mathematical procedures have been developed with a view to "recognizing" and eliminating the effects of parallel evolution but all are based upon the assumption that parsimony provides the best estimate of the course of evolution. Reconstruction of the most economical pathway for the evolution of a whole molecule often demonstrates that the assumption of parsimony is untrue for the changes at individual positions. Reconstruction of the most parsimonious solution for one molecule may demonstrate that the assumption of parsimony is untrue for another molecule. In contrast with earlier over-enthusiastic claims, Romero-Herrera, Lieska et al. (1979) concluded that: "the rigorous precision reflected in a parsimonious cladogram should not be confused with its correctness as a reconstruction. Such a cladogram provides us with a useful but nevertheless approximate picture of the past". In the Cambridge group we now doubt the usefulness of such a cladogram for phylogenetic reconstruction, but admit its usefulness for highlighting parallel evolution. Knowledge of this will contribute to our understanding of molecular evolution itself.

There is only one true pattern of phylogeny. It would be arrogant to believe that either biochemical or morphological data alone held the golden key to that pattern of evolution. I do not regard either molecular or morphological evidence as being a superior tool in this respect. I am convinced that we should be seeking a synthesis of diverse evidence rather than seeking to overthrow one line of evidence by another. Any such synthesis will inevitably lead to problems of assessing the relative weight to be given to different lines of evidence; in other words it will be a matter of opinion.

I do not believe that the study of myoglobin has made any substantial contribution to our knowledge of the pattern of evolution but I am convinced that knowledge of vertebrate phylogeny, based on the evidence of comparative anatomy and the fossil record, has made an enormous contribution to our understanding of the evolution of myoglobin. It seems possible that, in due course, fluctuations in the rate of fixation of mutations will be correlated with changes in adaptive zone. Future progress lies in a better understanding of the functional morphology of the molecule and in observations on the physiological properties of different myoglobins. It might then become possible to understand the functional significance of the numerous parallel substitutions which have helped to confound the study of phylogeny.

ACKNOWLEDGEMENT

I am grateful to my colleague, Dr A. E. Friday, for helpful criticism.

REFERENCES

Bruce, E. J., Castillo, O. & Lehmann, H. (1977). The myoglobin of Primates: *Symphalangus syndactylus* (Siamang). *FEBS Lett.* **78**: 113–118.
Farris, J. S. (1972). Estimating phylogenetic trees from distance matrices. *Am. Nat.* **106**: 645–668.
Fitch, W. M. & Langley, C. H. (1976a). Protein evolution and the molecular clock. *Fedn Proc. Fedn Am. Socs exp. Biol.* **35**: 2092–2097.
Fitch, W. M. & Langley, C. H. (1976b). Evolutionary rates in proteins: neutral mutations and the molecular clock. In *Molecular anthropology*: 197–219. Goodman, M. & Tashian, R. E. (Eds). New York: Plenum Press.
Friday, A. E. (1977). Evolution by numbers. In *Myoglobin*:142–166. Schnek, A. G. & Vandercasserie, C. (Eds). Brussels: Editions de l'Université.

Gardiner, B. G., Janvier, P., Patterson, C., Forey, P. L., Greenwood, P. H., Miles, R. S. & Jefferies, R. P. S. (1979). The salmon, the lungfish and the cow: a reply. *Nature, Lond.* 277: 175–176.

Goodman, M., Barnabas, J., Matsuda, G. & Moore, G. W. (1971). Molecular evolution in the descent of man. *Nature, Lond.* 233: 604–613.

Goodman, M. & Tashian, R. E. (1976). *Molecular anthropology.* New York: Plenum Press.

Halstead, L. B. (1978). The cladistic revolution – can it make the grade? *Nature, Lond.* 276: 759–760.

Halstead, L. B., White, E. I. & MacIntyre, G. T. (1979). Reply. *Nature, Lond.* 277: 176.

Joysey, K. A. (1978). An appraisal of molecular sequence data as a phylogenetic tool, based on the evidence of myoglobin. In *Recent advances in primatology* 3, *Evolution*: 57–67. Chivers, D. J. & Joysey, K. A. (Eds). London: Academic Press.

Joysey, K. A., Friday, A. E., Romero-Herrera, A. E. & Lehmann, H. (1977). Phylogenetic problems and parallel evolution in myoglobins. In *Myoglobin*: 167–178. Schnek, A. G. & Vandercasserie, C. (Eds). Brussels: Editions de l'Université.

Lehmann, H., Romero-Herrera, A. E., Joysey, K. A. & Friday, A. E. (1974). Comparative structure of myoglobin: Primates and tree-shrew. *Ann. N.Y. Acad. Sci.* 241: 380–391.

McKenna, M. C. (1969). The origin and early differentiation of Therian mammals. *Ann. N.Y. Acad. Sci.* 167: 217–240.

Moore, G. W., Barnabas, J. & Goodman, M. (1973). An iterative approach from the standpoint of the additive hypothesis to the dendrogram problem posed by molecular data sets. *J. theor. Biol.* 38: 423–457.

Moore, G. W., Goodman, M. & Barnabas, J. (1973). A method for constructing maximum parsimony ancestral amino acid sequences on a given network. *J. theor. Biol.* 38: 459–485.

Romero-Herrera, A. E., Lehmann, H. Castillo, O., Joysey, K. A. & Friday, A. E. (1976). Myoglobin of the orangutan as a phylogenetic enigma. *Nature, Lond.* 261: 162–164.

Romero-Herrera, A. E., Lehmann, H., Joysey, K. A. & Friday, A. E. (1973). Molecular evolution of myoglobin and the fossil record: a phylogenetic synthesis. *Nature, Lond.* 246: 389–395.

Romero-Herrera, A. E., Lehmann, H., Joysey, K. A. & Friday, A. E. (1976). Evolution of myoglobin amino acid sequences in Primates and other vertebrates. In *Molecular anthropology*: 289–300. Goodman, M. & Tashian, R. E. (Eds). New York: Plenum Press.

Romero-Herrera, A. E., Lehmann, H., Joysey, K. A. & Friday, A. E. (1978). On the evolution of myoglobin. *Phil. Trans. R. Soc.* (B) 283: 61–163.

Romero-Herrera, A. E., Lieska, N., Goodman, M. & Simons, E. L. (1979). The use of amino acid sequence analysis in assessing evolution. *Biochimie* 61: 767–779.

Simons, E. L. (1976). The fossil record of primate phylogeny. In *Molecular anthropology*: 35–62. Goodman, M. & Tashian, R. E. (Eds). New York: Plenum Press.

Watson, H. C. (1969). The stereochemistry of the protein myoglobin. *Prog. Stereochem.* 4: 299–333.

Reproductive Biology and
Neuroendocrinology

Chairman's Remarks

It is my pleasure to welcome you to the third section of this Symposium. It will not have escaped your notice that this morning we are venturing quite a distance from fields of primate biology. This is quite deliberate because it was felt proper that in this tribute we should give expression to what all of us have appreciated so much, the extraordinary breadth of interests of the one whom we are honouring in these two days. It is worth recalling that it was he who, as Secretary of the Zoological Society, initiated this series of Symposia which have been so successful over the years and in which he has taken a keen personal interest. I can assure you that at least the first two symposia were not in the field of primate biology; they dealt in fact with comparative endocrinology, and I had the pleasure of presenting one of them. A little bit before that time there had been initiated the still continuing series of International Symposia in comparative endocrinology, and well I recall his catalytic influence as Chairman in those early days. I have been checking my memory by looking at the records of the 1957 Symposium at Cold Spring Harbour. There indeed is the name Zuckerman, S., Birmingham, amongst the list of contributors. Zuckerman, S., Birmingham, did not actually contribute a paper, but I can assure you that he made a powerful mark on the conference with his extremely critical and perceptive concluding survey. What astonished me at the time, and what has continued to astonish me since, is his quite masterly ability to organize a mass of biological data to its maximum advantage. And so it is a particular pleasure to me, with those memories in my store, to preside at a session which will demonstrate, I hope, some of the perspectives which originate in the wider fields of interest of Lord Zuckerman. We are going to begin with a communication by one of his former Birmingham colleagues, Professor Holmes, now of the University of Leeds, who is going to talk to us about the pituitary gland.

We shall then pass to the insects. This is very proper, because neurosecretion is a field which has been advanced in a remarkable way along parallel and interlocking lines in invertebrate and vertebrate studies; it is a field which embraces the whole animal kingdom and, moreover, although it began as a primarily endocrine concept, we now know that the phenomenon of neurosecretion is a major element of total neurobiology and is by no means confined to a purely endocrine role. We shall then move up, if that is the correct word, to reptiles, to reproduction in the green sea turtle. It will be a particular pleasure to hear from Sir Alan Parkes for he was an early, if not the earliest, collaborator of Lord Zuckerman, drawn up into the collaboration by Lord Zuckerman's interest in the reproductive biology of primates.

E. J. W. BARRINGTON

Symp. zool. Soc. Lond. (1981) No. 46, 223–236

The Pars Intermedia of the Mammalian Pituitary Gland: Problems and Some Answers

R. L. HOLMES

Department of Anatomy, University of Leeds, UK

SYNOPSIS

The mammalian pars intermedia is very variable in relative size and structure in different species. Its development from the posterior wall of Rathke's pouch results in continuity with the pars distalis, which is associated with some inter-mingling of cells of these two parts of the adenohypophysis. A layer of flattened cleft cells separates the principal (secretory) cells from the intra-glandular cleft, or from the pars distalis when the cleft is not present. Cells of a similar type extend into the parenchyma of the pars intermedia and constitute a system of stellate cells closely associated with the secretory elements. Follicles are commonly found, and may have a specific, but unknown function.

The pars intermedia is innervated and undergoes structural changes if this nerve supply is interrupted. It is generally poorly vascularized, but preliminary studies have shown that horseradish peroxidase injected systemically diffuses rapidly and widely throughout the pars intermedia, suggesting that lack of a rich vasculature may not necessarily be associated with slow transport of secretion.

INTRODUCTION

The pars intermedia of the pituitary gland differs from the pars distalis, which constitutes a major component of the pituitary com-plex in cyclostomes, fishes, amphibians, reptiles, birds and mammals, in that it is not universally present as a distinct entity. In elasmo-branchs and teleosts it is closely associated with the neurohypophysis, forming the neurointermediate lobe. It is present in bony fishes, amphibians and reptiles, but absent in birds. Many species of mam-mals possess a pars intermedia, but it is lacking in some including Cetacea; it is well developed in some members of the primates which have so far been studied (for example the rhesus monkey) but poorly developed or vestigial in the mature chimpanzee, gorilla and human. Considerable problems of structure and function of this part of the mammalian pituitary gland remain to be solved.

The pars intermedia secretes a hormone, melanocyte stimulating hormone (MSH, Intermedin), which occurs in at least two chemical forms, alpha and beta. In amphibians, fishes and reptiles this hormone brings about darkening of the skin in response to variations in the background illumination, by causing dispersion of melanin granules in the melanophores. The hormone may also modify the degree of expansion of the pigment cells and, in the longer term, stimulate the synthesis of pigment. Chromatophores of other types, containing different pigments, may also be influenced.

In mammals, however, the cutaneous pigmentary system differs markedly from that of fishes, reptiles and amphibians. Chromatophores are lacking, and pigment is contained largely within the cells of the epidermis and the hairs derived from it. Melanin is synthesized by melanocytes lying mainly at the epidermo-dermal junction, which "inject" the pigment into the epidermal cells. Rapid changes in skin colour do not occur, although some species undergo seasonal variations in pigmentation of the coat, and ultra-violet light produces darkening of human skin. MSH is however present in the mammalian pars intermedia and its secretion appears to be controlled by MSH-releasing and MSH-release-inhibiting factors from the hypothalamus. It has been shown to promote the deposition of melanin in hair (Mistry & Weatherhead, 1976) and to exert a number of other effects (see review by Howe, 1973) but it is not clear which, if any, of these might be considered to illustrate the "normal" function of the mammalian gland.

ASPECTS OF DEVELOPMENT AND STRUCTURE

The pars intermedia develops from the posterior wall of Rathke's pouch, the upgrowth of epithelium from the future nasopharynx which gives rise to the adenohypophysis. This wall is from an early stage closely applied to the downgrowth from the neural tube which is to form the neurohypophysis. In many species of mammal, including some primates, the pars intermedia remains separated to a greater or lesser extent from the pars distalis by an intraglandular cleft which occupies the position of the original lumen of the pouch. The pars intermedia retains to a large extent in maturity its original relationships established both within Rathke's pouch and by the latter's association with the neural downgrowth, namely:

(1) the pars intermedia is directly continuous with the pars distalis superiorly and laterally;

(2) it may be associated with the pars tuberalis, which develops as lateral outgrowths from the upper part of Rathke's pouch, but it may be separated from this by part of the pars distalis;

(3) it commonly has a surface which forms part of the wall of the intraglandular cleft; but where this cleft is partially or wholly obliterated the pars intermedia is in direct contact with the pars distalis;

(4) it is closely applied to the infundibular process of the neuro-hypophysis and also, in many species, to the infundibular stem, particularly its lower, intraglandular part.

The varied form of the mammalian pars intermedia and its relation-ship to other parts of the pituitary are illustrated in Hanström's review (1966) and some of its particular features are discussed by Wingstrand (1966). These features are likely to have a bearing on the function of the pars intermedia.

The continuity of the pars intermedia with the pars distalis means that in areas where there is no intraglandular cleft it is often difficult to determine a precise boundary between the two parts of the gland. In the mouse and ferret, for example, there is an obvious extension of intermedia cells into the supero-rostral zone of the pars distalis, and a similar extension occurs laterally and inferiorly around the edge of the cleft. The extent of penetration of intermedia cells into the pars distalis is not entirely clear in optical microscopical studies. It has been suggested that in species such as birds and some mammals which do not appear to have a pars intermedia extensive penetration has occurred with intermingling of cells of the pars intermedia with those of the pars distalis. This could account for the presence of MSH activity in the pars distalis of those species so far examined.

A second consequence of the direct continuity of the pars inter-media with the pars distalis is that in these regions cells with the characteristics of those of the pars distalis occur within the pars intermedia (Fig. 1). These might either have developed *in situ* from stem cells lying in the marginal parts of Rathke's pouch, or have migrated in at a later stage of development. Whether or not such cells are functionally identical to similar ones lying within the pars distalis and controlled in the same way is at present a matter of speculation.

Penetration of intermedia cells into the neurohypophysis also occurs to a variable extent.

Cleft/Stellate and Principal Cells

The presence of an intraglandular cleft clearly determines a distinct separation of pars intermedia and pars distalis and serves as a barrier to vascular connexions between the two parts of the gland. The sur-faces of the cleft are lined by cells differing in form and structure from the glandular cells. In some species these are very flattened and most readily studied by electron microscopy. The author's obser-vations agree with those of Kagayama, Ando & Yamamoto (1969) that no basement membrane separates these cells from the rest of the pars intermedia. Tight junctions are present between adjacent

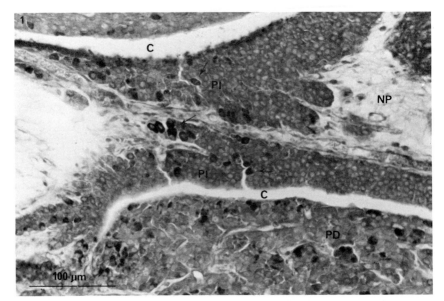

FIG. 1. Horizontal section through the upper part of the pituitary of a ferret. Dark granulated cells (arrows) typical of the pars distalis lie within the pars intermedia (PI). C, intraglandular cleft; NP, neural process; PD, pars distalis. (Stained by PAS-Alcian blue-haemalum-orange G.)

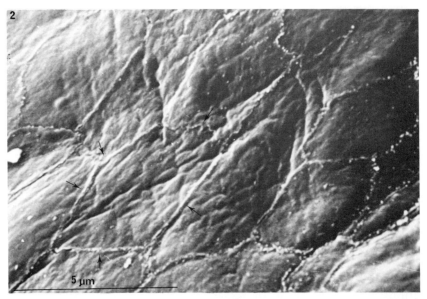

FIG. 2. Scanning electron micrograph of the surface of the pars intermedia bounding the intraglandular cleft (mouse) to show the outlines of the lining cells. The boundaries of one of these cells are indicated by arrows.

lining cells at their luminal ends, and scanning electron micrographs of the cleft surface show it to be completely covered by these thin plate-like cells (Fig. 2).

The cleft cells may perhaps be considered equivalent to the stellate cells of the pars distalis. Some of them extend long processes into the pars intermedia which come into contact with processes of similar cells lying amongst the principal secretory elements. Some of these deeply placed stellate cells are associated with the basement membrane of perivascular connective tissue spaces, where these occur. Tight junctions are seen between processes of stellate cells and also between these and glandular cells. In regions where the cleft is obliterated, cells with the characteristics of the lining cells of the two surfaces of the open cleft are closely apposed to each other.

There is no layer comparable to the cleft cells between the pars intermedia and the neurohypophysis. In some areas (in the mouse) cells of the pars intermedia are in close contact with nerve fibres with no intervening structures; in others dense basement membranes and blood vessels form a vascular lamina, from which some vessels extend into the intermedia. Nerve fibres also penetrate between the glandular cells, although there are considerable differences in the density of innervation in different species and even in different parts of the same specimen (see below).

The cleft or stellate cells probably correspond to the so-called ependymal type, with elongated slender nuclei, which was one of the two described in the mammalian pars intermedia by Lothringer (1886). He also noted the polygonal or rounded cells with oval nuclei which constitute the chief or principal components of the gland. Two forms of these were later described, a light and a dark, and it was suggested that these represented different states of activity of a single cell type. The distinction is particularly clear even by optical microscopy in some species such as the mole, but less marked in others.

Electron microscopy shows that the principal glandular cells have a characteristic content of granules, and that both dark (electron dense) and light (of variable electron-lucency) ones are usually present in the same cells. Other types of granulated cells have been observed, notably one resembling that thought to secrete ACTH in the pars distalis and which, it has been proposed, may secrete this hormone or one resembling it in the pars intermedia (Stoeckel, Dellmann, Porte & Gertner, 1971; Naik, 1972). Immunocytochemical studies by Naik (1973) have produced further evidence for the occurrence of ACTH-secreting cells, but he observed that MSH-secreting cells also reacted positively, although more weakly, for the presence of ACTH. The observations of Baker & Drummond (1972) suggested that while ACTH does occur in the pars intermedia, this part of the gland does not play a major role in the supply of the hormone in the same way as the corticotropic cells of the pars distalis.

Follicles

Follicular arrangements of cells, enclosing colloid or other material, constitute a common feature of the mammalian adenohypophysis. They occur to some extent in the pars distalis of most mammals so far studied, but in some species such as the potto they form a particularly prominent feature.

Follicles are commonly found in the pars intermedia. In the jird *Meriones unguiculatus* (Bhattacharjee, Chatterjee & Holmes, 1979) these are of two types. One is large and readily visible by optical microscopy; electron microscopy shows it to be lined by flattened cells whose somewhat electron-dense cytoplasm contains ribosomes, fine filaments, a compact Golgi complex and occasional lysosomes. The luminal borders of these cells bear microvilli; pinocytotic vesicles occur, and there are tight junctions between adjacent cells. The other type of follicle is smaller and only readily seen using the electron microscope. This is lined by low columnar or cuboidal cells which bear more microvilli than those of the other type of follicle and adjacent cells interdigitate extensively along their borders.

The lining cells of both types of follicle are surrounded by chief cells over much of their extent, and often show tight junctions with the latter. Elsewhere the follicular cells abut on the basal lamina of connective tissue (perivascular) space; some of the follicles extend to lie partly within the infundibular process. They may be empty or contain flocculent material. The second (smaller) type of follicle was found also to contain "crystalloid" strands, which show a pattern of linked hexagons (Chatterjee & Bhattacharjee, 1978). Similar material has been observed within the intraglandular cleft.

It is not known what function can be attributed to the follicles of the adenohypophysis. In the past they have been regarded as "remnants" of Rathke's pouch, devoid of any other significance. The structure of their constituent cells, however, is in keeping with some role involving transport and possible synthesis, and these possibilities should be borne in mind.

Endoplasmic Reticulum

The pars intermedia of the jird shows other features not so far described in the glands of other species of mammal. Many chief cells show the characteristics typical of pars intermedia cells, namely dense granules and granules of lower and variable electron density. Some of them, however, differ from the usual picture in that they contain particularly large amounts of rough endoplasmic reticulum. This occurs in several arrangements. Mostly it forms compact parallel cisternae, but it also occurs in concentric arrangements. In many of the chief cells lying close to the cleft and separated from its lumen only by the lining cells, the cisternae are greatly distended, so that only thin strands of cytoplasm remain between them. Yet another form occurs in cells lying adjacent to blood vessels or perivascular spaces, where the reticulum towards the vascular pole is arranged around a cytoplasmic vacuole which appears to have formed by distension within the endoplasmic reticulum. The luminal surface of the reticulum shows lack of ribosomes and blebbing. In some specimens indications that the lumen of the vacuoles might communicate with the extra-cellular space have been seen, but the possibility that this is an artefactual appearance cannot at present be excluded.

The functional significance of these unusual formations in secretory cells of the pars intermedia is not yet clear, although the great distension of the endoplasmic reticulum is reminiscent of that found in stimulated plasma cells, and probably indicates a high level of protein-synthetic activity.

INNERVATION

The pars intermedia contrasts markedly with the pars distalis as far as a nerve supply is concerned. The latter part of the adenohypophysis shows an almost total absence of nerve fibres, while the pars intermedia has a relatively rich innervation. Originally nerves were demonstrated by the classical metallic impregnation techniques but the lack of specificity of these methods, which may also reveal reticular fibres, gave rise to some erroneous conclusions. When stains for neurosecretory material were introduced, however, it became clear that neurosecretory nerve fibres were present in the pars intermedia, although the density of the innervation varied considerably in different species and in different parts of the tissue.

Electron microscopic studies were needed to demonstrate the details of the type and extent of the innervation. It is now clear that in a number of species amongst the rodents, carnivores and monkeys, typical neurosecretory fibres and also other types of nerve fibre enter the pars intermedia. Three types of fibre have been observed, namely type A (peptidergic, classical neurosecretory), type

B and type C (aminergic and cholinergic respectively). Principal (glandular) cells appear to be innervated by one or more types of these nerve fibres, but reports indicate that the extent varies from one in which almost every cell is innervated (e.g. ferret, Vincent & Anand Kumar, 1969) to a relatively sparse innervation in the mouse. Clearly there is a need for a much more precise assessment of the extent of the innervation in different species.

It seems probable that removal of the nerve supply to the pars intermedia, by cutting the pituitary stalk, changes the activity of its secretory cells. Such a lesion has been reported to be followed by an increase in cytoplasmic ribonucleoprotein (Holmes, 1961) in ferrets, and a true increase in volume of the intermedia in monkeys (Holmes, 1962). Increased secretion of MSH has also been found when the gland is separated from the hypothalamus (Kastin & Ross, 1964). These observations suggest that the nerves normally exert some inhibitory influence over the intermedia cells, although this would not rule out the existence of any excitatory influence as well.

VASCULARITY

Howe in his review (1973) draws attention to the variable vascularity of the mammalian pars intermedia; but although the density of blood vessels is low compared with other parts of the pituitary, it is certainly not negligible, as the work of Duvernoy & Koritké (1962) for example has shown. In regions where the intraglandular cleft separates the pars intermedia from the pars distalis, vessels enter the pars intermedia only from the neurohypophysis, and may form a series of loops extending into the intermedia and returning again to the infundibular process. Around the edges of the cleft, however, vessels can traverse the pars intermedia and communicate with those of both the pars distalis and the neurohypophysis. In rhesus monkeys it was found that vascular connections traversed the pars intermedia both in intact animals and also in those whose pituitary stalk had been divided (Holmes & Zuckerman, 1959). Suggestions that the pars intermedia of the laboratory rat is virtually avascular (for references see Howe, 1973) cannot be sustained (see Duvernoy & Koritké, 1962) and although the vascularity of the pars intermedia of the jird, for example, appears to be sparse, the tissue is certainly not devoid of vessels.

Horseradish Peroxidase Studies

The relatively poor vascularisation of this part of the pituitary gland in some species is surprising in view of its presumed endocrine function; but perhaps the nutrition of its cells and the passage of any secretory products into the systemic circulation depends less on the blood vessels than in other endocrine glands. The observations of Perryman (1976) and Perryman & Bagnara (1978) on frogs are of interest. They found that horseradish peroxidase (HRP) injected into the systemic circulation was rapidly distributed throughout the pars intermedia. Irregular extensions of perivascular spaces seemed to provide an efficient route for the transfer of material to and from cells, and possibly stellate cells played a part in such transfer.

Although in the mammalian pars intermedia perivascular spaces are not generally a marked feature, it seemed probable that the deep extensions of the cleft stellate cells might constitute the basis for a diffusion system. In order to test this hypothesis, a preliminary attempt has been made to apply the technique used by Perryman to the mouse.

Technique
Six adult male mice were used. They were anaesthetized with chloroform and injections made directly into the left ventricle. Four were injected with 20 mg HRP (Sigma Type II) dissolved in 1 ml normal saline; they were killed 2.5, 4.0, 4.5 and 5.0 minutes after the start of the injection. The other two animals were injected with normal saline alone and killed after 4.0 and 5.0 minutes.

After death the pituitary gland was rapidly exposed and flooded with 2% glutaraldehyde at a pH of 7.4 in phosphate buffer. The base of skull surrounding the sella turcica with the pituitary still in place was cut out and left in buffered glutaraldehyde in a refrigerator for four hours and transferred to buffer for overnight storage. The pituitary glands were then dissected free and small pieces (c. 0.5 mm) containing the pars intermedia were cut from them. These were incubated at room temperature in a solution containing 6 mg 3–3' diaminobenzine tetrahydrochloride and 0.1 ml 100 vol. hydrogen peroxide in 25 ml of 0.05 M Tris buffer (pH 7.6). The blocks were then rinsed in distilled water, post-fixed for two hours in cold 1% osmium tetroxide in 0.1 M phosphate buffer, dehydrated and embedded in the usual way using TAAB resin. Thin sections were later cut, stained by uranyl acetate followed by lead citrate and examined by electron microscopy.

Observations
The pars intermedia of the mice injected with saline alone showed no apparent differences from that of normal untreated mice.

The glands of each of the four animals injected with HRP showed extensive deposition of electron-dense reaction product. Near the surface of the cleft the reaction product was disposed in bands

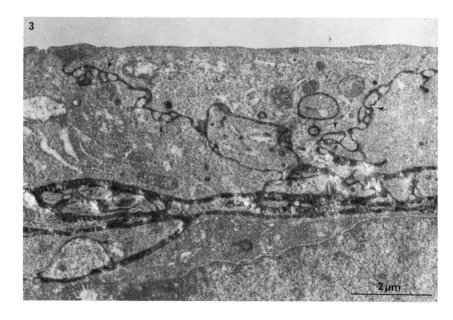

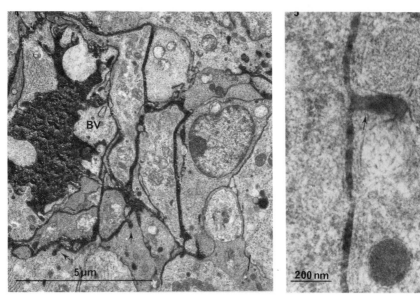

measuring up to *c*. 300 nm across running parallel to the surface. From these extended thinner bands of product which were continuous with complex formations outlining small, commonly circular areas of cytoplasm, which could in some areas be seen to be continuous with that of a cleft cell (Fig. 3). The appearance suggests that the reaction product was outlining extensive and complex interdigitations of the membranes of adjacent cells. Similar linear bands and complex patterns also occurred around blood vessels.

Towards the lumen of the cleft the dense material could be seen to extend only as far as the tight junctions between adjacent cells. In a few sites appearances suggestive of the passage of reaction product into the cells were seen.

More deeply in the pars intermedia, away from the cleft, reaction product outlined the secretory cells, and in sections in which a blood vessel was present there was clear continuity between the thin intercellular bands of the product with the thicker ones around the perivascular cells. The lines of reaction product between the secretory cells appeared to demonstrate intercellular spaces rather than processes of stellate cell cytoplasm (Fig. 4). At a number of sites (Fig. 5) it seemed that reaction product was being taken into the secretory cells as coated vesicles, and apparently isolated small aggregations of material lay free in the cytoplasm in some cells. Reaction product was also seen associated with dense bodies of the order of size of lysosomes.

The similarity of the pattern of distribution of reaction product in the pars intermedia in the mouse to that shown by Perryman (Perryman, 1976; Perryman & Bagnara, 1978) in the frog is striking, as is also the finding that reaction product was widely distributed by 2.5 minutes after the start of the injection of HRP, the shortest interval before killing of the four mice. It is possible that such distribution

---

FIGS 3—5. Transmission electron micrographs of the pars intermedia of a mouse injected with HRP four minutes before death.

(3) The intraglandular cleft can be seen at the top of the figure. Dense bands of HRP reaction product lie in the intercellular spaces between the lining cells. Towards the lumen this material reveals the complex interdigitations of adjacent cell surfaces (arrows). Note that in the top left hand corner the precipitate does not extend to the lumen, suggesting that diffusion of HRP was stopped here by a tight junction, although this is not clearly shown.

(4) A mass of precipitate lies within the lumen of a blood vessel (BV) within the pars intermedia. Outside the lumen it marks the spaces between perivascular cells and secretory glandular cells, and in places appears to be entering the cytoplasm (arrows).

(5) High power electron micrograph of HRP reaction product in the extracellular space between two secretory glandular cells of the pars intermedia. The arrow indicates uptake of material by one of the cells.

would be found after an even shorter interval; but in any event these preliminary observations suggest that the relative avascularity of the pars intermedia imposes little limitation on the rapid transport of material from the circulation throughout the pars intermedia, via the apparently small intercellular spaces between the secretory cells and the larger ones associated with perivascular elements and stellate cells. It appears likely that both the stellate and secretory cells can take up material from the extracellular space, but whether or not the stellate cells provide an intercellular route for interchange between blood and secretory tissue needs further study.

CONCLUSIONS

The mammalian pars intermedia secretes MSH, but the importance of this hormone in mammals is not clear. Although it has been shown to exert an effect on mammalian pigment cells, it is possible that, as suggested by Howe (1973) and others, its main activity may have become transferred either to other dermal structures such as sebaceous glands (Thody & Schuster, 1971), or to tissues having no close association with skin. The gland probably also secretes ACTH, and it has been proposed that it might constitute a source of this hormone supplementary to that of the pars distalis, controlled by a different release mechanism (Stoeckel et al., 1971). The source of ACTH in the pars intermedia could be from cells that originated in the pars distalis rather than from the posterior wall of Rathke's pouch. It should also be borne in mind that methods used to demonstrate the presence of ACTH might in effect be showing a precursor of MSH, since the two hormones have some basic molecular structure in common.

The great variations in relative size, cytology and innervation of the pars intermedia in different mammalian species suggest that its functional role, or at any rate its importance, varies from one species to another. Comparative studies, such as those of Legait (1963), need to be extended using the techniques now available to correlate structure with secretory activity as well as with habitat, and to examine the possibility of cyclical variations correlated with, for example, breeding seasons or hibernation.

The relatively poor vascularity of the pars intermedia, although again variable between species, is scarcely typical of endocrine organs. The preliminary studies reported above perhaps indicate that the vascular system is of less importance as a means of transport within the gland than might have been expected, and indicates another aspect of the gland in need of further study.

REFERENCES

Baker, B. L. & Drummond, T. (1972). The cellular origins of corticotropin and melanotropin as revealed by immunochemical staining. *Am. J. Anat.* 134: 395—410.

Bhattacharjee, D. K., Chatterjee, P. & Holmes, R. L. (1979). Follicles and related structures in the pars intermedia of the adenohypophysis of the jird (*Meriones unguiculatus*). *J. Anat.* 130: 63—67.

Chatterjee, P. & Bhattacharjee, D. (1978). Some ultrastructural observations on the follicles in the pars intermedia of the jird adenohypophysis. *J. Anat.* 127: 653.

Duvernoy, H. & Koritké, J.-G. (1962). Contribution à l'étude de la vascularisation du lobe intermédiaire de l'hypophyse. *Bull. Ass. Anat., Paris* 48: 547—561.

Hanström, B. (1966). Gross anatomy of the hypophysis in mammals. In *The pituitary gland* 1: 1—57. Harris, G. W. & Donovan, B. T. (Eds). London: Butterworths.

Holmes, R. L. (1961). Changes in the pituitary gland of the ferret following stalk section. *J. Endocr.* 22: 7—14.

Holmes, R. L. (1962). The pituitary gland of normal and stalk-sectioned monkeys with particular reference to the pars intermedia. *J. Endocr.* 24: 53—58.

Holmes, R. L. & Zuckerman, S. (1959). The blood supply of the hypophysis in *Macaca mulatta*. *J. Anat.* 93: 1—8.

Howe, A. (1973). The mammalian pars intermedia: a review of its structure and function. *J. Endocr.* 59: 385—409.

Kagayama, M., Ando, A. & Yamamoto, T. Y. (1969). On the epithelial lining of the cleft between pars distalis and pars intermedia in the mouse adenohypophysis. *Gunma Symp. Endocr.* 6: 125—135.

Kastin, A. J. & Ross, G. T. (1964). Melanocyte-stimulating hormone (MSH) and ACTH activities of pituitary homografts in albino rats. *Endocrinology* 75: 187—191.

Legait, E. (1963). Cytophysiologie du lobe intermédiaire de l'hypophyse des mammifères. In *Cytologie de l'adénohypophyse*: 215—230. Benoit, J. & Da Lage, C. (Eds). Paris: C.N.R.S.

Lothringer, S. (1886). Untersuchungen an der Hypophyse einiger Säugetiere und des Menschen. *Arch. mikrosk. Anat.* 28: 257—292.

Mistry, U. & Weatherhead, B. (1976). Effects of α-melanocyte-stimulating hormone on the melanin content of the hair of the Siberian hamster (*Phodopus sungorus*). *Gen. comp. Endocr.* 29: 267.

Naik, D. V. (1972). Electron microscopic studies on the pars intermedia in normal and in mice with hereditary nephrogenic diabetes insipidus. *Z. Zellforsch. mikrosk. Anat.* 133: 415—434.

Naik, D. V. (1973). Electron microscopic-immunocytochemical localization of adrenocorticotropin and melanocyte stimulating hormone in the pars intermedia cells of rats and mice. *Z. Zellforsch. mikrosk. Anat.* 142: 305—328.

Perryman, E. K. (1976). Permeability of the amphibian pars intermedia to peroxidase injected intravascularly. *Cell Tiss. Res.* 173: 401—406.

Perryman, E. & Bagnara, J. T. (1978). Extravascular transfer within the anuran pars intermedia. *Cell Tiss. Res.* 193: 297—313.

Stoeckel, M. E., Dellmann, H. D., Porte, A. & Gertner, C. (1971). The rostral zone of the intermediate lobe of the mouse hypophysis, a zone of particular concentration of corticotrophic cells. *Z. Zellforsch. mikrosk. Anat.* 122: 310–322.
Thody, A. J. & Schuster, S. (1971). Pituitary control of sebum secretion in the rat. *J. Endocr.* 51: vi–vii.
Vincent, D. S. & Anand Kumar, T. C. (1969). Electron microscopic studies on the pars intermedia of the ferret. *Z. Zellforsch. mikrosk. Anat.* 99: 185–197.
Wingstrand, K. G. (1966). Microscopic anatomy, nerve supply and blood supply of the pars intermedia. In *The pituitary gland* 3: 1–27. Harris, G. W. & Donovan, B. T. (Eds). London: Butterworths.

Symp. zool. Soc. Lond. (1981) No. 46, 237—251

Form, Activity and Function in Insect Neurosecretion

L. H. FINLAYSON

Department of Zoology and Comparative Physiology, University of Birmingham, Birmingham, UK

SYNOPSIS

A precise definition of neurosecretion is not possible although "typical" neurosecretory neurones are easily recognized. The first proof of an endocrine function for neurosecretory cells was obtained for insects.

Neurosecretory cells occur in every ganglion of the central and visceral nervous systems of insects. There is also a system of extraganglionic neurosecretory neurones with granule-laden fibres running in all directions in thorax and abdomen. One such neurone has been found in close association with the ecdysial gland. Another system of neurosecretory fibres runs from neurones in the central nervous system. Fibres of both central and peripheral origins terminate in neurohaemal endings or in close association with a wide range of tissues.

Action potentials generated by neurosecretory neurones of invertebrates and lower vertebrates are longer in duration than those of "ordinary" neurones. In the stick insect the neurosecretory action potential is propagated by the movements of calcium ions in both soma and fibres.

At least ten peptidergic neurohormones have been extracted from insects and there may be a similar number of other active peptides. Immunochemical analyses have revealed a number of peptides which appear to be similar to mammalian peptides and in some cases physiological mimicry has been demonstrated.

Innervation of target tissues may be aminergic. Several amines have been found in insect tissues and one neurone of the metathoracic ganglion has been shown to secrete octopamine.

INTRODUCTION

The study of neurosecretion in lower vertebrates and invertebrates at Birmingham began with the appointment in 1959 by Lord Zuckerman of the late Sir Francis Knowles to a lectureship in endocrinology in the Department of Anatomy in Birmingham. He made notable contributions to the subject and all of us who have followed him have been influenced by his work. My own excursions into neurosecretion began by accident when Dr. M. P. Osborne and I discovered that some of the neurones we were studying were not sensory but neurosecretory (Finlayson & Osborne, 1968). Everyone who looks at neurosecretory cells by whatever technique they employ cannot fail to be struck by their distinctive appearance. Under the light microscope, using dark-field illumination, they are a delicate blue, in contrast to the rest of the nervous tissue which is quite transparent, or stained by a variety of methods they stand out in strong contrast to the ordinary neurones. Under the electron microscope the secretory inclusions are typically so numerous and so characteristic in shape and structure that once more the specific distinctiveness of these cells is obvious. As in all biological phenomena, however, we soon discover that distinctions are not always perfectly clear and definitions rear their ugly heads. Like many others, Knowles became involved in arguments over the definition of neurosecretion and at first tried to limit it to the confines of endocrinology (Knowles & Bern, 1966). The phenomenon, unfortunately, refuses to be so limited (Knowles, 1974) and a precise definition is not yet agreed, nor do I see why it ever should be. There is no doubt that many neurones can be properly described as neurosecretory because they are so distinct in appearance, but when the neurones in question are not so distinct, or if definitions in terms of function are attempted, difficulties arise.

The study of neurosecretion in invertebrates and in insects in particular has a long history and it was in insects that an endocrine function for neurosecretion was first demonstrated. Kopeč showed in 1917 that the brain of the gipsy moth caterpillar was essential for moulting and Wigglesworth in 1939 localized the source of the brain hormone in two groups of neurosecretory cells in the medial region of the forebrain of the blood-sucking bug *Rhodnius*. Since then great advances have been made in studies of neurosecretion in insects and in mammals. This review will present a brief survey of the neurosecretory system of insects, with particular emphasis on the peripheral nervous system, and make some comparisons with that of the mammal.

FORM

There are many descriptions of the appearance and distribution of neurosecretory cells in the central nervous system of insects, but there is considerable variation in these descriptions, no doubt reflecting differences between insects but also, no doubt, reflecting differences in techniques. The use of stains for light microscopy has produced a large literature but a confusing picture. Clearly there are within the ganglia of the central nervous system of insects a variety of neurosecretory neurones, judged by their reactions to stains, and also by their appearance under the electron microscope. Neurosecretory neurones have been found in every ganglion of the central nervous system and of the visceral (stomatogastric) system.

In addition to the neurosecretory cells in ganglia there are others in a peripheral system which have received much less attention generally, but which have been investigated in the Department of Zoology and Comparative Physiology in Birmingham. Not only are there peripheral neurosecretory neurones, but there is widespread distribution of neurosecretory material by nerve fibres from both central and peripheral neurones. The peripheral system has been studied in most detail in the stick insect *Carausius morosus* (Orchard & Finlayson, 1977c; Fifield & Finlayson, 1978).

Peripheral Neurones

Outside the central nervous system of insects there are many neurones and, until comparatively recently, they were all considered to be sensory in function. There are two basic types, one associated with setae and other cuticular structures and the other more deeply seated, in a few cases attached to a special strand of connective tissue but more often without any auxiliary apparatus (Finlayson, 1976). It is the second type which is of interest because it includes the neurosecretory cells. They are typically multipolar and are found under the epidermis, on muscle fibres, on connective tissue strands and on nerves in many parts of the body. Those of the stick insect which have been investigated in detail have been shown to be either mechanoreceptors that respond to being stretched by an increase in their discharge frequency, or to be neurosecretory neurones (Orchard & Finlayson, 1976). A similar distinction can be found in peripheral neurones of blowflies (Gelperin, 1971) and of tsetse flies (Anderson & Finlayson, 1978). The neurosecretory neurones belong to the series that are situated on or in peripheral nerves (Fifield & Finlayson, 1978). In each abdominal segment of the stick insect there are four neurosecretory neurones on each link nerve which connects the transverse nerve of the median (unpaired) system to the main segmental nerve, making a total of eight per segment. In the thorax there is a similar arrangement but with two neurosecretory neurones in the prothorax, 12 in the mesothorax and eight in the metathorax (Orchard & Finlayson, 1977c).

Under the electron microscope these link nerve neurones (LNNs) are seen to be packed with spheroidal, electron-dense granules. They are membrane-bound when recently secreted by the Golgi bodies but when fully formed they are so densely packed that the membrane is invisible. The occurrence of "omega" profiles where granules are being released makes it unlikely that the membrane disappears before release of the granule because the "omega" profile is probably produced by the fusion of the membrane of the granule with the membrane of the nerve fibre.

From the cell bodies of the LNNs we traced dendritic processes running in every direction possible and terminating over a wide area of the peripheral nervous system (Fifield & Finlayson, 1978). At that time we did not find LNN fibres terminating in close proximity to any tissues, but since then Dr Ian Findlay has traced small processes from the LNNs to the spiracular muscle. He also found, accompanying these processes, a neurosecretory fibre from the central nervous system which contains granules of a different form (Fifield & Finlayson, 1978). Recently, Mrs Anne Griffiths, a research student in my laboratory, has found a typical peripheral neurosecretory neurone in close association with the endocrine ecdysial (prothoracic) gland of caterpillars of the moths *Mamestra*, *Agrotis* and *Manduca*. This is a very important discovery and rather surprising in view of the immense amount of work which has been carried out on the ecdysial gland.

Although we have not looked for neurosecretory neurones on the gut of the stick insect, they have been found on the gut of other insects. In the cockroach *Leucophaea* (Miller, 1975) the larva of the rhinoceros beetle *Oryctes* (Nagy, 1978) and in the larva of the moth *Manduca* (Reinecke, Gerst, O'Gara & Adams, 1978) there are neurosecretory neurones at the junction of the intestine and the rectum. Processes from the neurones in *Manduca* innervate muscles of the hindgut. A remarkable feature of the two pairs of proctodaeal neurones in *Manduca* is that they are multinucleate, with 25–28 nuclei per neurone in the final instar larva. Mrs Griffiths has also found multinucleate neurosecretory neurones in the lateral regions of the cutworm larva of the moth *Agrotis*. The original description of these neurones by Hinks (1975) stated that there were between one and 14 neurones in a pair of small peripheral ganglia in each segment of the body, but in the light of Mrs Griffiths' findings each "ganglion" may be a single neurone. The neurones in the thoracic segments and in the first and last abdominal segments are uninucleate. According to Hinks (1975) there are between two and 14 neurones in abdominal segments 3–7. The number of cells varies within the species of Lepidoptera he examined but he described an increase in number from segments 2–5 and then in segment 8 but a single neurone. It is likely that, as in *Agrotis*, these numbers represent multiple nuclei within a single cell and not multiples of cell somata, but we have not examined a range of species to check this possibility.

The increase in number of nuclei is correlated with an increase in the size of the soma of the neurone, and presumably also with an increase in the amount of neurosecretion produced. It would appear, therefore, that there is a need for a higher output from the neurones in certain abdominal segments.

Neurosecretory Fibres of Peripheral Nervous System

In addition to the processes from the LNNs there is another extensive system of peripheral neurosecretory fibres. In several species of insects they have been found on muscles, salivary glands, ecdysial glands, heart, reproductive organs, alimentary canal — in fact, most tissues and organs of the insect. Although it has generally been assumed that these fibres originate in the central nervous system, some of them may be processes from peripheral neurones such as the LNNs of the stick insect and the ecdysial gland neurone of moth larvae.

ACTIVITY

Neurosecretory neurones propagate action potentials like "ordinary" neurones but in the invertebrates and in the lower vertebrates they differ from those of ordinary neurones in several respects. The duration of the action potential is longer, sometimes as much as 20 times longer, than action potentials of motor or sensory neurones in the same animal, e.g. the medial neurosecretory cells of the brain of the flesh fly *Sarcophaga* (Wilkens & Mote, 1970). Commonly they are from two to 10 times as long, as in the LNNs of the stick insect (Orchard & Finlayson, 1977a, b). A further difference between these neurones and other peripheral neurones (Orchard & Finlayson, 1977a) and non-neurosecretory central neurones (Pitman, Tweedle & Cohen, 1972) is that the cell body of the neurosecretory neurone is electrically excitable and can generate an action potential, whereas the cell body of the other neurones is inexcitable. Another difference between the LNNs and other neurones, central and peripheral, of the stick insect is that the action current is propagated by movements of calcium ions and not by sodium ions. The LNNs are exposed to the haemolymph of the stick insect, not enclosed in a well-developed glial sheath as are other peripheral and central neurones. The ionic dependence of the LNNs upon calcium may thus be correlated with the particular ionic composition of the haemolymph of the stick insect, which like most herbivorous insects has a very low concentration of sodium, and may not be specifically correlated with the neurosecretory function of the LNNs. Because calcium is so intimately concerned with the process of secretion in general, the calcium dependence of the neurosecretory link nerve cells is particularly interesting. Neurosecretory neurones in some molluscs have also been shown to use calcium in the propagation of the action potential, but only in the cell soma; in the axon the current is carried by sodium ions (Wald, 1972; Junge & Miller, 1974).

It has been suggested that the influx of calcium triggers the release of neurosecretory material and that the prolonged action potential allows a longer time for secretion to take place (Gainer, 1978). This hypothesis assumes that the function of the action potential is to promote release of secretion. While this assumption is very likely to be correct it is, at present, based on circumstantial evidence, mainly from work on the hypothalamus and on the crustacean eye-stalk, in which the frequency of firing of neurones can be correlated with the release of their hormones (see Yagi & Iwasaki, 1977). Similar evidence is available in the insects. In the stick insect, for example, there is a diurnal cycle of secretory activity in the cell soma, as revealed by the fluorescent dye acridine orange, and also a similar cycle of electrical activity (Finlayson & Orchard, 1978). The two are not precisely synchronized, but this is not surprising because release does not occur at the soma; the dye is monitoring changes in the appearance of the region where neurosecretory granules are produced, whereas electrical activity is recorded from peripheral projections of the neurones where presumably release of material is occurring.

FUNCTION

Neurohormones are probably predominant in the endocrine system of insects. At least nine peptidergic neurohormones have been extracted (Table I) and there is physiological evidence for about a dozen more (Goldsworthy & Mordue, 1974; Maddrell, 1974; Normann, 1974; Steele, 1976).

Comparison with Vertebrate Hormones

Normann (1974) discovered a hypoglycaemic factor produced in the medial neurosecretory cells of the brain in blowflies. It is similar to insulin in its physiological effect which can be mimicked by insulin (Normann & Duve, 1978). Furthermore, the effects of the hyperglycaemic and adipokinetic hormones are mimicked by the vertebrate pancreatic hormone glucagon, which depresses the level of glucose

TABLE I

Peptidergic neurohormones extracted from insects

Source	Hormone	Function
Brain + corpus cardiacum	Ecdysiotropic	Activates ecdysial gland
	Diuretic ⎫ Antidiuretic ⎭	Control excretion
	Adipokinetic	Mobilizes fat
	Eclosion	Regulates time of adult emergence
	Bursicon	Controls tanning of cuticle
	Hypoglycaemic ⎫ Hyperglycaemic ⎭	Regulate blood sugar
Sub-oesophageal ganglion	Diapause	Controls embryonic diapause
All ventral ganglia	Bursicon	Controls tanning of cuticle

References: Goldsworthy & Mordue, 1974; Maddrell, 1974; Steele, 1976; Berlind, 1977; Bodnaryk, 1978; Normann & Duve, 1978.

and raises the levels of trehalose and fat in the blood. It appears that these molecules of the insulin and glucagon type were used by invertebrate animals to regulate their blood sugar levels long before the vertebrates appeared on the scene.

Recent studies, using the powerful technique of immunochemical analysis, have shown that insect tissues contain several substances which must be closely similar to mammalian peptides (Table II) in structure.

The parts of the nervous system which contain vasopressin-like substance have an antidiuretic effect on the Malpighian tubules and on the rectum. Conversely, an antidiuretic hormone (neurohormone D) has been shown to influence the flow of water through the isolated urinary bladder of the toad (Gersch, Richter, Stürzebecher & Fabian, 1969; Gersch, Richter & Stürzebecher, 1971).

Neurosecretory Innervation of Non-Neural Endocrine Glands

The ecdysial gland produces α-ecdysone under the influence of a tropic hormone from the brain. α-Ecdysone is not the moulting hormone but a precursor or prohormone. The control of moulting has in historical sequence been attributed to the brain (Kopeč, 1917), then to the medial neurosecretory cells of the brain (Wigglesworth, 1939), then to the ecdysial (prothoracic) gland (Fukuda, 1940), then to α-ecdysone produced by the ecdysial gland, then to β-ecdysone produced from α-ecdysone by other tissues or synthesised by tissues other than the ecdysial glands (King *et al.*, 1974).

TABLE II

Mammalian peptidergic substances found in insects

Immunoreaction to:	Tissue	Animal	Reference
Vasopressin	(a) Axons of brain (b) 2 cells of sub- oesophageal ganglion	Cricket Cricket	Strambi, Strambi *et al.* (1978); Strambi, Rougon-Rapuzzi *et al.* (1979)
α-endorphin	Sub-oesophageal ganglion	Silkworm Pine processionary caterpillar	Rémy, Girardie & Dubois (1978)
Enkephalins	Optic lobes etc.	Locust	Gros, Lafon-Cazal & Dray (1978)
Somatostatin	Medial n.s. cells of brain	Locust	Doerr-Schott, Joly & Dubois (1978)
Glucagon	Corpus cardiacum	Blowfly	Normann & Duve (1978)
Insulin	Medial n.s. cells of brain	Blowfly	Normann & Duve (1978)

The ecdysial gland has long been known to have neurosecretory innervation (Scharrer, 1964; Herman & Gilbert, 1966) which has been presumed to come from central neurones via the median (unpaired) nervous system. This neurosecretory innervation is not required for the functioning of the ecdysial gland at moulting and at metamorphosis. A transplanted brain without nervous connections to the ecdysial gland can trigger the gland by releasing ecdysiotropic hormone into the blood. What then is the function of the neurosecretory innervation?

The problem is further complicated by Mrs Griffiths' discovery of a peripheral neurosecretory neurone closely associated with the ecdysial gland in the caterpillar larvae of moths. Granule-laden processes from the neurone make intimate contact with the ecdysial gland and appear to release their secretion on to the surface of the gland cells. We have no evidence yet to indicate that there is also neurosecretory innervation of the gland from the central nervous system, so the original descriptions of neurosecretory innervation may refer to processes from the ecdysial gland neurone and not to fibres originating in the central nervous system. However, Normann (1965) found two types of fibre in the neurosecretory innervation of the ecdysial gland of the blowfly, one containing granules of the type found in the medial cells of the brain or the intrinsic cells of the corpus cardiacum, but the other of unknown origin and containing granules of a different type. Perhaps the second type of granule is produced by a peripheral neurone like that in moth larvae.

Mrs Griffiths has also found that the ecdysial gland neurone in *Agrotis* degenerates just before the ecdysial gland itself degenerates during the metamorphosis of the pupa into the adult. Neurosecretory neurones of the corresponding series in the abdomen do not degenerate. This is further evidence that the ecdysial gland neurone is likely to be closely bound to the ecdysial gland both functionally and anatomically. Such complexity illustrates the main theme of this review — that neurosecretory innervation in insects is widespread throughout the body and varied in its cytological structure. Differences in the cytological appearance of the neurosecretory material must surely reflect differences in physiological functions. If this is so the ecdysial glands of insects may be influenced by three types of neurosecretion: (1) a tropic blood-borne neurohormone from the brain; (2) direct neurosecretory innervation from brain or corpus cardiacum or ventral ganglion; and (3) innervation from a peripheral neurosecretory neurone.

Neurosecretory innervation is also found in the corpus allatum, a classical endocrine gland of ectodermal origin, which produces the hormone unique to insects, juvenile hormone or neotenin. This hormone has the remarkable property of maintaining an insect in a juvenile form during its period of growth. It is also used in the adult, when youth has been lost forever, as a gonadotropic hormone concerned with the control of egg development.

Aminergic Innervation of Tissues

In the examples of neurosecretory innervation of endocrine glands, as in most examples of neurosecretory innervation of tissues, the function of the innervation is not known. It has been suggested by Miller (1975) that direct "neurosecretomotor" innervation of tissues may always be aminergic (B-fibres of Knowles, 1965). If all neurosecretomotor innervation is non-peptidergic and release of secretion takes place only at synapses or very close to the target tissue, there might be an argument for reviving the definition of neurosecretion by Knowles and others (Knowles & Bern, 1966) and confining it to peptidergic neurohormones forming the "final common pathway of neuroendocrine integration". However, there are still problems which make this definition unsatisfactory. It is only in the case of one of the dorsal unpaired median neurones of the locust metathoracic ganglion (DUMETi) that the identity of the active principle is known with certainty. This neurone innervates the extensor-tibiae muscles of the metathoracic (jumping) legs and produces an amine (octopamine) which is apparently released in the close vicinity of extensor-tibia muscle fibres and has an inhibitory effect upon an intrinsic rhythm generated by part of this muscle (Hoyle, 1974, 1975; Hoyle & Barker, 1975). The electrical activity of DUMETi or the application of octopamine also have a facilitating influence on synaptic transmission in the control of "slow" extensor-tibiae muscle fibres (Evans & O'Shea, 1977). The granules of DUMETi are electron-dense and measure 60–190 nm in diameter (Hoyle, Dagan, Moberly & Colquhoun, 1974; Hoyle, 1975) and therefore they fall into the B category which is appropriate for aminergic fibres.

Other biogenic amines have been found in insect tissues. Dopamine has been found in extracts of medial neurosecretory cells of the brain (Houk & Beck, 1977) and in other tissues (see Walker & Kerkut, 1978). In the ingluvial (visceral) ganglion of the cockroach *Blaberus* there is ultrastructural evidence of possible peptidergic activity (dense and clear granules 80–200 nm in diameter) as well as smaller (40 nm in diameter) inclusions which are clear and resemble synaptic vesicles, and a few small dense core vesicles about 60 nm in diameter. Cytochemical studies indicate that the small clear vesicles may be storing an amine, but if that is so what is the role of the large granules? If they are not precursors of the smaller vesicles it would appear that these neurones are secreting at least two substances (Chanussot, 1972; Chanussot, Dando, Moulins & Laverack, 1969).

Dr Ian Findlay, in my laboratory, has been making extracts of the neurohaemal tissue of the transverse nerve of the stick insect, including the "perisympathetic organ" of Raabe (1967). In this region of the transverse nerve three different types of neurosecretory fibre can be recognized according to the appearance of the granules they contain. Only extracts from the distal region of the nerve have been found by Dr Findlay to have an effect on the activity of the excretory organs, the Malpighian tubules. The extract causes a brief rapid output of urine.

The problem of identifying active substances from the peripheral neurosecretory system of insects is difficult because of the very small amounts of tissue involved, and because several secretory products are often present in the same nerve or neurohaemal tissue. The finding that nerves containing different types of granules are sited in different regions of the peripheral nervous system may be helpful, as in Dr Findlay's work, in the separation of the various active principles they contain. The product of the LNNs can be obtained from the somata themselves, but extract uncontaminated by other neurosecretory granules should be obtainable from the long nerve that runs to the dorsal region of each abdominal segment and which carries only LNN neurosecretion (Fifield & Finlayson, 1978).

CONCLUSIONS

The neurosecretory system of insects is more diffuse than that of mammals. There is anatomical and physiological evidence for local control by neurocrine secretions of tissues and organs. The variety of types of neurosecretory granules and the variety of release points indicate that many active substances have yet to be isolated and identified. Insect endocrinology was simple 20 years ago, but with each discovery of another hormone or of a neurosecretory cell that produces an unidentified substance, the picture becomes increasingly complex.

The analogue of the hypothalamus—pituitary system of mammals is not confined to the brain—corpus cardiacum complex but extends to neurosecretory innervation of target tissues throughout the body including the corpus allatum and the ecdysial gland. The circulatory system of the insect is open and so the method of delivery of a high concentration of neurosecretion to a target tissue by blood vessels as in the hypothalamic—anterior pituitary system, is not practicable.

Evidence is beginning to be found for similarities in peptide molecules used by insects in their neuroendocrine systems and those of vertebrates.

With the refined microtechniques now available, particularly those utilizing immunology, and with further advances in biochemical methods, it is likely that the next decade will see the isolation and synthesis of a whole range of substances which make up the pharmacopoeia of insects and other invertebrates as well as those of vertebrates. The neurosecretory cells which stand out so dramatically in light and electron microscopy are part of the chemical factory which provides the minute quantities of highly active substances that regulate the metabolic and neuromuscular activities of all animals.

Insects are of great economic and medical importance and so there is considerable interest in the nervous system and neuroendocrine systems of insects because these appear to be the site of action of most insecticides. Research on mammals is aimed, in the long term, at improving the health of man and domestic animals by manipulation of their chemical messengers, but the corresponding research on insects is aimed at their destruction. Research in this field may help to bring a rational approach to the production of new insecticides.

ACKNOWLEDGEMENT

I am grateful to Dr Ian Findlay for reading the manuscript and making helpful criticisms and suggestions.

REFERENCES

Anderson, M. & Finlayson, L. H. (1978). Topography and electrical activity of peripheral neurons in the abdomen of the tsetse fly (*Glossina*) in relation to abdominal distension. *Physiol. Ent.* 3: 157—167.

Berlind, A. (1977). Cellular dynamics in invertebrate neurosecretory systems. *Int. Rev. Cytol.* 49: 171—251.

Bodnaryk, R. P. (1978). Structure and function of insect peptides. *Adv. Insect Physiol.* 13: 69—132.

Chanussot, B. (1972). Étude histologique et ultrastructurale du ganglion ingluvial de *Blabera craniifer* Burm. (Insecte, Dictyoptère). *Tissue Cell.* 4: 85—97.

Chanussot, B., Dando, J., Moulins, M. & Laverack, M. S. (1969). Mise en évidence d'une amine biogène dans le système nerveux stomatogastrique des insectes: Étude histochimique et ultrastructurale. *C. r. hebd. Séanc. Acad. Sci., Paris* 268: 2101—2104.

Doerr-Schott, J., Joly, L. & Dubois, M. P. (1978). Sur l'existence dans la pars intercerebralis d'un insecte (Locusta migratoria R. et F.) de cellules neurosécrétrices fixant un antisérum antisomatostatine. C. r. hebd. Séanc. Acad. Sci., Paris 286: 93–95.

Evans, P. D. & O'Shea, M. (1977). An octopaminergic neurone modulates neuromuscular transmission in the locust. Nature, Lond. 270: 257–259.

Fifield, S. M. & Finlayson, L. H. (1978). Peripheral neurons and peripheral neurosecretion in the stick insect, Carausius morosus. Proc. R. Soc. (B) 200: 63–85.

Finlayson, L. H. (1976). Abdominal and thoracic receptors in insects, centipedes and scorpions. In Structure and function of proprioceptors in the invertebrates: 153–211. Mill, P. J. (Ed.). London: Chapman & Hall.

Finlayson, L. H. & Orchard, I. (1978). Neurosecretory and electrical activity of extra-ganglionic neurons in the stick insect. In Comparative endocrinology: 323–326. Gaillard, P. J. & Boer, H. H. (Eds). Amsterdam: Elsevier/North Holland Biomedical Press.

Finlayson, L. H. & Osborne, M. P. (1968). Peripheral neurosecretory cells in the stick insect (Carausius morosus) and the blowfly larva (Phormia terraenovae). J. Insect Physiol. 14: 1793–1801.

Fukuda, S. (1940). Induction of pupation in silkworm by transplanting the prothoracic gland. Proc. Imp. Acad. Tokyo 16: 414–416.

Gainer, H. (1978). Input–output relations of neurosecretory cells. In Comparative endocrinology: 293–304. Gaillard, P. J. & Boer, H. H. (Eds). Amsterdam: Elsevier/North Holland Biomedical Press.

Gelperin, A. (1971). Abdominal sensory neurons providing negative feedback to the feeding behaviour of the blowfly. Z. vergl. Physiol. 72: 17–31.

Gersch, M., Richter, K. & Stürzebecher, J. (1971). Weitere experimentelle Untersuchungen zum Wirkungsmechanismus des Neurohormons D von Periplaneta americana (L.) am "biologischen Modell" der Harnblase von Bufo bufo (L.): Beeinflussung des Na+ -Transportes. Gen. comp. Endocr. 17: 281–286.

Gersch, M., Richter, K., Stürzebecher, J. & Fabian, B. (1969). Experimentelle Untersuchungen zum Wirkungsmechanismus des Neurohormons D von Periplaneta americana (L.) am "biologischen Modell" der Harnblase von Bufo bufo (L.). Gen. comp. Endocr. 12: 40–50.

Goldsworthy, G. J. & Mordue, W. (1974). Neurosecretory hormones in insects. J. Endocr. 60: 529–558.

Gros, C., Lafon-Cazal, M. & Dray, F. (1978). Présence de substances immunoréactivement apparentées aux enképhalines chez un insecte, Locusta migratoria. C. r. hebd. Séanc. Acad. Sci., Paris 287: 647–650.

Herman, W. S. & Gilbert, L. I. (1966). The neuroendocrine system of Hyalophora cecropia (L.) (Lepidoptera: Saturniidae). I. The anatomy and histology of the ecdysial glands. Gen. comp. Endocr. 7: 275–291.

Hinks, C. F. (1975). Peripheral neurosecretory cells in some Lepidoptera. Can. J. Zool. 53: 1035–1038.

Houk, E. J. & Beck, S. D. (1977). Distribution of putative neurotransmitters in the brain of the European corn borer, Ostrinia nubilalis. J. Insect Physiol. 23: 1209–1217.

Hoyle, G. (1974). A function for neurons (DUM) neurosecretory on skeletal muscle of insects. J. exp. Zool. 189: 401–406.

Hoyle, G. (1975). Evidence that insect dorsal unpaired median (DUM) neurons are octopaminergic. J. exp. Zool. 193: 425–431.

Hoyle, G. & Barker, D. L. (1975). Synthesis of octopamine by insect dorsal median unpaired neurons. *J. exp. Zool.* 193: 433–439.

Hoyle, G., Dagan, D., Moberly, B. & Colquhoun, W. (1974). Dorsal unpaired median insect neurons make neurosecretory endings on skeletal muscle. *J. exp. Zool.* 187: 159–165.

Junge, D. & Miller, J. (1974). Different spike mechanisms in axon and soma of molluscan neurone. *Nature, Lond.* 252: 155–156.

King, D. S., Bollenbacher, W. E., Borst, D. W., Vedeckis, W. V., O'Connor, J.D., Ittycheriah, P. I. & Gilbert, L. I. (1974). The secretion of α-ecdysone by the prothoracic glands of *Manduca sexta in vitro. Proc. natn. Acad. Sci. U.S.A.* 71: 793–796.

Knowles, F. G. W. (1965). Neuroendocrine correlations at the level of ultrastructure. *Arch. Anat. microsc.* 54: 343–357.

Knowles, F. G. W. (1974). Twenty years of neurosecretion. In *Neurosecretion – the final neuroendocrine pathway*: 3–11. Knowles, F. G. W. & Vollrath, L. (Eds). Berlin, New York: Springer-Verlag.

Knowles, F. G. W. & Bern, H. A. (1966). Function of neurosecretion in endocrine regulation. *Nature, Lond.* 210: 271–272.

Kopeč, S. (1917). Experiments on metamorphosis of insects. *Bull. int. Acad. Sci. Cracovie* (B) 1917: 57–60.

Maddrell, S. H. P. (1974). Neurosecretion. In *Insect neurobiology*: 307–357. Treherne, J. E. (Ed.). Amsterdam: North-Holland.

Miller, T. A. (1975). Neurosecretion and the control of visceral organs in insects. *A. Rev. Ent.* 20: 133–149.

Nagy, F. (1978). Ultrastructure of a peripheral neurosecretory cell in the proctodaeal nerve of the larva of *Oryctes nasicornis* L. (Coleoptera: Scarabaeidae) *Int. J. Insect Morph. Embryol.* 7: 325–336.

Normann, T. C. (1965). The neurosecretory system of the adult *Calliphora erythrocephala* I. The fine structure of the *corpus cardiacum* with some observations on adjacent organs. *Z. Zellforsch. mikrosk. Anat.* 67: 461–501.

Normann, T. C. (1974). Neurosecretory cells in insect brain and production of hypoglycaemic hormone. *Nature, Lond.* 254: 259–261.

Normann, T. C. & Duve, H. (1978). Neurohormones regulating haemolymph levels of trehalose, glucose and lipids in blowflies. In *Comparative endocrinology*: 441–444. Gaillard, P. J. & Boer, H. H. (Eds). Amsterdam: Elsevier/North-Holland Biomedical Press.

Orchard, I. & Finlayson, L. H. (1976). The electrical activity of mechanoreceptive and neurosecretory neurons in the stick insect *Carausius morosus. J. comp. Physiol.* 107: 327–338.

Orchard, I. & Finlayson, L. H. (1977a). Electrically excitable neurosecretory cell bodies in the periphery of the stick insect, *Carausius morosus. Experientia* 33: 226–228.

Orchard, I. & Finlayson, L. H. (1977b). Electrical properties of identified neurosecretory cells in the stick insect. *Comp. Biochem. Physiol.* 58A: 87–91.

Orchard, I. & Finlayson, L. H. (1977c). Studies on peripheral neurons and neurohaemal tissue in the thorax of the stick insect (*Carausius morosus*). *Experientia* 33: 1440–1442.

Pitman, R. M., Tweedle, C. D. & Cohen, M. J. (1972). Electrical responses of insect central neurons: augmentation by nerve section or colchicine. *Science, N.Y.* 178: 507–509.

Raabe, M. (1967). Récherches récentes sur la neurosécrétion dans la chaîne nerveuse ventrale des insectes. *Bull. Soc. zool. Fr.* **92**: 67–71.

Reinecke, J. P., Gerst, J., O'Gara, B. & Adams, T. S. (1978). Innervation of hindgut muscle of larval *Manduca sexta* (L.) (Lepidoptera: Sphingidae) by a peripheral multinucleate neuroscretory neuron. *Int. J. Insect Morphol. Embryol.* **7**: 435–453.

Rémy, C., Girardie, J. & Dubois, M. P. (1978). Présence dans le ganglion sous-oesophagien de la chenille processionnaire du pin (*Thaumetopoea pityocampa* Schiff) de cellules révélées en immunofluorescence par un anticorps antiendorphine. *C. r. hebd. Séanc. Acad. Sci., Paris* **286**: 651–653.

Scharrer, B. (1964). The fine structure of blattarian prothoracic glands. *Z. Zellforsch. mikrosk. Anat.* **64**: 301–326.

Steele, J. E. (1976). Hormonal control of metabolism in insects. *Adv. Insect Physiol.* **12**: 239–323.

Strambi, C., Rougon-Rapuzzi, G., Cupo, A., Marin, N. & Strambi, A. (1979). Mise en évidence immunocytologique d'un compose apparenté à la vasopressine dans le système nerveux du grillon *Acheta domesticus. C. r. hebd. Séanc. Acad. Sci., Paris* **288**: 131–133.

Strambi, C., Strambi, A., Cupo, A., Rougon-Rapuzzi, G. & Martin, N. (1978). Étude des taux d'une substance apparentée à la vasopressine dans le système nerveux de grillons soumis à différentes conditions hygrométriques. *C. r. hebd. Séanc. Acad. Sci. Paris* **287**: 1227–12230.

Wald, F. (1972). Ionic differences between somatic and axonal action potentials in snail giant neurons. *J. Physiol.* **220**: 267–281.

Walker, R. J. & Kerkut, G. A. (1978). The first family (adrenaline, noradrenaline, dopamine, octopamine, tyramine, phenylethanolamine and phenylethylamine). *Comp. Biochem. Physiol.* **61C**: 261–266.

Wigglesworth, V. B. (1939). Source of the moulting hormone in *Rhodnius. Nature, Lond.* **144**: 953.

Wilkens, J. L. & Mote, M. I. (1970). Neuronal properties of the neurosecretory cells in the fly *Sarcophaga bullata. Experientia* **26**: 275–276.

Yagi, K. & Iwasaki, S. (1977). Electrophysiology of the neurosecretory cell. *Int. Rev. Cytol.* **48**: 141–186.

Symp. zool. Soc. Lond. (1981) No. 46, 253–265

Reproduction in the Green Sea Turtle, *Chelonia midas*

ALAN S. PARKES[*]

Christ's College, Cambridge, UK

SYNOPSIS

The relative inaccessibility of the sea turtles has retarded research but consider-able information about their biology in general and their reproduction in par-ticular has been obtained by tagging females on nesting beaches and observing the nature and contents of the nest and the behaviour of the hatchlings. This is especially true of *Chelonia mydas*, for which much additional information has also been yielded from successful captive breeding on a land-based farm on Grand Cayman Island, British West Indies. Most of the parameters of reproduc-tive activity on the farm are similar to those found in the wild, and a low hatch rate of farm-laid eggs is offset by a greater frequency of nesting.

INTRODUCTION

Chelonia mydas is widely distributed between 26°N and 26°S latitude, roughly between the tropics, and breeds freely within that range. The species is thought to be of the order of 100 million years old in biological history and has probably not changed much in that long period. It may truly be said, therefore, to have stood the test of time. It is a remarkable animal. It weighs up to 230 kg; it can survive for several weeks immobilized on its back (literally turned turtle) and for this reason was for centuries the main source of fresh meat for ships' crews in the Caribbean; it can swim 5000 km and lay a thousand eggs, apparently without feeding; it can develop anaerobic metabolism and thus sleep under water; it can produce up to 30% of its body weight as eggs in 10 weeks and it can survive massive injuries.

A remarkable animal, indeed, but not one easy to study in detail. Mating takes place at sea, the male rarely leaves the water, the female comes ashore only to nest and the hatchlings rush for the water as soon as they leave the nest. In these circumstances, we are ignorant of many basic facts of the life history of *Chelonia*. We do not know

[*]Correspondence to: Sir Alan Parkes, 1 The Bramleys, Shepreth, Royston, Herts SG8 6PY, UK.

where the hatchlings spend their first year, the age at sexual maturity, whether the male is fertile throughout the year, the length of reproductive life or how long individuals live. Until a few years ago, our knowledge of reproduction in wild *Chelonia* was limited to what could be obtained from tagging females on nesting beaches and observing the frequency of nesting and the nature and contents of the nests. In this area, the paper by Hendrickson (1958) on the Sarawak turtles and the review by Hirth (1971) are outstanding, but tribute must also be paid to the pioneer work of Moorhouse (1933), one of the few biologists to make use of the material available from turtles slaughtered at soup factories. More general accounts of *Chelonia* have been given by Carr (1968) in his classic book *The turtle*, and by Bustard (1972), Pritchard (1967) and Rebel (1974). Recently, however, more detailed information has become available from the captive breeding of *Chelonia* on a land-based "farm" on Grand Cayman Island, British West Indies. This operation started in 1968 and mating in the artificial "lagoon" followed by nesting on the artificial beach started in 1973. Results for 1973 were recorded by Ulrich & Owens (1976) and Simon, Ulrich & Parkes (1975); for 1974 by Ulrich & Parkes (1978) and a summary of those for 1975 and 1976 by Wood & Wood (1978). For the 1977—79 results on the farm, see Addendum, p. 264.

BREEDING SEASON

Chelonia has a restricted or clearly marked breeding season which varies with different geographical races in different parts of the world. In the Western Atlantic there is a decided east to west time shift. On Ascension Island nesting is at its height in February, in Surinam during March and in Costa Rica in August (Simon, 1975). For the Far East, Hendrickson (1958) records that in Malaysia and Sarawak nesting showed very clear peaks during June—September, but that a few turtles nested at all times of year, possibly because of the proximity to the equator and the relatively constant temperature. The peak of nesting corresponded to the dry season in May to October, roughly as elsewhere in the northern hemisphere (Fig. 1). In the southern hemisphere, as with many other animals, the seasonal breeding of *Chelonia* is reversed. Peak nesting on Heron Island on the Great Barrier Reef was observed by Moorhouse (1933) to occur during the warm season October to March, when the mean air temperature is above 24°C (Fig. 2): it would be intriguing to find out whether northern hemisphere turtles transported to the southern hemisphere

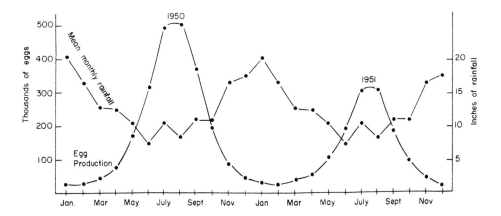

FIG. 1. Egg production and rainfall on the Sarawak turtle islands. From Hendrickson (1958). Egg production data from Harrisson (1951, 1952). Rainfall data from records kept for Talang Talang Besar Island by the Sarawak Dept. of Civil Aviation.

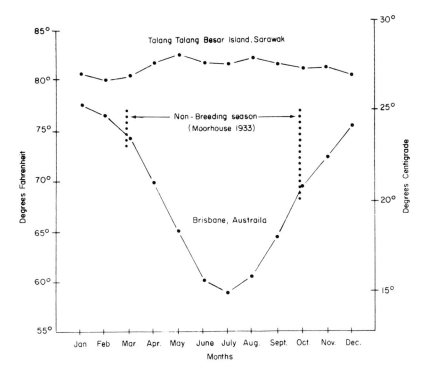

FIG. 2. Mean monthly air temperatures for Brisbane, Australia (nearest available station to Heron Island, Great Barrier Reef), and Talang Talang Besar Island, Sarawak. Heron Island turtles show a definite non-breeding season; in Sarawak, at least some nesting occurs in all months. From Hendrickson (1958).

reverse their nesting season in accordance with the climate. In the meantime, one may note that on the green sea turtle farm on Grand Cayman Island, turtles of different origins share a peak of nesting in July—August corresponding probably to the breeding season of wild *Chelonia* when the island was a major breeding ground (Ulrich & Owens, 1976). It is not known whether the male is fertile all year round or whether, like the female, it has a limited season.

Breeding in the wild is often associated with a migratory pattern. For instance, turtles feed off the north-east coast of Brazil and nest on the beaches of Ascension Island, a dichotomy probably resulting from giant earth movements over millions of years having gradually separated the breeding and feeding grounds (Carr & Coleman, 1974). The captive breeding on Grand Cayman Island, however, shows that migration is not a necessary antecedent to breeding. Rather the contrary, because on the farm breeding is more intensive (see p. 263).

SEXUAL ACTIVITY

Mating takes place at sea off the nesting beaches, often at night. The mating couple stay on the surface so far as possible and are often accompanied by other males trying to displace the mounted male. The females may be receptive briefly or for two weeks or more, during which time their sexual activity waxes and wanes and they may mate with several different males. The process has been described by many authors, notably Hendrickson (1958) and Booth & Peters (1972). The males rarely, if ever, leave the water. On the Cayman farm events follow a similar pattern but the special conditions permit most interesting observations as to factors affecting sexual behaviour.

By the beginning of 1972 there were 64 ex-wild adult females and 10 ex-wild males in the breeding pool at the Cayman farm; some had been there for 30 months, some for only 12 months. No sexual activity had been observed, and the introduction of a large male from Ascension Island in 1972 did not alter the situation. Then in April 1973 two large males from Surinam were added to the pool; one of these started to copulate immediately and the other within a few hours. Later, males which had been in the pool for a year or more, including the Ascension male, also became sexually active (Ulrich & Owens, 1976). Three conclusions seem to be permissible: (1) that in 1972 the females were still suffering from captivity effects and were not sexually receptive even to a new male; (2) that by the following year some at least of the females had become sexually receptive but that by that time a "familiarity effect", such as has been described for other animals, had arisen due to the close proximity in which the males and females had been kept for at least a year; (3) that the introduction of the new Surinam males to the now receptive females had led to sexual activity and stimulated the previously quiescent males in the pool. As a result of these observations, the males were isolated from the females between November 1973 and April 1974.

A similar effect of competition was seen in 1974. A male and female, both of which had been active in the previous year, were isolated in a large holding tank for several months in the early part of the season without showing sexual activity. They were then transferred to the breeding pool where the female was immediately pursued by males already there. This roused the original male, which was very soon copulating with the female with which it had previously been isolated without result. The mating couple were attended by several pool males waiting their turn (Ulrich & Parkes, 1978).

These observations are, of course, similar to those made for many other species, including primates, but, again as with other species, they pose the whole question of the social organization of the species and of the nature of the medium by which the organization is achieved. Of the three obvious possibilities, sight, sound and smell, smell seems, as in many other animals, to be the most likely. *Chelonia* has a large and complicated area of olfactory epithelium and a large olfactory bulb in the brain. Moreover it can distinguish between at least certain different smells (Manton, Carr & Ehrenfeld, 1972). It has complicated axillary and inguinal glands, first described by Rathke in 1848, and ultrastructurally by Ehrenfeld & Ehrenfeld (1973), which produce a whitish secretion — a few drops can sometimes be expressed by hand — which has a musky smell. The use of the odoriferous substances of animals in the perfume trade, and my previous experience in this field (see Parkes, 1966), led me to think that the assistance of professional perfumers might be useful in typing the odorous substance or substances in this secretion, but it was not found practicable to obtain enough for either olfactory or chemical analysis.

The role of Rathke's glands in the sexual life of *Chelonia* is quite uncertain, especially as they are well developed in the immature animal and show no seasonal variation. Rathke's glands do not, however, exhaust the possibilities — glands, presumably exocrine, have been described in other sites, and olfaction remains the most likely means of social and sexual integration in *Chelonia*. It is probably also involved in the fact that the young adult usually, but not always, returns for the first breeding season, and thereafter for later ones, to the beach on which it had left the nest as a hatchling years before (cf. the homing instinct of salmon, eels and other fish).

A. S. Parkes

NESTING

At some stage after mating the female crawls up a beach and selects a nesting site. There she digs a body pit with an egg chamber at the rear end. She deposits her soft shelled eggs, fills up the hole and crawls back to the water. There is no other maternal care. Unlike the case of, say, the crocodile, no effort is made to guard the nest. Incubation is relatively simple. The eggs are laid at a depth of 50–60 cm at which temperature fluctuations in the sand are largely eliminated, giving a fairly constant level, in Hendrickson's experience, of about 38°C. Here, with suitable humidity and oxygenation, hatching takes place in 50–55 days during which time the temperature of the egg mass is considerably increased by metabolic heat. The tremendous mortality from predators incurred by hatchlings, in spite of their frenzied activity, as they break the surface of the sand and rush to the sea and in the shallow water, has often been described.

The interval between mating and nesting is of great interest. For obvious reasons, such information is difficult to obtain for wild populations, but captive breeding on the Cayman farm has given a clear answer. In 1974, eight of the ten females observed to mate nested 21–39 days later, the highest hatch rate occurring in the four females nesting 30–34 days after mating. No doubt further data will increase the spread, but the most frequent interval is likely to be three to five weeks.

Back in the water, the female in the wild may or may not mate again. Authorities disagree. But on the Cayman farm, where the animals are under close observation during the breeding season, the great majority of laying females have not been observed to mate again on returning to the water and some have actually been seen to fight off importunate males. A very few only allow brief and probably ineffectual mounting on returning to the water after laying. Nevertheless the females nest again after an interval of 9–14 days, the most frequent interval on the farm being 10 days, which agrees very closely with that found by Hendrickson (1958) for the Sarawak turtles (Fig. 3) but is rather shorter than the 12–14 days given by Bustard (1972) for Heron Island.

The process is repeated until anything up to seven, eight or nine clutches of eggs have been laid, totalling perhaps 1000 or more eggs, in 10–12 weeks. Hendrickson (1958) found that a substantial proportion of Sarawak turtles laid five or more times in a season (Fig. 4). On the Cayman farm in 1973 and 1974, 21 of the 33 laying females laid four, five or six clutches, and eight laid seven or eight. Larger turtles lay more eggs (Bustard, 1972; Simon & Parkes, 1976). There

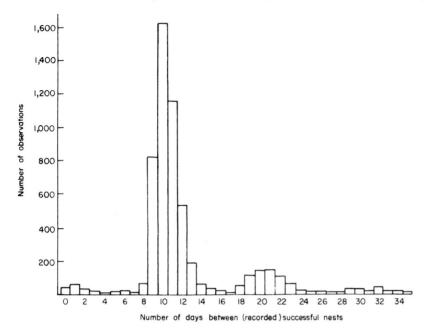

FIG. 3. Number of days interval between nestings by marked female *Chelonia mydas*. The preponderance of nine to 12 day intervals is clearly shown. The minor peaks around 20–21 days and 29–33 days probably represent double or treble intervals caused by nestings having been missed. The few very short intervals result from some residual eggs being dropped one or two days after the main clutch. From Hendrickson (1958).

TABLE I

Hatch rate and clutch sequence (all females)

Clutch sequence	No. of clutches	No. of eggs	No. eggs per clutch	Viable hatchlings	Viable hatch (%)
1st	14	1585	133	559	35.2
2nd	14	1798	128	770	42.8
3rd	14	1814	129	824	45.4
4th	13	1530	118	728	47.6
5th	10	1234	123	629	50.9
6th	8	978	122	470	48.0
7th	5	543	109	253	46.5
8th	2	182	91	97	53.2

From Ulrich & Parkes (1978).

is no solid evidence that the fertility of the eggs decreases with the later clutches (Table I). Little is known about the performance over a period of years of females in the wild.

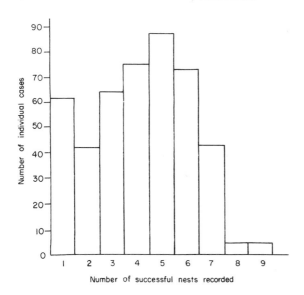

FIG. 4. Numbers of nests recorded for individual marked female turtles. From Hendrickson (1958).

The nesting sequence poses two intriguing problems: (1) what happens in the female during the mating–nesting interval and why is the mating–nesting interval so much longer than the inter-nesting interval? (2) Are spermatozoa stored during the course of one nesting season, say three to four months, and if so must one consider the possibility that they can also survive from one year to the next?

OVULATION

The ovary at the beginning of the breeding season consists of a much folded, almost diaphanous ligament fringed with ovarian cortex in which are embedded two or three sets of yolks of clearly different sizes, presumably representing the first two or three prospective ovulations, backed up by a mass of minute oocytes with little or no yolk formation (see Fig. 5). The disappearance of the set of largest yolks by ovulation permits a wave of development among the other oocytes.

It is virtually certain that normally the first ovulation and oviposition of the season is coitus-induced and, in the absence of further mating, that the successive ovulations during the season result from the same stimulus. Therefore it seems that, as in many other animals, coitus in the turtle has two functions, to inseminate the female with spermatozoa and to trigger a neuroendocrine

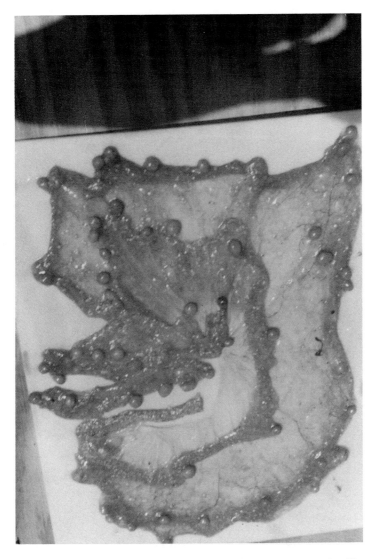

FIG. 5. Ovary of Nicaraguan female mid-March 1973, showing ligament and string–like ovarian cortex in which are visible the most advanced follicles with yolks about 1 cm in diameter. Follicles of the next wave, barely distinguishable in the photograph, have yolks about 2 mm in diameter (photograph by A.S.P.).

Facing p. 260.

mechanism leading to ovulation, but that in the turtle the reaction is elaborated in that coitus leads to a succession of ovulations at about 10-day intervals over a period of two or three months. The difference between the mating–nesting interval (three to five weeks) and the interval between successive ovulations (10 days) is presumably accounted for by the time taken for the activation of the ovulation-producing mechanism. As to the mechanism, *Chelonia* has a well developed hypophysis, which is known to contain gonadotrophic substances, and it is reasonable to suppose that the first ovulation of the season is induced by the same hypothalamic–hypophysial–gonadal mechanism as is found in other types in which ovulation is coitus-dependent. It remains to be seen how, in the absence of further mating, the production of the later ovulations is mediated — it is highly unlikely on mechanical if on no other grounds that successive crops of eggs co-exist in the oviduct from a massive single ovulation.

STORAGE OF SPERMATOZOA IN FEMALE *CHELONIA*

On the Cayman farm close observation in 1973–75 provided no evidence that the majority of females mated again after the start of nesting, but in 1976 this possibility was precluded by the isolation of the females after the initial mating (Wood & Wood, 1978). It is thus clear that the eggs produced at successive nestings through the season are fertilized by spermatozoa inseminated before the first nesting. It is virtually certain, therefore, that spermatozoa survive in the oviduct through the nesting season so that the eggs of each successive wave are fertilized at ovulation. Considering the tremendous activity of the oviduct during the nesting season and the lack of evidence of the existence of spermathecae, such survival of spermatozoa is remarkable. It is more credible, however, than the idea that the whole season's supply of eggs, mature and immature, are all fertilized in the ovary, or in the oviduct after a mass ovulation, before the first nesting. Moreover, in view of the lack of evidence that fertility of the eggs decreases with later clutches (Table I) it must be supposed that spermatozoa are still present in the oviduct in substantial numbers at the end of the nesting season. This raises the question as to whether they can survive in a functional state to the next season and then effect fertilization of the eggs of a currently unmated female.

The laying of fertile eggs by long-isolated females has been reported for some of the terrestrial Testudinae, but the weight of evidence is

against it for *Chelonia*. Firstly, the long-term survival of spermatozoa
would do nothing to promote ovulation so that the laying of eggs by
an unmated female *Chelonia* would imply spontaneous ovulation or
the retention of an ovulation-producing stimulus from mating in a
previous season. As to the latter, it is relevant that when the Cayman
farm was first stocked several females taken from nesting beaches
laid one or even two residual clutches of eggs on the farm beach, but
showed no sign of reproductive activity in the following year. As to
spontaneous ovulation: in 1973 observation of mating on the farm
was not continuous, and later, the difficulty of watching a large
number of animals in nearly half an acre of water up to ten feet
deep, especially at night, makes the completeness of the mating
records less than certain. In 1974 four females laid without observed
matings but three of them, laying a total of more than 2000 eggs,
all had good hatch rates. Records for 1975 are not available but in
1976 four animals were recorded as laying without having been
observed to mate; all four were, however, credited with some early
embryonic death, and one with a 39.0% hatch, implying fertilization
of the eggs (Wood & Wood, 1978). There is no hard evidence here of
spontaneous ovulation as a normal occurrence or of storage of
spermatozoa from one season to another, proof of which would require
the prolonged isolation of a female after an initial mating, or, better
still, her mating in isolation with a sterile male to stimulate ovulation.
There is thus no basis for the contention that the mating of the ex-
wild females on the farm has been ineffective and that their successful nesting has been due to the storage of spermatozoa derived from
matings in the wild before capture. The mating and successful nesting
by farm-reared females finally disposes of what was essentially an
attempt to throw doubt, by any means, on the results of captive
breeding of *Chelonia* on the Cayman farm.

EFFECTS OF DOMESTICATION

Many of the criteria of reproductive activity are remarkably similar
on the Cayman farm to those reported for wild populations of
Chelonia, including breeding season, number of clutches and eggs
laid by a female in a season, the inter-nesting interval, number of
eggs per clutch and incubation time.

In two important particulars, however, the captive stock differs
markedly from wild stocks. *Chelonia* females, in the wild, do not
normally breed every year: some breed every second year, the
majority every third year and some no doubt at longer intervals.

On the farm, the average frequency is much greater. Up to the end of 1976, 12 females had nested one year out of two and seven, two years out of three or three years out of four. Two had laid in two, two in three, and four in four successive years. Such high figures have never been reported for the wild and probably stem from the abundant food and hence the diversion of metabolic effort from food-gathering and migration to reproduction.

By contrast, the hatch rate of farm-laid eggs is disappointing compared, if not with that in the wild, at least with eggs brought from natural rookeries and hatched on the farm. It is difficult to arrive at a representative figure for the hatch rate in the wild. A concerted effort by a large number of hatchlings is required to break out of the nest through the sand, so that a small number of hatchlings from a small clutch of eggs or from a bad hatch may never see the light of day. Hence, only the marking of a large number of nests at the time of laying, their excavation when hatching has or should have taken place and the counting of dead hatchlings, dead in shell, addled eggs and empty shells will give an adequate figure for the hatch rate in the wild.

Moorhouse as long ago as 1933, working on Heron Island in the Great Barrier Reef, found hatch rates in 11 nests of between 41% and 85%, but most were between 50% and 59%. Pritchard (1967) gives 30%, as an overall figure. In contrast, by 1973 a figure of 80.4% had been achieved with the 14 803 eggs obtained from a beach in Costa Rica and incubated artificially at the Cayman farm (Simon, 1975). In the same year, 11 268 farm-laid eggs were incubated under similar conditions, but later in the year gave a hatch rate of 42.3%. This rate was raised to 44.8% for 9752 eggs in 1974 (Ulrich & Parkes, 1978), but fell to 20.4% for 17 427 eggs in 1975 and to 34.8% for 15 186 eggs in 1976 (Wood & Wood, 1978). The reason for the low hatchability of the farm-laid eggs is not known but it is curious that four nests not detected at the time of laying on the Cayman farm beach when excavated after hatching were found to have had a hatch rate of 87.7% of 442 eggs.

The ex-wild females, therefore, breed more frequently but produce eggs with a lower hatch rate than females in the wild, otherwise their performance is remarkably similar. Additionally, farm-reared turtles are easier to handle than the ex-wild stock. If the problems of rearing the young stock, always likely to be severe in the early days of intensive farming, can be solved satisfactorily, *Chelonia mydas* will be well on the way to becoming a domestic animal.

ACKNOWLEDGEMENTS

I am much indebted to Professor J. R. Hendrickson for permission to reproduce Figs 5, 6, 8 and 10 of his 1958 paper, and to Dr J. R. Wood for sending me a copy of the script of his 1980 paper.

ADDENDUM AUGUST 1980

Since this paper was written, the 1977—79 results on the farm have been recorded by Wood & Wood (1980). Broadly, they confirm the findings in previous years as to sexual behaviour, mating—nesting interval, inter-nesting interval, size of clutch and number of clutches laid in the captive wild stock. The frequency of breeding by the captive females continued to be much greater than has been reported for females in the wild — three females have nested in seven consecutive seasons, four in six, and six in five out of seven. Increased duration of mating up to 300—400 min increased the incidence of nesting and the hatchability of the eggs, but overall hatchability continued to be low.

Among the females reared from eggs taken in the wild and hatched on the farm, six nested at eight years old, 11 at nine years and five at ten years, but many within these age groups have not started to breed. Performance improved with those laying in more than one season, but was still much below that of the captive wild females. Stock raised from eggs laid on the farm are not yet of breeding age.

REFERENCES

Booth, J. & Peters, J. A. (1972). Behavioural studies on the green turtle (*Chelonia mydas*) in the sea. *Anim. Behav.* **20**: 808—812.
Bustard, R. (1972). *Australian sea turtles*: Their natural history and conservation. London & Sydney: Collins.
Carr, Archie (1968). *The turtle*. London: Cassell.
Carr, A. & Coleman, P. J. (1974). Seafloor spreading theory and the odyssey of the green turtle. *Nature, Lond.* **249**: 128—130.
Ehrenfeld, J. G. & Ehrenfeld, D. W. (1973). Externally secreting glands of freshwater and sea turtles. *Copeia* **1973**: 305—314.
Harrisson, Tom (1951). The edible turtle (*Chelonia mydas*) in Borneo. 1. Breeding season. *Sarawak Mus. J.* **3**: 592—596.
Harrisson, Tom (1952). Breeding of the edible turtle. *Nature, Lond.* **169**: 198.
Hendrickson, J. R. (1958). The green sea turtle, *Chelonia mydas* (Linn.) in Malaya and Sarawak. *Proc. zool. Soc. Lond.* **130**: 455—535.
Hirth, H. F. (1971). Synopsis of biological data on the green turtle *Chelonia mydas* (Linnaeus). *FAO Fish. Biol. Synopsis* No. 85.

Manton, M., Carr, A. & Ehrenfeld, D. (1972). An operant method for the study of chemoreception in the green turtle, *Chelonia mydas*. *Brain, Behav. Evolut.* **5**: 188—201.

Moorhouse, F. W. (1933). Notes on the green turtle (*Chelonia mydas*). *Rep. Gr. Barrier Reef Comm.* **4** (1): 1—22.

Parkes, Alan S. (1966). *Sex, science and society*, Newcastle upon Tyne: Oriel Press.

Pritchard, P. C. H. (1967). *Living turtles of the world*. Jersey City, NJ: TFH Publications.

Rathke, H. (1848). *Ueber die Entwickelung der Schildkröten*. Braunschweig: F. Vieweg und Sohn.

Rebel, T. P. (1974). *Sea turtles and the turtle industry of the West Indies, Florida and the Gulf of Mexico*. Florida: University of Miami Press.

Simon, M. (1975). The Green sea turtle (*Chelonia mydas*): Collection, incubation and hatching of eggs from natural rookeries. *J. Zool., Lond.* **176**: 39—48.

Simon, Martin H. & Parkes, Alan S. (1976). The Green sea turtle (*Chelonia mydas*): nesting on Ascension Island, 1973—1974. *J. Zool., Lond.* **179**: 153—163.

Simon, Martin H., Ulrich, Glenn F. & Parkes, Alan S. (1975). The Green sea turtle (*Chelonia mydas*): mating, nesting and hatching on a farm. *J. Zool., Lond.* **177**: 411—423.

Ulrich, G. F. & Owens, D. W. (1976). Preliminary note on reproduction of *Chelonia mydas* under farm conditions. *Proc. World Maricult. Soc.* **5**: 205—212.

Ulrich, G. F. & Parkes, Alan S. (1978). The Green sea turtle (*Chelonia mydas*): further observations on breeding in captivity. *J. Zool., Lond.* **185**: 237—251.

Wood, James R. & Wood, Fern E. (1978). Captive breeding of the green sea turtle (*Chelonia mydas*). *Proc. World Maricult. Soc.* **7**: 533—541.

Wood, J. R. & Wood, F. E. (1980). Reproductive biology of captive green sea turtles *Chelonia mydas*. *Am. Zool.* **20**: 499—505.

Symp. zool. Soc. Lond. (1981) No. 46, 267—281

Sexual Physiology of the Brain

B. A. CROSS

ARC Institute of Animal Physiology, Babraham, Cambridge, UK

SYNOPSIS

Many reproductive functions are now known to be dependent upon central nervous mechanisms. Morphological differences have been observed in the brains of males and females at both light and electron microscope levels and can be reversed experimentally by suitable manipulation of perinatal sex steroids. The masculine pattern of development which depends upon aromatisation of testosterone from the foetal or newborn testis is associated with non-cyclic release of luteinizing hormone (LH) from the pituitary gland and a characteristic pattern of synaptic inputs from the amygdala to the medial preoptic area. Unit recording in the hypothalamus of the female rat has revealed cyclic neuronal changes and clarified the mechanism of neuronal excitation of ovulation. Simultaneous radioimmunoassay of LH and prolactin in rats after suckling or stimulation of the medial preoptic area together with unit studies of tuberoinfundibular neurones yield evidence of separate cell populations that may regulate discharge of the two hormones. Further insight into the coding of nervous impulses in these neuroendocrine systems is provided by studies on the oxytocin and vasopressin cells of the magnocellular nuclei which suggest an underlying mechanism for pulsatile release of pituitary hormones.

INTRODUCTION

Not very long ago sexual physiology was mainly confined to the workings of the gonads and reproductive tracts of the male and female. Production of germ cells, the migration, maturation and ejaculation of spermatozoa, seminal transport, implantation of the conceptus, pregnancy, parturition, and lactation were all processes which it was thought could occur without the obligatory participation of the brain. We now know that scarcely a single role in the reproductive repertoire does not engage central nervous mechanisms either directly, as in the lordosis response of oestrous rodents, or indirectly, e.g. via hypothalamic control of gonadotrophic secretion.

In this chapter we shall mention some neural phenomena that depend upon sex and some sexual processes that depend upon neural mechanisms. The dividing line between the two categories is not always clear.

SEXUAL DIMORPHISM IN THE BRAIN

Sex chromatin (Barr bodies) was first found in the nuclei of brain cells in female cats, and may now be regarded as the definitive marker of genetic brain sex. There is a lengthening list of other sexual dimorphisms in the brain, from gross weight and size, through relative prominence of certain brain nuclei or their constituent cell bodies to the content of monoamines or other transmitter substances (see Glucksmann, 1974).

Many of the morphological differences are probably related to sexual or neuroendocrine functions. A striking example is the work of Gorski, Gordon, Shryne & Southam (1978) at the level of the light microscope which showed that the size of the medial preoptic nucleus is consistently larger in male rats than in females. The dimorphism is sex hormone dependent, as evidenced by its reversal in genetic females treated in the critical neonatal period with testosterone or in males castrated at birth. Somewhat similar sexually dimorphic effects were earlier described by Dörner & Standt (1968). The first quantitative ultrastructural studies were those of Raisman & Field (1971, 1973) who described a sexually dimorphic difference in the number of afferent synapses on dendritic spines in a medial preoptic zone in rats. The spine synapses derived from non-strial sources (i.e. not amygdala) were twice as abundant in females as in males. Neonatal exposure to androgen was shown to be crucial to the occurrence of the male dimorphic pattern. For the preoptic area of the hamster Greenough, Carter, Steerman & De Voogd (1977) analysed dendritic patterns in Golgi-stained material and discovered sex differences which they believed were related to functionally different afferent inputs. It is of considerable interest in connection with these micromorphological phenomena that selective effects of sex hormones on growth of neurites in cultures of newborn mouse preoptic/hypothalamic tissue have been demonstrated by Toran-Allerand (1976).

Because so many classical studies implicate the medial preoptic area in sexual behaviour and reproductive cycles most attention has been focused on dimorphic patterns in that area. Doubtless other morphological differences would be revealed by application of suitable quantitative techniques elsewhere. To give but three instances: the nuclei of cells in the medial amygdaloid nucleus of the male squirrel monkey are significantly larger than those of the female (Bubenik & Brown, 1973); there are differences in the structure of the ependyma lining the third ventricle in male and female rhesus monkey and cyclical changes occur only in the latter (Anand Kumar,

1968); and Réthelyi (1979) reports that female rats possess a neuro-vascular contact surface in the median eminence and pituitary stalk that is 26% larger than that of males. Whether the many psychological differences in preference and aptitude often attributed to sex have any morphological correlate in the brain remains to be determined.

SEXUAL DIFFERENTIATION OF THE HYPOTHALAMUS

One of the most compelling doctrines to emerge in reproductive biology in the last two decades is that of the organizing action of sex steroids upon functional hypothalamic systems during a critical period in late foetal or neonatal life. A wealth of data from the effects of testosterone treatment in newborn females, castration or anti-androgen treatment in newborn males and direct implantation of sex steroids in the neonatal hypothalamus has proved beyond cavil that adult males and females are possessed of functionally differentiated hypothalamic systems (Gorski, 1966; Neumann, Steinbeck & Hahn, 1970; Brown-Grant, 1973). Normal males or "masculinized" females do not exhibit a cyclic release of gonado-trophic hormone (LH) as in the normal ovulating female and their sexual behaviour is typically male. "Feminized" males on the other hand, if provided with ovarian grafts, experience a cyclic release of gonadotrophin and receptive behaviour (lordosis) characteristic of oestrus.

It is now widely accepted that the masculinizing effect of testosterone during the critical period of hypothalamic differentiation occurs paradoxically after its aromatization to oestrogen (McDonald & Doughty, 1974; Naftolin & Ryan, 1975). The mechanism of action of the steroid on developing brain is not fully understood but probably involves as an essential ingredient an effect upon neuronal protein synthesis (Kobayashi & Gorski, 1970). In all mature female mammals so far tested, including primates, the hormonal signal to the brain for triggering the discharge of an ovulatory dose of LH from the adenohypophysis is a rise in plasma oestrogen from the ovaries bearing ripening follicles. The ability to respond to this stimulus depends upon the presence of oestrogen receptors in the preoptic/anterior hypothalamic area, and it has been proposed that neonatal androgen causes sterility of the adult by blocking the subsequent formation of the receptors (see Zigmond, 1975).

The full scope of sexually differentiated systems is by no means certain, but it is not confined to mating behaviour and LH secretion.

Neill (1972) has shown, for example, that the surge of prolactin secretion that occurs in the cyclic female rat is a feature of the sexually differentiated hypothalamus, for the prolactin response to oestrogen is present in intact females and neonatally castrated males but absent in normal males or androgen sterilized females. Probably the regulation of other metabolic hormones does not entirely escape the organizing influence of neonatal sex steroids. Ingestive behaviour at all events seems to be almost as dependent as mating behaviour on hormonal gender (Nance, Bromley, Barnard & Gorski, 1977).

Concepts of hormonal feedback preceded present knowledge of sexual differentiation but it is often difficult to know whether a given response to hormone treatment is sex-dependent or not. Mating behaviour is quite well understood, however, and though mounting responses occur in individuals of either sex the typical lordosis response of receptive females is not present in sexually differentiated males. The extensive researches of Pfaff and his group (Pfaff, Lewis, Diakow & Keiner, 1973; Pfaff, Montgomery & Lewis, 1977) have analysed the neuro-locomotor reflexes involved in lordosis and correlated the sites of the motor neurone pools in the brain that mediate the response with those in which uptake of radio-labelled oestrogen occurs. We are worse informed about central nervous mechanisms that operate in such reproductive activities as penile erection, male copulatory thrusts or moderating abdominal contractions in labour. All these have spinal reflex components, presumably modified by forebrain (including hypothalamic) influences to which neonatal and adult hormone states may make a contribution.

Two caveats seem desirable at this point. Not all nervous effects of hormones result from central actions, e.g. Komisaruk, Adler & Hutchison (1972) have shown that oestrogen enlarges the sensory field of the pudendal nerve in spayed rats. Secondly, psychosocial stimuli can upset the preprogrammed hypothalamus e.g. by interfering with the aromatization of androgens (Di Prisco, Lucarini & Dessi-Fulgheri, 1978) and the triggering action of oestrogen on LH in talapoin monkeys (Bowman, Dilley & Keverne, 1978).

ELECTROPHYSIOLOGICAL STUDIES IN NEURAL SEX

Hypothalamus and Ovulation

The traditional methods of studying brain mechanisms have been electrical stimulation, ablation and recording. The first two reveal mechanisms by exaggerating or deranging normal function, whereas recording aims to detect neural processes in normal operation. All three approaches provide evidence in the living animal but post-mortem study of the brains is necessary to reveal lesion or electrode sites, and the distribution by immunofluorescent tags of various functional types of neurone. Though we know a great deal about the sexually differentiated ovulation mechanism involving oestrogen triggering of medial preoptic/anterior hypthalamic (PO/AH) neurones and the excitation of LH–RH cells in the arcuate/median eminence area (Sawyer, 1975), no one so far has confidently recorded specifically the electrical activity of this neuronal system.

Unit recording has been the technique favoured in the author's laboratory, alone or in combination with direct or indirect measures of pituitary hormone secretion (Cross, 1973a, b; Cross, Dyball et al., 1975). Studies of the firing rates of single neurones in the PO/AH area on female rats showed a peak of activity on the day of vaginal pro-oestrus. This cyclic event was observed both in intact rats under urethane anaesthesia and in acutely decerebrated rats with hypothalamic islands. Ovariectomy a day earlier prevented the rise in discharge activity so it would seem we were recording the neural response to endogenous hormonal (oestrogen) feedback from the ovary acting directly upon the hypothalamus (Cross, 1973b). However, inasmuch as the connexions of the recorded neurones were unknown they could as likely be concerned with sexual behaviour or feeding as with gonadotrophin secretion. Dyer (1973) improved upon this situation by distinguishing three sub-populations of PO/AH neurones on the basis of their response to shocks delivered to the ventromedial/arcuate region of the hypothalamus. Type A cells were antidromically excited and had low resting discharge rates and mean conduction rate of $0.32\,\mathrm{m\,s^{-1}}$. They were candidates for LH–RH neurones. Type B cells were synaptically excited or inhibited by ventromedial/arcuate shocks but were not antidromically activated. Type C cells were unaffected by the test stimuli indicating that they possessed neither afferent nor efferent connexions to the ventromedial/arcuate region. Dyer was able to conclude that the pro-oestrous peak of discharge activity previously observed must reside in the B and C neurones.

In a further set of experiments Dyer, MacLeod & Ellendorff (1976) looked for sexually dimorphic differences in the Type A population of PO/AH neurones. They found that electrical shocks delivered to the medial amygdala synaptically influenced the firing rate of these cells much more frequently in males than in females. Moreover neonatal castration resulted in the development of the female pattern of amygdalo-preoptic connexions whereas females neonatally treated with testosterone showed an intermediate position. It is tempting to compare this neurophysiological evidence of hormone mediated sexual dimorphism with the sexually dimorphic zone delineated by Raisman & Field (1973) on ultrastructural criteria. However, the Type A cells were not restricted to this zone, and in any case the sex-differentiated spine synapses seen by electron microscopy did not originate from the amygdala.

The existence of direct neural connexions between the medial preoptic area and the arcuate nucleus shown in the preceding experiments has been confirmed by degeneration studies using the light and electron microscope (Köves & Réthelyi, 1976). We are probably dealing with two separate pools of neurones in the sexually differentiated ovulatory mechanism, viz. the preopticoarcuate neurones and the tuberoinfundibular (LH–RH) neurones. Can we give an educated guess as to the patterning of ovulatory signals at these two sites? Stimulation experiments in pro-oestrous rats with ovulation blocked by pentobarbitone have recently offered a clue. It would seem that the duration of the signal rather than the actual number of impulses or their frequency is the critical event. Thus a 45 min period of stimulation of the arcuate nucleus at $10 \, s^{-1}$ was as effective in eliciting ovulation as stimulation at $100 \, s^{-1}$, and shorter periods at high frequency were less effective. However, three 5 min periods of stimulation at $100 \, s^{-1}$ spread over 45 min were as effective as continuous stimulation (Dyer & Mayes, 1978). Evidently all that is necessary to evoke ovulation is a series of short spurts of LH–RH, utilizing the "priming" effect on the adenohypophysis that was described by Fink, Chiappa & Aiyer (1976). In the Babraham experiments (Dyer & Mayes, 1978), preoptic stimulation by contrast had to be sustained at $100 \, s^{-1}$ for the full 45 min to elicit subsequent ovulation, so the relationship between the two postulated neuronal pools is not a simple one.

Sadly, although it is becoming a routine procedure to monitor the electrical discharges of tuberoinfundibular neurones by antidromic stimulation from the median eminence (e.g. Sawaki & Yagi, 1976), there is still no way in life of identifying LH–RH neurones. Nevertheless we have good reason to suppose that the tuberoinfundibular system contains peptidergic cells and aminergic neurones. Among the latter a prominent group consists of the dopamine-containing cells, and dopamine is believed to be identical with the so-called prolactin-inhibiting factor (PIF, see Tindal, 1978). Now recent experiments at Babraham have shown not only that arcuate/median eminence stimulation at $10 \, s^{-1}$ could release an ovulatory quota of LH in pro-oestrous rats, but also that stimulation at $7.5\,s^{-1}$ over a three-hour period during lactation blocks the normal suckling-induced release of pro-lactin (Wakerley & ter Haar, 1978). Is there a chance that we might learn to distinguish in life the neurones mediating these very differ-ent effects?

A problem of great importance for the control of fertility in man and animals is the nature of the reciprocal relationship between ovulation and suckling anoestrus (amenorrhoea). Plasma levels of LH in lactating anoestrous rats are basal and even ovariectomy has no significant effect in elevating plasma LH (Smith & Neill, 1977). Nevertheless recent experiments in our laboratory (Isherwood & Cross, 1980) with lactating rats bearing indwelling catheters in their right atria have shown that removal of the pups for six hours on the ninth day *post-partum* elicits a five-fold increase in plasma LH con-centration which is rapidly reversed when the litter is replaced. In this situation we are witnessing acutely the differential effect of the suckling stimulus on prolactin and LH (Fig. 1). Before long it should be possible to record the neuronal basis of this reciprocal mechanism which is presumably a property of the female rather than the male hypothalamus.

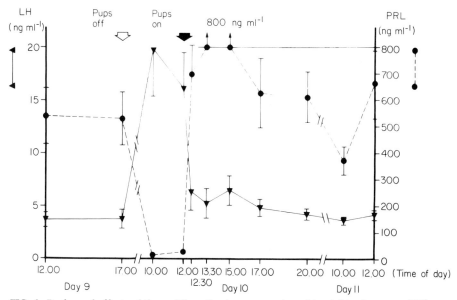

FIG. 1. Reciprocal effects of the suckling stimulus on secretion of luteinising hormone (LH) and prolactin (PRL). Graphs show plasma concentrations of the two hormones (means and standard error) collected from indwelling atrial cannulae over days 9–11 in six lactating rats. Note the low plasma LH in suckled rats and the five-fold increase after removing the litter. Restoration of suckling reversed this effect and provoked a large increase of prolactin within 30 min.

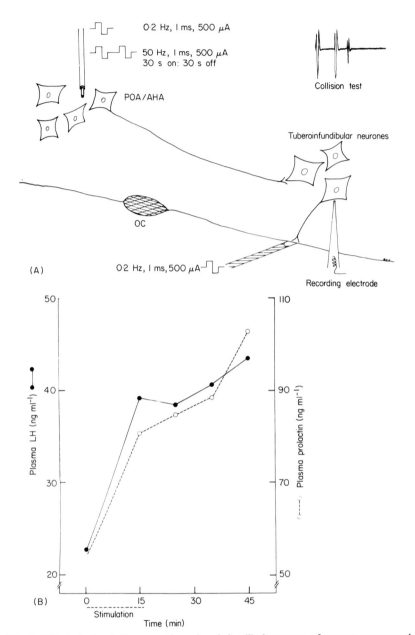

FIG. 2. Effect of stimulating preoptico-tuberoinfundibular system of neurones on secretion of LH and prolactin. (A) Diagram of experimental arrangement for recording tuberoinfundibular neurones identified antidromically by the collision test from stimulating electrodes on the median eminence. A second stimulating electrode in the preoptic area delivers either single shocks (0.2 Hz) or 50 Hz stimulation; OC, optic chiasma. (B) Curves showing the mean increases in LH and prolactin from a 15 min period of 50 Hz stimulation in the preoptic area in 33 rats.

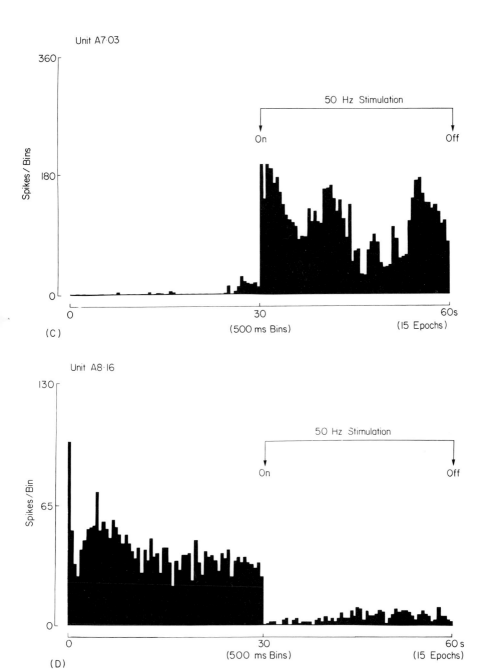

FIG. 2. (*cont.*) (C) A tuberoinfundibular neurone greatly accelerated in firing rate by the preoptic stimulation, possibly a LH–RH neurone. (D) A tuberoinfundibular neurone inhibited by the preoptic stimulation, possibly a dopamine (PIF) neurone.

In the meantime there is another prospect of dissecting out *in vivo* the dopamine (PIF) from the LH—RH neurones in the tuberoinfundibular system. Electrical stimulation of the medial preoptic area evokes a dual release of LH and prolactin measured by radioimmunoassay. We should expect then to find two populations of neurones in the tuberoinfundibular system, one which is accelerated by the preoptic stimulation (LH—RH cells) and one which is inhibited (dopamine or PIF cells). Dyer & Saphier (in press) have recently found just this situation. Of 44 tuberoinfundibular neurones identified by antidromic shocks to the median eminence (23% of all recorded cells) 14 were excited by preoptic stimulation and 12 inhibited. The remaining 18 were unaffected (Fig. 2). Further experiments are in progress to repeat the observations after selective destruction of the dopamine neurones.

Hypothalamus and Neurohypophysis

It will be apparent that in the study of the neuronal systems involved in gonadotrophin and prolactin secretion a grave disadvantage has been our inability to specify in advance the nature of the neurones being recorded. Undoubtedly our confidence in identifying neuroendocrine cells in life is greatest for the magnocellular system of oxytocin and vasopressin neurones whose cell bodies occupy the supraoptic and paraventricular nuclei and whose axons terminate in the neural lobe (Cross & Wakerley, 1977).

These endocrine neurones can be detected by antidromic activation from the neural lobe and distinguished from each other by their characteristic responses to various physiological stimuli. Oxytocin cells are characterized by their explosive volley of discharges (20—$80 s^{-1}$) before each reflex milk ejection in suckled lactating rats. Between each burst the oxytocin neurones fire at low irregular rates and the few seconds of frenzied activity occur synchronously in all the cells, i.e. about half the neurones in both magnocellular nuclei. Their combined effort in delivering something like half a million impulses to their endings releases 0.5—1.0 mU oxytocin from the neurohypophysis into the bloodstream, and during suckling sessions of an hour or more the process may recur at intervals of 5—15 min (Wakerley & Lincoln, 1973).

Vasopressin cells in the resting state exhibit low firing activity indistinguishable from oxytocin cells but they do not participate in the barrage of firing at milk ejection. However, if dehydration occurs they develop phasic activity with short bursts of accelerated firing ($5—20 s^{-1}$) interspersed by quiescent periods. The onset and duration of the bursts is not synchronized and both the duration and intra-burst frequency increases with increasing plasma molarity (Wakerley, Poulain & Brown, 1978). Similar activation of the vasopressin cells can be provoked by haemorrhage (Poulain, Wakerley & Dyball, 1977).

Obviously there is a pronounced coding of impulses in the two populations of magnocellular neurones. The synchronous bursts of the oxytocin cells provoke a tidal wave of oxytocin in the blood-stream to stimulate the comparatively insensitive mammary myo-epithelium, whereas the asynchronous phasic discharges of the vasopressin cells elicit a slow trickle of antidiuretic hormone to restrain renal loss of water. There is no sound reason to suppose that vasopressin cells have any sexual role and it is of interest that their characteristic phasic activity is seen in animals of both sexes, and also in superfused hypothalamic slices (Haller, Brimble & Wakerley, 1978), implying an intrinsic property of the vasopressin neurones. On the other hand the characteristic bursts of the oxytocin cells have not been observed in males and their synchronous occurrence sug-gests a well-organized afferent excitatory mechanism. Whether this is a sexual specialization of the female is not clear. Apart from milk ejection the most probable function of oxytocin is in parturition when second-stage contractions of the myometrium are augmented by the hormone released in response to distension of the birth canal ("Ferguson reflex"). In this connexion Dreifuss, Tribollet & Baertschi (1976) have reported that vaginal distension of lactating rats pro-duced reflex milk ejection responses, though in these experiments some vasopressin neurones (phasically active) appeared to be excited. Also it has been demonstrated clearly that oxytocin neurones can be stirred into greater activity by stimuli with no sexual significance, e.g. haemorrhage or dehydration. The sustained moderate acceleration of firing rate then observed is easily distinguishable both from the phasic activity of vasopressin cells and from their own synchronous milk ejection bursts.

Perhaps the main value of our present understanding of magno-cellular neuroendocrine cells is in offering models for the possible modes of action of other peptidergic neurones. For example Dutton & Dyball (1979) have shown that stimulation of incubated rat neural lobes with pulse patterns derived from tape recordings of phasically firing supraoptic neurones in dehydrated rats released more vaso-pressin than the same number of pulses regularly spaced in time. Optimal inter-spike interval appears to be shorter for oxytocin cells than for vasopressin cells. Some comparable observations have been made for LH—RH released from the mediobasal hypothalamus incubated *in vitro* (Dyer, Mansfield & Yates, 1979). These indicate that the optimal interspike interval for hormone release is affected not only by the length of the stimulation but also by the gonadal status of the animal. This type of study becomes increasingly urgent as more evidence accumulates for the pulsatile release of gonadotrophic hormone (Midgley & Jaffe, 1971; Blake & Sawyer, 1974).

CONCLUSIONS

There is no longer any need to doubt that female brains differ from male brains in morphological detail and functional capacity. The physiological differences seem more concerned with vegetative capabilities than cognitive and pertain chiefly to reproductive cycles and the associated behaviour patterns. Genetic sex determines gonadal development but the evidence suggests that the gonadal hormones are the effective determinant of brain sex, and experimental manipulation of these in perinatal animals can produce lasting developmental changes in brain function. In the adult animal the effects of sex hormones on the brain are likely to be short-term and reversible, though the precise response in each case can be expected to depend upon both brain gender and current hormonal status.

Cases are known where the neural mechanism of a sexual function (e.g. ovulation or milk ejection) expressed in neurophysiological terms appears to differ radically in females from that in males. We cannot always be sure, however, that it would be impossible to duplicate the neural response in males if suitable conditions were found. There is a continuing need for basic studies on the behaviour of neurones involved in the mediation of sexual responses. Already a convincing account can be given of the activity of the magnocellular neurones that release oxytocin in suckling-induced milk ejection and there are good prospects in the near future of unravelling the neuronal systems that control the release of luteinizing hormone and prolactin from the pituitary gland.

REFERENCES

Anand Kumar, T. C. (1968). Sexual differences in the ependyma lining the third ventricle of the anterior hypothalamus of adult rhesus monkeys. *Z. Zellforsch. mikrosk. Anat.* **90**: 28–36.

Blake, C. A. & Sawyer, C. H. (1974). Effects of hypothalamic deafferentation on the pulsatile rhythm in plasma concentrations of luteinizing hormone in ovariectomised rats. *Endocrinology* **94**: 730–736.

Bowman, L. A., Dilley, S. R. & Keverne, E. B. (1978). Suppression of oestrogen-induced LH surges by social subordination in talapoin monkeys. *Nature, Lond.* 275: 56—58.

Brown-Grant, K. (1973). Recent studies on the sexual differentiation of the brain. In *Foetal and neonatal physiology*: 527—545. Comline, R. S., Cross, K. W., Dawes, G. S. & Nathanielsz, P. W. (Eds). Cambridge: University Press.

Bubenik, G. A. & Brown, G. M. (1973). Morphologic sex differences in primate brain areas involved in regulation of reproductive activity. *Experientia* 15: 619—621.

Cross, B. A. (1973a). Unit responses in the hypothalamus. In *Frontiers in neuroendocrinology*: 133—171. Ganong, W. F. & Martini, L. (Eds). New York: Oxford University Press.

Cross, B. A. (1973b). Towards a neurophysiological basis for ovulation. *J. Reprod. Fert.* (Suppl.) 20: 97—117.

Cross, B. A., Dyball, R. E. J., Dyer, R. G., Jones, C. W., Lincoln, D. W., Morris, J. F. & Pickering, B. T. (1975). Endocrine neurons. *Rec. Progr. Hormone Res.* 31: 243—294.

Cross, B. A. & Wakerley, J. B. (1977). The neurohypophysis. In *International review of physiology, Endocrine Physiology II*, 16: 1—34. McCann, S. M. (Ed.). Baltimore: University Park Press.

Di Prisco, C. L., Lucarini, N. & Dessi-Fulgheri, F. (1978). Testosterone aromatisation in rat brain is modulated by social environment. *Physiol. Behav.* 20: 345—348.

Dörner, G. & Standt, J. (1968). Structural changes in the preoptic anterior hypothalamic area of the male rat following neonatal castration and androgen substitution. *Neuroendocrinology* 3: 136—140.

Dreifuss, J. J., Tribollet, E. & Baertschi, A. J. (1976). Excitation of supraoptic neurone by vaginal distention in lactating rats; correlation with neurohypophysial hormone release. *Brain Res.* 113: 600—605.

Dutton, A. & Dyball, R. E. J. (1979). Phasic firing enhances vasopressin release from the rat neurohypophysis. *J. Physiol.* 290: 433—440.

Dyer, R. G. (1973). An electrophysiological dissection of the hypothalamic regions which regulate the pre-ovulatory secretion of luteinizing hormone in the rat. *J. Physiol.* 234: 421—442.

Dyer, R. G., MacLeod, N. K. & Ellendorff, F. (1976). Electrophysiological evidence for sexual dimorphism and synaptic convergence in the preoptic and anterior hypothalamic areas of the rat. *Proc. R. Soc. Lond.* (B) 193: 421—440.

Dyer, R. G., Mansfield, S. & Yates, J. O. (1979). Release of luteinising hormone releasing hormone from hypothalamic slices: evidence that gonadectomy reduces the response to electrical stimulation. *J. Endocr.* 80: 31—32P.

Dyer, R. G. & Mayes, L. C. (1978). Electrical stimulation of the hypothalamus: new observations on the parameters necessary for ovulation in rats anaesthetised with pentobarbitone during the pro-oestrous "critical period". *Expl Brain Res.* 33: 583—592.

Dyer, R. G. & Saphier, D. J. (1981). Electrical activity of antidromically identified tuberoinfundibular neurones during stimulated release of luteinising hormone and prolactin in pre-oestrus rats. *J. Endocr.* 89: In press.

Fink, G., Chiappa, S. A. & Aiyer, M. S. (1976). Priming effect of luteinising hormone releasing factor elicited by preoptic stimulation and multiple injections of synthetic decapeptide. *J. Endocr.* **69**: 359–372.

Glucksmann, A. (1974). Sexual dimorphism in mammals. *Biol. Rev.* **49**: 423–475.

Gorski, R. A. (1966). Localization and sexual differentiation of nervous structures which regulate ovulation. *J. Reprod. Fert.* (Suppl.) **1**: 67–88.

Gorski, R. A., Gordon, J. H., Shryne, J. E. & Southam, A. M. (1978). Evidence for a morphological sex difference within the medial preoptic area of the rat brain. *Brain Res.* **148**: 333–346.

Greenough, W. T., Carter, C. S., Steerman, C. & De Voogd, T. J. (1977). Sex differences in dendritic patterns in hamster preoptic area. *Brain Res.* **126**: 63–72.

Haller, E. W., Brimble, M. J. & Wakerley, J. B. (1978). Phasic discharge of supraoptic neurones recorded from hypothalamic slices. *Expl Brain Res.* **33**: 131–134.

Isherwood, K. M. & Cross, B. A. (1980). Effect of the suckling stimulus on secretion of prolactin and luteinising hormone in conscious and anaesthetised rats. *J. Endocr.* **87**: 437–444.

Kobayashi, F. & Gorski, R. A. (1970). Effects of antibiotics on androgenization of the neonatal female rat. *Endocrinology* **86**: 285–289.

Komisaruk, B. R., Adler, N. & Hutchison, J. (1972). Genital sensory field: enlargement by estrogen treatment in female rats. *Science, Wash.* **178**: 1295–1298.

Köves, K. & Réthelyi, M. (1976). Direct neural connection from the medial preoptic area to the hypothalamic arcuate nucleus of the rat. *Expl Brain Res.* **25**: 529–539.

McDonald, P. G. & Doughty, C. (1974). Effect of neonatal administration of different androgens in the female rat: correlation between aromatization and the induction of sterilization. *J. Endocr.* **61**: 95–103.

Midgley, A. R. & Jaffe, R. B. (1971). Regulation of human gonadotropins: X Episodic fluctuation of LH during the menstrual cycle. *J. clin. Endocr.* **33**: 962–969.

Naftolin, F. & Ryan, K. J. (1975). The metabolism of androgens in central neuroendocrine tissues. *J. Steroid Biochem.* **6**: 993–997.

Nance, D. N., Bromley, B., Barnard, R. J. & Gorski, R. A. (1977). Sexually dimorphic effects of forced exercise on food intake and body weight in the rat. *Physiol. Behav.* **19**: 155–158.

Neill, J. D. (1972). Sexual differences in the hypothalamic regulation of prolactin secretion. *Endocrinology* **90**: 1154–1159.

Neumann, F., Steinbeck, H. & Hahn, J. D. (1970). Hormones and brain differentiation. In *The hypothalamus*: 569–603. Martini, L., Motta, M. & Fraschini, F. (Eds). New York: Academic Press.

Pfaff, D., Lewis, C., Diakow, C. & Keiner, M. (1973). Neurophysiological analysis of mating behaviour responses as hormone-sensitive reflexes. In *Progress in physiological psychology* **5**: 253–297. New York & London: Academic Press.

Pfaff, D., Montgomery, M. & Lewis, C. (1977). Somatosensory determinants of lordosis in female rats: behavioral definition of estrogen effect. *J. comp. Physiol. Psychol.* **91**: 134–145.

Poulain, D. A., Wakerley, J. B. & Dyball, R. E. J. (1977). Electrophysiological differentiation of oxytocin- and vasopressin-secreting neurones. *Proc. R. Soc. Lond.* (B) 196: 367–384.

Raisman, G. & Field, P. M. (1971). Sexual dimorphism in the preoptic area of the rat. *Science, Wash.* 173: 731–733.

Raisman, G. & Field, P. M. (1973). Sexual dimorphism in the neuropil of the preoptic area of the rat and its dependence on neonatal androgen. *Brain Res.* 54: 1–29.

Réthelyi, M. (1979). Regional and sexual differences in the size of the neuro-vascular contact surface of the rat median eminence and pituitary stalk. *Neuroendocrinology* 28: 82–91.

Sawaki, Y. & Yagi, K. (1976). Inhibition and facilitation of antidromically identified tuberoinfundibular neurones following stimulation of the median eminence in the rat. *J. Physiol.* 260: 447–460.

Sawyer, C. H. (1975). Some recent developments in brain-pituitary-ovarian physiology. *Neuroendocrinology* 17: 97–124.

Smith, M. S. & Neill, J. D. (1977). Inhibition of gonadotrophin secretion during lactation in the rat: relative contribution of suckling and ovarian steroids. *Biol. Reprod.* 17: 255–261.

Tindal, J. S. (1978). Neuroendocrine control of lactation. In *Lactation: a comprehensive treatise* 4: 67–114. Larson, B. L. (Ed.). New York & London: Academic Press.

Toran-Allerand, C. D. (1976). Sex steroids and the development of the newborn mouse hypothalamus and preoptic area *in vitro*: implications for sexual differentiation. *Brain Res.* 106: 407–412.

Wakerley, J. B. & Lincoln, D. W. (1973). The milk-ejection reflex of the rat: a 20–40-fold acceleration in the firing during oxytocin release. *J. Endocr.* 57: 477–493.

Wakerley, J. B., Poulain, D. A. & Brown, D. (1978). Comparison of firing patterns in oxytocin- and vasopressin-releasing neurones during progressive dehydration. *Brain Res.* 148: 425–440.

Wakerley, J. B. & ter Haar, M. B. (1978). Plasma concentrations of prolactin and thyrotrophin during suckling: effects of stimulation of the median eminence. *J. Endocr.* 76: 557–558.

Zigmond, R. E. (1975). Binding, metabolism, and action of steroid hormones in the central nervous system. In *Handbook of psychopharmacology* 5: 239–328. Iversen, L. L., Iversen, S. D. & Snyder, S. H. (Eds). New York: Plenum Press.

Primate Sociology and Behaviour

Chairman's Remarks

There are at least two good reasons why a session on primate behaviour should be included in this symposium. The first, and more general one, reflects the current interest in studying behaviour of primates in a variety of free living habitats as well as in the laboratory. A meeting hoping to survey primate biology would be unthinkable without some reference to work in this area. Indeed, it is surprising that it should have taken so long for this kind of work to attract the attention of zoologists, particularly since there had existed, since the 1930s and 1940s, several publications which ought to have alerted us to the fascination inherent in such studies. And this, of course, brings me to my second — and more particular — reason. Almost 50 years ago now, there first appeared a quite remarkable book: *The social life of monkeys and apes*. Despite the enormous volume of primate behaviour in the years since then this book is still widely read, quoted, discussed and criticized, as it was when it appeared, to judge by contemporary accounts. It was the first attempt to treat the study of primate behaviour seriously and critically, a most necessary task in view of the anecdotes and folklore which provided most of the so-called information which was then available. But it did more than that, for in reporting the tumultuous events following attempts to establish a baboon colony in the London Zoo, it provided a discussion and critique of such factors as the formation of dominance hierarchies, the evolution and use of sexual signals, and the correlation between an animal's internal reproductive state and its overt behaviour: all topics which continue to preoccupy us to this day. Questions were also raised, some of which (for example the function of altruistic behaviour) have only recently been pursued with any vigour, while others (whether animals distinguish between the living and the dead is one) are still largely unexplored.

This part of the meeting therefore consists of three papers which attempt to discuss our present state of knowledge in three of the areas which form part of the framework of *The social life of monkeys and apes* as it was written half a century ago. In the first paper, R. D. Martin describes some current concepts concerning the behaviour of primates in the wild, paying particular attention to the interplay between a species' ecological environment and its physical and social adaptations. J. Herbert discusses the way that hormones may be used to modify or determine the behaviour within social groups of monkeys, and how different hormonal mechanisms may play distinct biological roles. R. E. Passingham deals with selected topics of the primates' behavioural repertoire from the point of view of a physiological psychologist, and attempts to relate behaviour to brain function.

J. HERBERT

Symp. zool. Soc. Lond. (1981) No. 46, 287–336

Field Studies of Primate Behaviour

R. D. MARTIN

Department of Anthropology, University College London, London, UK

SYNOPSIS

The scientific study of primate behaviour under natural conditions dates back approximately 50 years, one of the first contributions being Zuckerman's *The social life of monkeys and apes* (1932). Rapid expansion of primate field studies since 1930 has now yielded sufficient data to permit provisional attempts to produce a synthetic theory of primate social organization. Major themes in this synthesis are ecological parameters (especially feeding ecology), reproductive strategies and scaling to body size, with particular emphasis on bias-free quantification of both behaviour and ecology. Future developments in this area will be characterized by an increasing need for long-term monitoring of natural primate populations and by greater integration of field and laboratory studies (notably in the areas of nutrition, reproduction, genetics and parasitology).

Studies on prosimian primates, which have been slow to develop and have generally been somewhat under-represented, are particularly significant for a variety of reasons. Most importantly, studies of nocturnal prosimian species can throw light on nocturnal:diurnal distinctions in general and on the ancestral primate condition (probably characterized by nocturnal adaptation – Martin, 1979). Since nocturnal prosimians are typically small-bodied, it is easier to apply broad population survey techniques and methods such as radio-tracking which have been developed with small mammals. Furthermore, the apparently independent evolution of diurnal lemurs on Madagascar permits a test of hypotheses concerning the evolution of social organization in diurnal monkeys and apes. The "solitary" pattern typical of nocturnal prosimians contrasts with the gregarious habit typical of diurnal lemurs, monkeys and apes, but *all* primates exhibit well-developed social networks and the usual pattern for nocturnal primates is a dispersed "harem" system.

Major classifications of patterns of primate social organization are reviewed (Carpenter, 1954; Crook & Gartlan, 1966; Eisenberg, Muckenhirn & Rudran, 1972), leading to the recognition of five basic types: "solitary"; monogamous family groups; one-male groups ("harem" groups); age-graded-male groups; multi-male groups. Contrary to previous conclusions, it is suggested that the evolution of primate social organization has passed from dispersed harem systems in "solitary" nocturnal forms to gregarious "harem" systems in diurnal forms, with multi-male groups or monogamous family groups appearing as specialized (and relatively rare) patterns under certain conditions. This agrees with Zuckerman's (1932) early identification of the "harem" group as fundamental to primates.

The differential scaling of characters to body size (allometry) is discussed with particular reference to Kleiber's (1932) rule relating basal metabolic rate to body weight. Studies analysing parameters of primate behaviour (e.g. foraging group size; ranging area; day range) with respect to body size are reviewed. Empirical rules derived from such studies, for example the observation that folivorous primates typically have smaller home ranges and shorter day ranges than frugivorous primates, are supported by comparisons of pairs of sympatric primate species matched for body size. Inclusion of new data on nocturnal pro-simians and re-analysis of some categories of data show that nocturnal insectivor-ous/frugivorous forms require even larger home ranges at any given body size than diurnal frugivores or folivores, that the degree of range coverage per day is not clearly related to territorial behaviour, and that (surprisingly) day range length is not related to body size when arboreal and terrestrial primates are con-sidered separately (the latter having longer day ranges than the former). The in-crease in ranging area requirements with body size does not appear to be dictated by metabolic requirements alone, since the increase exceeds the expected level. Future work in this area will require detailed analysis of energy budgets for individual species.

Reproductive strategies are briefly discussed, with special reference to the concept of r- and K-selection (MacArthur & Wilson, 1967; Pianka, 1970). The concept is illustrated by a comparison between *Galago alleni* (rainforest species; relatively K-selected) and *G. senegalensis moholi* (wooded savannah species; relatively r-selected). Consideration of reproductive strategies confirms the view that monogamous family groups can only arise under very special conditions.

Although primate behaviour can exhibit wide ranges of variation, this does not invalidate conclusions drawn from analysis of the average conditions for a large sample of primate species. The conclusions reached can be integrated into a general framework enhancing our understanding of primate social organization and providing a useful support for vital conservation measures.

INTRODUCTION

This is a very appropriate time for a review of field studies of primate behaviour, since it is approximately 50 years since such studies first began to take shape in a genuinely scientific framework. Two major contributions can be cited as marking the beginning of this period: Zuckerman's *The social life of monkeys and apes* (1932) and Carpen-ter's monograph (1934) on the behaviour and ecology of the howler monkey (*Alouatta "palliata"*, now referred to as *Alouatta villosa*), conducted at the Smithsonian Tropical Research Institute in Panama. Mention should also be made of Nissen's (1931) study of the chim-panzee (*Pan troglodytes*) in West Africa, which preceded both of these publications and should certainly be regarded as a noteworthy early attempt to collect systematic field data on a primate species. Significantly, both Nissen and Carpenter were inspired by Robert Yerkes, whose influence on the early development of this subject cannot be overstated. It can truly be said that the early 1930s saw the emergence of detailed and systematic observations on primates under natural conditions. Since that time, such studies have gathered in momentum to such an extent that the latest guide to primate field studies circulated by the Primate Society of Great Britain listed 92 field studies planned (12%), in progress (70%) or recently completed (18%) in various parts of the world. As a result of the mass of data generated by such studies, quite sophisticated analyses of natural primate behaviour are now possible (given critical selection of the data to be used) and in the last seven years four major textbooks on the subject have appeared (A. Jolly, 1972; Clutton-Brock, 1977a; Bramblett, 1976; Sussman, 1979). An attempt will be made in this chapter to review some of the major trends in primate field studies which have developed over the past 50 years and to trace the outlines of some of the associated theoretical avenues which are now being explored with some success.

Zuckerman's *The social life of monkeys and apes* was an important landmark for primate field studies not only in its timing but also in the identification of themes which have remained as central issues to the present day. Before discussing these themes, however, it is instructive to mention some aspects of historical interest. In the preface to his book, Zuckerman traced its origin to a paper he had given in February 1929 to the Anthropological Society of University College London. He had been encouraged by a number of people present at his lecture to publish his material in an expanded form, and one of those people was the social anthropologist Malinowski. This underlines the fact that the study of primate behaviour is of considerable relevance to anthropology generally, though there is perhaps cause for disappointment that this area of interdisciplinary contact has not yielded more concrete results than it has over the past half-century of primate field studies. It is also noteworthy that Zuckerman's early work on primates involved a close association with the Zoological Society of London. Zoological collections still have a prominent part to play in the development of our understanding of primate behaviour and it is likely, for reasons outlined below, that an increasing call will be made on the facilities available in zoos and elsewhere for the study of primates in captivity. Finally, it should be noted that Zuckerman's central concern in *The social life of monkeys and apes* was that of "treating overt behaviour as the result or expression of physiological events which have been made obvious through experimental analysis". Although this viewpoint has not been shared by many who have embarked on primate field studies since 1932, it has become increasingly obvious over recent years that a genuine understanding of primate behaviour under natural conditions demands (among other things) attention to physiology and to other aspects which can only be studied in captivity. By the same token, a genuine understanding of primate behaviour is utterly dependent upon field studies conducted in the natural environmental setting where the various primate species have evolved. One of the main aims of this review is to demonstrate the intrinsic value of primate field studies while at the same time emphasizing the need for increasing integration between field and laboratory investigations.

Despite the fact that Zuckerman's book was written almost 50 years ago, and despite the fact that he had himself conducted only brief field investigations of a single primate species (the Chacma baboon, *Papio ursinus*) in South Africa, the following points which he made have stood the test of time and remain highly relevant to primate field studies today:

(1) There was a great need for accurate field data on primate behaviour under natural conditions. Information available in 1930 was essentially anecdotal and precise accounts of ecological aspects were lacking.

(2) The interpretation of primate social behaviour was dependent, first and foremost, on study of ecology, reproductive physiology and individual variability.

(3) For mammals generally, the following three categories of behaviour were particularly important with respect to social organization:

    (i)   the search for food;
    (ii) the search for mates;
    (iii) avoidance of "enemies" (= predators).

(4) It was possible to identify for each primate species broad patterns of social organization typical for that species, which would be reflected in behaviour observed in captivity. (Hamadryas baboons *(Papio hamadryas)* had been seen to exhibit a polygynous form of social organization in several zoological collections and recent field studies (e.g. Kummer, 1968) have confirmed that this pattern is characteristic in the wild. Reports cited by Zuckerman of family groups among wild gibbons have been repeatedly confirmed in subsequent field studies (Carpenter, 1940; Ellefson, 1968; Chivers, 1972) and his description of the limited social groupings of orang utans has been consistently borne out by recent systematic observations under natural conditions (MacKinnon, 1974; Horr, 1975; Rijksen, 1978).

(5) The basic pattern of primate social organization seemed to be polygynous, in some cases at least accompanied by the occurrence of solitary males.

In view of recent interest in the possible genetic basis of "altruistic behaviour" of primates (Wilson, 1975), it is also noteworthy that Zuckerman included an entire chapter on "Altruism and Society". Although he was more concerned with the presence or absence of *conscious* altruism in non-human primates, it is striking that he should have recognized this as a major theme at such an early stage in the development of studies of primate social organization. Equally striking are the following two statements, which aptly set the scene for any modern review of primate field studies:

> It is conceivable that the day may come when it will be possible to discuss adequately all forms of overt behaviour, human or animal, in the same terms.   (Zuckerman, 1932: 20.)

> Analogy, as Hogben has implied, will have to give way to analysis if there is to be an end to irrelevant and anthropomorphic classifications of animal society.   (Zuckerman, 1932: 22.)

REQUIREMENTS FOR A SYNTHETIC THEORY

It is probably true to say that developments over the last decade have at last heralded the emergence of a genuinely synthetic theory of primate behaviour and ecology. Ten years ago, it was possible to trace only the barest outlines of such a synthesis. In the 50 years which have elapsed since Zuckerman predicted the emergence of a synthetic approach, field studies of primate behaviour have passed (remarkably rapidly, under the circumstances) through the successive stages of description, classification, quantification, analysis, prediction and testing. It is a mark of the sophistication of present analyses of primate behaviour that a genuinely predictive basis is now discernible and the coming years should see increasing testing of hypotheses and predictions both in the field and in the laboratory.

Since this review must necessarily be limited in its scope, it should at once be emphasized that there are two fairly distinct (though obviously interdependent) approaches which have advanced hand-in-hand. One can consider overall patterns of primate social organization and their relationship to natural environmental conditions and one can consider the fine details of social relationships between individuals. Although the latter aspect must of course be included in an overall synthetic theory, there is simply not enough space to do justice to it in this review. The reader is referred, instead, to two excellent accounts by Hinde (1974, 1977) for discussion of inter-individual relationships within primate societies. This present review will concentrate instead on the analysis of broad relationships between primate behaviour and ecology.

It is, by now, well established that field studies of primate behaviour must include ecological investigations to provide any worthwhile basis for interpretation. There has been an increasing trend towards quantification of the behaviour of primates under natural conditions (see Clutton-Brock, 1977b) and this trend has been accompanied by increasing attention to the quantification of ecological aspects. One corollary of this is the requirement that any primate field study should, where possible, cover at least 12 months, since annual climatic variations, and their repercussions on ecological conditions, must be encompassed by patterns of primate social organization. Feeding ecology has come to be recognized as a prime factor relevant not only to the social organization of primate species, but also to that of other mammal groups (e.g. ungulates: Geist, 1974; Jarman, 1974) and to that of birds (see Crook, 1965). Quantification of ecological parameters in primate field studies has thus been focused mainly on climatic factors and on specific aspects of major food sources. In some cases where individual primate species also feed on animal prey (especially arthropods), ecological quantification has been extended to the study of variations in animal food supply. Exact quantification of plant food availability is difficult, but much valuable information can be obtained by investigation of the changing states of individual plant species of known importance in primate diets (phenological studies). Quantification of behaviour, on the other hand, has been focused on the estimation of broad parameters of social organization (social group size; home range area; population density; socionomic sex ratio), on measuring aspects of feeding behaviour (day range length; daily activity budget; foraging behaviour; feeding activities), and on quantifying inter-individual relationships. With all these behavioural categories, there has been an increasing emphasis on the use of methods which minimize sources of bias (e.g. arising from differential recording of conspicuous behaviour patterns or of the behaviour of conspicuous individuals) and which guarantee an effective sampling of various activities observed under relatively difficult field conditions. These methods have been effectively reviewed by J. Altmann (1974) and by Clutton-Brock (1977b). This increasing attention to quantification of both behaviour and ecology has greatly increased the value of primate field studies with respect to attempted syntheses. However, further developments in this direction will be required as more and more factors are taken into account, such as parasites (Freeland, 1976, 1977), nutrient content of foodstuffs (Hladik & Hladik, 1972) and plant toxins

(Glander, 1975; Freeland & Janzen, 1974). One area in which quantification remains pitifully inadequate is that of reproduction; accurate demographic data from natural primate populations are essential for a valid synthesis. With many of the parameters which remain to be quantified in future primate field studies, it is obvious that data will have to be collected over a period of several years to provide an accurate reflection of the natural situation. (For example, Pollock (1975a) has suggested for the indri of Madagascar (*Indri indri*) that the population he studied may typically produce infants in synchrony only once every three years.) This is but one indication of the fact that the kinds of questions now being raised by primatologists will require really extensive field-studies in the future. Hence the value of establishments such as the Smithsonian Tropical Research Institute in Panama which have permitted data collection on individual primate species in a series of studies conducted over a considerable number of years (e.g. see Carpenter, 1962, and Smith, 1977 for a review of information gathered on howler monkeys, *A. villosa*).

Given the variability of good quantitative data derived from primate field studies, it is possible to proceed to preliminary interpretation and prediction, as will be shown below for a number of specific features of primate social organization. Before considering the actual data, however, it is necessary to identify some of the main issues which have been important in the development of primate field studies. In the first place, it must be emphasized that the order Primates contains numerous prosimian species (lemurs, lorises and tarsiers) as well as the better-known simians (monkeys, apes and man). Many studies of "primate social organization" in fact deal only with the monkeys and the apes. Indeed, field studies of prosimian primates were relatively neglected until the 1960s and modest expansion in such studies began a full decade after the dramatic expansion of field studies of simian primates (particularly Old World Monkeys) which took place in the mid-1950s (see S. A. Altmann, 1967). For whatever reason, there has been a general tendency to overlook the prosimian primates in analyses of primate social organization, but there are numerous advantages in including the prosimians in overall theoretical approaches, as will be shown.

In the development of any science, classification of the data emerges as an important stage prerequisite to analysis. It is therefore essential to consider the emergence and consolidation of schemes of classification of patterns of social organization, since they necessarily influenced the subsequent development of analytical frameworks. Having established some kind of classificatory basis, it is possible to proceed to synthetic approaches, and here special mention must be made of the significance of body size. It has been accepted for some time in the field of primate morphological evolution that the expression of particular characters is constrained by, among other things, body size and associated metabolic requirements. But even primate morphologists have classically considered body size as a factor only in certain specific areas (e.g. relative brain size: Stephan & Bauchot, 1965; Bauchot & Stephan, 1969; Jerison, 1973; Martin, 1973a) and have been slow to realize its pervasive importance in other areas (e.g. in locomotor adaptation). In studies of primate social organization, explicit consideration of body size as one determinant has only very recently been articulated and there is still much to be done. Allometry, or the differential scaling of specific characteristics to body size, is now becoming recognized as an essential component of primate adaptation and is relevant to patterns of primate social organization no less than to morphological aspects. In particular, since feeding ecology has been empirically recognized as a major factor in the determination of the form of primate societies, physiological investigations into allometric variation in metabolic requirements must also be taken into account.

It has already been pointed out that Zuckerman (1932) placed particular emphasis on reproduction as a factor influencing primate social organization. It is generally true to say that since 1932 field studies of primates have (largely because of their short-term nature) not paid adequate attention to reproductive biology, though there have been a few notable exceptions (e.g. Rowell, 1967; Goss-Custard, Dunbar & Aldrich-Blake, 1972; Crook, Ellis & Goss-Custard, 1976). It has now become quite apparent from ecological approaches to "reproductive strategies" that a synthetic theory of primate social organization must include detailed analysis of reproductive aspects as well as of other features, such as feeding ecology, which have been more widely accepted. In fact, since natural selection can be virtually equated with differential reproductive success, it is difficult to see why primate reproductive strategies have received so little attention in many of the attempted syntheses which have been published to date. Now that "sociobiological" approaches are becoming fashionable (Wilson, 1975), it is rapidly becoming clear that detailed attention to "reproductive strategies" of individuals within primate societies is required. From the evidence now available, it would seem that feeding ecology, reproductive adaptations and allometric scaling of these functions to body size are all major topics requiring investigation for the successful interpretation of primate social organization.

Another recent trend identifiable in primate field studies generally allied to increasing quantification of behaviour and ecology, is the integration of such studies with laboratory-based research on primates. Many of the questions raised by field studies, such as those relating to nutrition and metabolism, to reproduction, to genetics and to parasitism, call for detailed laboratory investigations. This trend can only increase in the years to come as more fundamental questions are asked about primate physiology in relation to primate behaviour under natural conditions.

As one final point, before turning to actual examples, something should be said about *intraspecific variability* in primate behaviour. This aspect has been neglected by some and over-emphasized by others in the literature. It must at once be said that variation in individual features of primate social organization (e.g. social group size and composition; home range size; population density) can be considerable, particularly where there is substantial variation in habitat conditions. It is therefore necessary to consider the range of variation within any primate species as well as establishing the average condition for that species. However, over-emphasis of the variability of primate societies can, just as in any other area of primate biology, obscure the significance of the average condition for each species, and it is overly pessimistic to rule out any firm conclusions about the relationship between primate behaviour and ecology merely because variability exists. In the sections that follow, the data considered represent *modal values* for each species (Clutton-Brock, 1974) and it is felt that the conclusions drawn are valid despite the considerable variability which has been described for particular species. Nevertheless, the wide range of variability in social organization of certain primate species does obviously limit the precision of predictions which can be made from overall analyses, and this must be remembered when such predictions are "tested". In addition, something can be learned about primate social organization from the distribution of such intraspecific variability as exists. It is, for example, striking that primate species which live in true monogamous family groups (single breeding pairs with their offspring) typically exhibit considerably less variability in numerous parameters of social organization than do species with other forms of social organization. This point has been independently recognized by Clutton-Brock & Harvey (1977a: 8):

> For example, monogamous breeding seems to be an extremely stable characteristic and in those primate genera where it is found, it apparently occurs in all congeneric species. Moreover, facultative polygyny is not known to occur in any of these species. It seems likely that the switch from monogamy to polygyny is not easily achieved and that phylogenetic inertia is probably important.

(It should be noted, however, that this statement was not meant to convey the impression that any evolutionary progression would have been from monogamy to polygyny — see later.)

Thus we must keep the question of intraspecific variability in mind whilst not allowing it to obscure the search for unifying principles broadly applying to the social organization of primates generally.

THE IMPORTANCE OF FIELD STUDIES OF PROSIMIAN PRIMATES

It is evident from the modern consensus of opinion regarding evolutionary relationships among the living primates that man's closest zoological relatives are the great apes, followed (in order of decreasing affinity) by the lesser apes, the Old World monkeys, the New World monkeys and the prosimians. At first sight, therefore, it might seem reasonable to conclude that we can learn most about human evolutionary origins by studying the apes and the Old World monkeys, and least by studying the prosimians, and that this perspective should determine priorities for research. But there are several compelling reasons why this narrow view should be rejected. In fact, it now seems likely that a sound theoretical basis for interpreting the evolution of primate social behaviour will depend not only upon comprehensive study of prosimians and simians but also upon a synthetic view of *mammalian* social organization, with due reference to avian systems and perhaps to the social organization of other vertebrate (and even invertebrate) groups. In addition to this general requirement for a comprehensive theory of animal social organization, there are also a number of specific reasons why prosimian field studies have a vital role to play.

There is good reason to believe that the common ancestors of the living primates were nocturnal in habit (Martin, 1973a, 1979) and that diurnal life has accordingly emerged as a secondary phenomenon during primate evolution. Nocturnal habits have been retained by the majority of living prosimian species, whereas only one simian primate species (the owl monkey, *Aotus trivirgatus*) is nocturnal. Thus, if one wishes to consider the differential effects of diurnal and nocturnal habits on the evolution of primate social organization, some information from prosimian field studies is required. It is also true that one can only really begin to interpret early primate developments in the light of information derived from modern prosimians — their most primitive modern descendants.

The prosimians also offer a specific opportunity for testing certain hypotheses about the relationship between behaviour and ecology, in that a minority of modern Madagascar lemur species exhibit diurnal habits. It seems likely (Martin, 1972) that the lemurs of Madagascar, in the absence of competition from simian primates, have radiated to fill ecological niches parallel to those characteristically occupied by monkeys or apes in Africa, Asia and South America. For example, the indri (*Indri indri*) exhibits many parallels to the Asiatic gibbons and siamangs (Hylobatidae) with its monogamous habit and territoriality associated with loud calls, while the ringtail lemur (*Lemur catta*) is semi-terrestrial and exhibits parallels in group size and habits to macaques. Since this evolutionary development was certainly quite independent of the adaptive radiation of monkeys and apes, it is particularly valuable to establish to what extent diurnal lemurs comply with empirical rules established for the relationships between behaviour and ecology applying to simian primates (see later — Fig. 4).

It is also noteworthy that among the primates, nocturnal habits are associated with relatively small body size, while diurnal habits are typically associated with larger body size (modal body size for nocturnal primates = 500 g; modal body size for diurnal primates = 5 kg — see also Martin, 1979). This is a rule which applies to mammals generally (Charles-Dominique, 1975). For present purposes, one significant implication of this fact is that inclusion of prosimians in quantitative analyses of the relationship between body size and specific behavioural characteristics will extend the range of body weights covered and thus enhance any conclusions which are drawn. Furthermore, by virtue of their typically smaller body sizes, prosimians offer a number of advantages in field studies compared to simians. With the smallest, nocturnal prosimian species it is relatively easy to combine the standard techniques of studying diurnal primates with established techniques utilised for small mammals generally (e.g. trapping, marking, release and recapture; analysis of stomach contents). This is particularly valuable since it is becoming increasingly apparent that questions about primate social organization involve factors operating at the level of the *population* as well as those affecting the individual *social group*. For example, reproductive "strategies" require investigation at both levels. Given the fact that the number of primates in a given area accessible to a single field research worker broadly increases as the body size of the species decreases (see later), it is clear that it will be easier to answer certain questions about primate social organization by studying prosimians, rather than simians. It is no accident that certain new techniques, such as radio-tracking, have been applied more readily and more successfully in the study of small-bodied nocturnal prosimians (see Charles-Dominique, 1977a; Bearder & Martin, 1979), rather than in field studies of large-bodied diurnal simian species. There can, therefore, be little doubt that field studies of prosimian primate species are considerably more relevant to a synthetic theory of primate social behaviour than a simple consideration of zoological proximity to man would imply.

CLASSIFICATIONS OF PRIMATE SOCIAL ORGANIZATION

As explained above, the classification of patterns of primate social organization is an essential preliminary to conducting an analysis of relationships to other factors (e.g. to feeding ecology). It should be made clear at the outset that such classifications must necessarily be arbitrary in some respects, just as with animal taxonomy, and there will always be borderline cases which create problems. But it is unproductive to criticize classifications of primate social organization merely because of these drawbacks; even crude generalizations can lead us along the road to a deeper understanding of the factors involved. Attempts at classification of primate social organization can be traced back at least as far as Carpenter (1954), but the really seminal publication is that of Crook & Gartlan (1966). This publication was particularly important in that it included reference to some of the first real field studies of prosimian behaviour and ecology and because a clear attempt was made to relate the different categories of social organization to different ecological features across the order Primates. Naturally, the publication of this paper was followed by a host of publications pointing out apparent exceptions to the "rules" which had been identified. Yet enough material contained in the Crook & Gartlan paper is of lasting value to make this a convenient starting-point for any further classificatory exercise. Five different categories of primate social organization were recognized, as shown in Table I, and Crook & Gartlan noted the following as specific features probably associated with the distinctions between the categories: (1) nocturnal vs. diurnal habits; (2) forest environment vs. more open country environment; (3) broad dietary types (omnivorous vs. frugivorous vs. folivorous). These major distinctions remain valid in a general sense, despite the fact that clear exceptions have been identified in subsequent field studies. For instance, it has been found that many arboreal forest-living colobine and cercopithecine species exhibit "one-male-groups" (supposedly an adaptation to arid conditions) and that quite large social groups containing several adult males can be found in forest habitats (e.g. *Miopithecus talapoin* — Gautier-Hion, 1973) as well as in more open-country

TABLE I

Classifications of primate social organization

1. *"Adaptive grades" of Crook & Gartlan (1966):*

   I Nocturnal; pair-living; "solitary"; probably territorial; mainly insectivorous

   II Crepuscular or diurnal; small family parties based on a single male; very small groups; territorial with display marking; frugivorous or folivorous

   III Diurnal; multi-male groups; small to occasionally large parties; may be territorial or simply exhibit mutual avoidance; frugivorous or frugivorous—folivorous

   IV Diurnal; multi-male groups; medium to large parties; may be territorial or simply exhibit mutual avoidance; vegetarian-omnivorous

   V Diurnal; one-male groups; medium to large parties; large feeding congregations; vegetarian-omnivorous

2. *Categories of Eisenberg, Muckenhirn & Rudran (1972):*

   I "Solitary". Insectivore-frugivores and folivores

   II Parental family. Frugivore—insectivores and folivore—frugivores

   III One-male troop. Arboreal folivores; arboreal frugivores and semi-terrestrial frugivores

   IV Age-graded-male troop. Arboreal folivores, arboreal frugivores, semi-terrestrial frugivore-omnivores and terrestrial folivore—frugivores

   V Multi-male troop. Arboreal frugivores, semi-terrestrial frugivore—omnivores

habitats. Two more fundamental drawbacks of this early classificatory scheme can be noted. Firstly, the term "solitary" applied to certain small-bodied, nocturnal prosimian species can be misleading if it is taken to be the opposite of "social". It has been repeatedly shown over the last decade (Martin, 1973b; Charles-Dominique, 1977b; Bearder & Martin, 1979) that nocturnal prosimians can exhibit quite complex systems of social organization (Fig. 1). As Charles-Dominique (1977b, 1978a) has pointed out, the opposite of "solitary" is "gregarious" in this context. Both solitary and gregarious primates exhibit well-defined systems of social organization, but only the latter actually move around as immediately recognizable social groups. The importance of this observation is accentuated by the fact that the orang-utan (*Pongo pygmaeus*), one of man's closest zoological relatives, exhibits a pattern of social organization virtually identical to the "solitary" pattern identified for several nocturnal prosimian species. (Incidentally, the fact that the orang-utan exhibits such a social system, falling within Crook & Gartlan's (1966) first category, is itself at variance with the ecological details of their

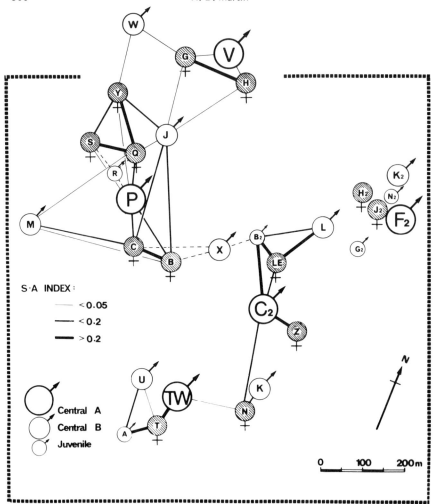

FIG. 1. Pattern of social relationships among lesser bushbabies (*Galago senegalensis moholi*) established on the basis of associations of individuals at diurnal sleeping sites (Bearder & Martin, 1979). This pattern of social contacts through sharing of sleeping sites is related to overlap between home ranges and to encounters between the bushbabies during nocturnal activity. It can be seen that single males with conspicuous physical development and terri-torial activity ("Central A" males) have associations with a number of females (2-5) and tolerate the presence of subordinate males ("Central B" males and juveniles). N.B. The area enclosed by the box represents the main study area of 1 km².

classification.) This, yet again, illustrates the value of studies of nocturnal prosimians in enlarging our perspective of primate social systems. The second major drawback with the Crook & Gartlan (1966) classification resides in their use of the term "grade" to describe their five categories. This term has a long-established usage in evolutionary biology in the definition of successive categories of increasing complexity which are taken to approximate to an evolutionary sequence (Martin, 1973a). Indeed Crook & Gartlan (1966) indicated fairly clearly that they regarded their categories of primate social organization as representing an evolutionary sequence of this kind:

> . . . we attempt to allocate species recently investigated to a series of "Grades" representing "levels" of adaptation in forest, tree savannah, grassland and arid environments respectively. Anatomical investigations of fossil and living material reveal a progressive adaptive radiation from forest-dwelling insectivorous primates to larger open country animals predominantly vegetarian. It is not surprising, therefore, to find correlated trends in the behavioural data.

It is unfortunate that their basic classification, which still has much to commend it, should have been linked with evolutionary implications which have not been supported by subsequent work. Six years later, another key paper appeared (Eisenberg et al., 1972) which introduced an important modification into the Crook & Gartlan classificatory scheme, while at the same time radically altering the evolutionary interpretation (Table I). Crook & Gartlan had designated the one-male social unit (= "harem group") as the "most advanced" of primate social systems, attributing its appearance to adaptation to arid, open country conditions. As Eisenberg et al. (1972) correctly point out, one of the most unusual forms of social organization among the primates is in fact the so-called "multi-male group" (i.e. a group containing more than one sexually and socially mature adult male, in addition to adult females and immature animals of both sexes). This form of social organization is unusual among mammals and it is actually relatively rare even among the primates. For this and other reasons, Eisenberg et al. (1972) recognized the multi-male system as the "most advanced" of primate social patterns and they recognized a new category intermediate between the classical one-male system and the full multi-male system: the so-called age-graded-male system. Here, there are a number of sexually mature, adult males present, but there is a hierarchy among these males based on age and only the alpha male is presumed to be actively involved in breeding. This is a useful additional category and the very fact that it has been recognized has helped to clarify the likely course of evolution of primate social organization. Eisenberg et al. (1972) also emphasized that a "solitary" feeding pattern does not necessarily imply a lack of social organization. If we forget, for a moment, the specific question of whether primates move together in obvious

foraging parties (gregarious pattern) and simply consider the relationships of individuals to one another, it can at once be seen that
nocturnal prosimians exhibit social networks comparable to those
characterizing the one-male units and age-graded-male units of
diurnal primates. In terms of dominance relationships and likely
priorities of male breeding access to females, to the social system of
Galago alleni (Charles-Dominique, 1977a) is comparable to the
harem system described for numerous diurnal cercopithecine monkey species, such as *Cercopithecus mitis*, while the social system of
Galago senegalensis (Bearder & Martin, 1979) is comparable to the
age-graded-male system recognized for the vervet monkey (*Cercopithecus aethiops*). Interestingly, both sets of examples chosen here
show that harem systems and age-graded-male systems can occur in
different species belonging to the same genus. This, in itself, indicates
that it is a relatively simple matter for species to shift from one system to another in evolution. This suggestion is confirmed by the fact
that individual species (e.g. the howler monkey, *Alouatta villosa*, and
the Hanuman langur, *Presbytis entellus*) may exhibit either of these
two systems, depending on local habitat conditions. Such lability
contrasts markedly with the consistency typically exhibited by
species living in true monogamous family groups.

Taking the five categories of primate social organization defined
by Eisenberg *et al.* (1972), one can re-examine hypotheses regarding
the likely sequence of evolution. Crook & Gartlan (1966) regarded
the "solitary" pattern of nocturnal prosimians as primitive and
implied that the evolution of primate societies took place first
through the formation of family groups, with this leading on to
multi-male groups and then to one-male groups as a specialization in
certain circumstances. Eisenberg *et al.* (1972) were less explicit about
the likely evolutionary sequence, and they avoided using the term
"grade" for their categories of primate social organization. However,
they did explicitly refer to the multi-male groups as the most advanced conditions and they also implied that the family group might,
at least in some cases, be an intermediate stage in the evolution of
one-male groups. But all the work published to date on the behaviour
of nocturnal lemurs and lorises indicates that their social systems,
although involving solitary foraging, are closest — in terms of male/
female relationships at least — to the polygynous systems of gregarious diurnal primates (Fig. 2). This is also true of studies of various
other nocturnal placental mammals (e.g. the palm civet, *Nandinia
binotata* — Charles-Dominique, 1978b) and even of nocturnal marsupials (e.g. the brush-tailed opossum, *Trichosurus vulpecula* —
Winter, 1975) and the one-male system is prevalent among diurnal

ungulate species (Jarman, 1974). There is thus a considerable amount of comparative information indicating a solitary polygynous system, based on overlapping male and female home ranges, in early (probably nocturnal) mammals, which was retained by the early (probably nocturnal) primates and translated into a gregarious polygynous system among many diurnal primates. This is in agreement with Zuckerman's (1932) early identification of the harem system as a basic pattern among monkeys and apes. It is accordingly likely that Eisenberg *et al.* (1972) are correct in identifying true multi-male social systems in primates as rather unusual products of further evolution (via the age-graded-male system) from a more typical polygynous system.

In most previous discussions, however, the distinctiveness of the true monogamous family group system has been largely overlooked. Crook & Gartlan (1966) and Eisenberg *et al.* (1972) have concurred in implying that the family group could be a precursor in the formation of one-male or multi-male groups. Yet, since nocturnal lemurs and lorises generally exhibit a dispersed form of the polygynous gregarious group (Fig. 2), such a precursor is unnecessary. In addition, it should be recognized that the true family group, as a breeding system, differs significantly from *all* other patterns of primate social organization. There are relatively few primate species with true family groups, many of which have now been carefully studied in the field (*Tarsius* spp., Niemitz, 1979; J. MacKinnon, personal communication; *Indri indri*, Pollock, 1975b; *Lemur mongoz*, Sussman & Tattersall, 1976; *Saguinus oedipus*, Dawson, 1977; *Callicebus* spp., Mason, 1968; Kinzey *et al.*, 1977; *Aotus trivirgatus*, Wright, 1978; *Hylobates lar*, Ellefson, 1968; *Symphalangus syndactylus*, Chivers, 1974). In all cases, available evidence indicates that once an adult male and an adult female have formed a pair, they remain together for life. What is unusual about this system, compared to other primate social systems, is that the male's breeding access is restricted to a single female (thus confining his breeding output at any point in time) and there is therefore limited potential genetic variability among his offspring (whatever the implications of this may be). Female primates generally are restricted in their breeding access to males by virtue of the fact that they are viviparous and have long gestation periods, but pair-living similarly restricts the potential genetic variability among their offspring in that they cannot mate with a sequence of different males. It therefore seems likely that rather special conditions must exist for the emergence of true monogamous family groups in primate social life. Empirically, this is borne out by the fact that family groups occur only among fully

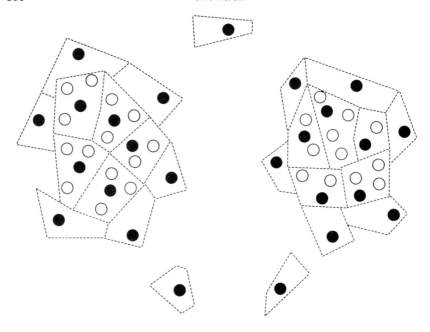

FIG. 2. Diagram illustrating the typical pattern of social relationships found in nocturnal prosimians (from Martin, 1972, reprinted by permission of the Royal Society of London). The figure illustrates two localized "population nuclei", each containing a core of central males and associated females, surrounded by a fringe of peripheral males. Dotted lines indicate patterns of association through home range overlap. Peripheral males can migrate between population nuclei and replace central males under certain conditions. (Black circles = males; open circles = females.) Although variations on the base pattern can occur (e.g. in degree of tolerance of peripheral males and in the average number of females associated with each central male), the system is typically polygynous.

arboreal, tropical forest-living primates; they have never been re-corded for primate species inhabiting relatively open country, where terrestrial adaptation is common. This suggests that relative environmental stability may be an important factor favouring monogamous systems in primates.

In summary, one can envisage (at least as a viable alternative) the evolution of primate social systems from a dispersed "solitary" polygynous pattern present among the nocturnal ancestors of modern primates (Fig. 2). This basic pattern (with minor variations) would have been retained to the present day by most of those lemur and loris species which have retained nocturnal habits (Charles-Dominique, 1978a). With the shift to diurnal life, among the lemurs of Madagascar and among the simians of the Old and New Worlds, this system would have coalesced to produce gregarious foraging parties with a polygynous structure (one-male groups or age-graded-male groups).

Over and above this, in relatively rare cases, there would have been further evolution to produce *either* multi-male systems *or* true monogamous family groups. The latter, once they have appeared, remain remarkably consistent and stable.

SCALING, METABOLIC RATE AND FEEDING ECOLOGY

The classifications set out by Crook & Gartlan (1966) and Eisenberg *et al.* (1972) make no explicit reference to body size as a factor influencing primate social organization. Yet it is obvious that since body size affects (among other things) food requirements and locomotor capacities, any social system must be adapted to take account of the body size factor as well as complying with ecological requirements and reproductive patterns. In order to appreciate the potential influence of body size on social organization, it is necessary to go back to first principles. The concept of *allometry* (non-linear change in specific parameters with changing body size) is now well established in the literature on primate morphological evolution and an excellent review of the subject in general has been provided by Gould (1966). For present purposes, it should be pointed out that we are concerned mainly with *interspecific allometry*, involving examination of average conditions for different species to determine the manner in which given parameters are related to body size. However, the general approach to allometry is essentially the same whether we are dealing with intraspecific or interspecific comparisons. The approach is usually empirical in kind, with the observed values for a given parameter plotted against different body sizes (in the case of interspecific allometry, the average body sizes of individual species), in order to determine the basic pattern of scaling of that parameter to body size. It has generally been found that the following expression can be used to express the relationship between a given parameter (P) and body size (W):

$$P = k \times W^{\alpha}.$$

This empirically determined formula can be converted to logarithmic form as follows:

$$\log P = \alpha \log W + \log k.$$

Such logarithmic conversion has the advantage of producing a linear expression. A best-fit line drawn through paired logarithmic values of P and W will allow immediate determination of the power function α (the slope of the line) and of the constant k ($\log k =$ the intercept on the ordinate). The allometric formula thus determined can then be used as a relatively simple indicator of the overall relationship between the selected parameter and body size, and the power function itself may provide clues to the underlying mechanisms. Where α has a value greater than unity, we can talk of *positive allometry*; where α has a value less than unity (the more usual situation), we can refer to *negative allometry*. In the (relatively rare) cases where α *is* unity, we have a condition of *isometry* and a straight-line relationship can be determined without resorting to the use of logarithms.

It is necessary here to comment on the calculation of a best-fit line for an allometric relationship. The simplest and most widely used technique is to calculate a *linear regression*, as with previous analyses of the relationships between social organization parameters and body size in primates (Milton & May, 1976; Clutton-Brock & Harvey, 1977a, b). However, it has been suggested (Kermack & Haldane, 1950) that use of the linear regression is not really appropriate for biological data in general, since it implies that one variable in the bivariate plot (viz. body size in allometric plots) is independent and can be measured with negligible error, while the other parameter is dependent and subject to measurement errors (see also Sokal & Rohlf, 1969). With biological systems, such a clear-cut distinction between independent and dependent variables is rarely justifiable, and it is probably better to use a line-fitting procedure which does not imply that this distinction exists and which assumes that errors are likely to be present in both variables. One may use either the *reduced major axis* (Kermack & Haldane, 1950) or, preferably, the *major* (= principal) *axis* (Sokal & Rohlf, 1969) to satisfy this requirement. Of course, there are questionable assumptions inherent in the use of *any* statistical line-fitting technique (e.g. relating to normality of distribution), so it is impossible to lay down hard-and-fast rules. However, from a purely pragmatic viewpoint it is desirable to use a line-fitting procedure which indicates the overall trend in a bivariate plot without unduly emphasizing one of the variables, and that function is best served by the major axis (Sokal & Rohlf, 1969; M. Hills, personal communication).

From some points of view, it may not matter which technique is used to determine a best-fit line. If there is little scatter of points around the fitted line, all three techniques (linear regression; reduced major axis; major axis) will be in good agreement. However, as the scatter of points around the line increases, so the discrepancy between the allometric formulae yielded by the three techniques increases. Since logarithmic plots of primate social organization parameters against body size characteristically exhibit marked scatter (see later), the problem is particularly acute in this context. For any given set of data, the value of the correlation coefficient (r) remains the same regardless of the line-fitting procedure, but if any importance is attached to the actual value determined for the allometric power function (i.e. the slope on a double logarithmic plot) the allometric formula derived with a linear regression may be misleading. The discrepancy between the value of α determined with a linear regression and that determined with the reduced major axis or major axis increases as the scatter of the data around the fitted line increases. In fact, this discrepancy is also systematic in that the regression always yields a *lower* slope than the reduced major axis or the major axis (e.g. see Table III, p. 316). One practical implication of this is that any prediction of the value of a particular parameter (e.g. home range size) from an empirical allometric formula based on a regression will involve overestimation where body size is small and underestimation where body size is large. In view of these considerations, all three statistical procedures are employed below in order to provide a comprehensive picture and to highlight discrepancies which arise.

Whatever statistical technique one uses, the main aim of the exercise in studying allometric relationships is usually that of recognizing and hence eliminating the effect of body size in comparisons as a step preliminary to the investigation of other factors. One way of doing this is to use the best-fit expression to calculate for each species a theoretical value for the parameter concerned at a standard body size. This is equivalent to each point up or down the slope prescribed by the expression sliding to the standard body size selected. In studies of primate brain size, this technique has been successfully used in the calculation of "encephalization indices" (Stephan, 1972) or "indices of cephalization" (Jerison, 1973) to permit meaningful comparison between individual species of differing body size. Such an approach could also be used with profit in the comparison of primate species of differing body size with respect to quantitative features of social organization. But it must be noted that determination of an index in which the effect of body size has been "removed" depends upon the slope of the best-fit line. It is therefore essential to ensure that the slope prescribed should be as meaningful as possible in both biological and statistical terms.

The study of allometric relationships can also permit a quantitative approach to the concept of the "grade". For example, Bauchot & Stephan (1969) have shown that the following four mammalian categories exhibit similar allometric relationships between brain weight and body weight, with a slope close to 0.67 but with increasing values of the constant k: "basal" insectivores; "advanced" insectivores; prosimians; simians (see also Martin, 1973a). One can treat these as four "grades" of relative brain size, taking the statistical best-fit condition of each group as representative. It seems likely that a similar treatment on primate social organization could be equally profitable, though a more complex set of relationships is involved.

Allometric studies are generally empirical in nature and it is often felt that they go no farther than providing a quantitative description of relationships between specific parameters and body size. This is not necessarily the case, though. The value of the slope determined may in itself be significant and, given a knowledge of allometric relationships previously determined, it may be possible to predict the slope or at least to comment on departures from the slope which might have been expected. At this point, it is necessary to introduce one of the most fundamental allometric relationships that has been established for mammals generally: that existing between basal metabolic rate (resting energy utilization in kcal per day) and body size. As it happens, one of the most influential papers on this subject (by Kleiber, 1932) was published in the same year as *The social life of monkeys and apes*, followed by further contributions from Brody, Proctor & Ashworth (1934) and from Benedict (1938). The subject has recently been effectively reviewed by Hemmingsen (1950), by Kleiber (1961) and by Schmidt-Nielsen (1972). From all these studies, it has emerged unequivocally that the relationship between basal metabolic rate (MR) and body size (W) in mammals (and, indeed, in other vertebrate and even invertebrate groups) is best expressed by the following formula:

$$MR = k \times W^{\frac{3}{4}}$$

As can be seen, this is a negative allometric relationship: large-bodied mammals use less energy per unit of body weight in a standard time than do small-bodied mammals. The relationship is so well established that it can be accepted as a biological principle which should be taken into account in any situation where the energy requirements of species of different body sizes are being considered. The consistency of this relationship is reflected by the closeness of fit of the data points to the empirically determined allometric expression, reflected in the high value of the correlation coefficient obtained with Kleiber's (1961) data (see Table II). (Incidentally, these data provide a good illustration of the fact that the slope determined with a linear regression differs only slightly from that determined with a reduced major axis or major axis when there is little scatter on the bivariate plot. Hence, all of these statistical techniques are practically equivalent for calculations scaling metabolic rate to body size.)

TABLE II

Relationship between basal metabolic rate and body size (data for 26 selected mammalian taxa from Kleiber, 1961)

Linear regression	$\log MR = 0.756 \log BW - 0.442$
Reduced major axis	$\log MR = 0.757 \log BW - 0.446$
Major axis	$\log MR = 0.757 \log BW - 0.446$

MR = basal metabolic rate; BW = body weight.
Correlation coefficient $(r) = 0.998$.
Coefficient of determination $(r^2) = 0.997$.
95% Confidence limits of slope of principal axis = $0.740 - 0.774$.

One can proceed from the standard equation given above and refer to the *metabolic body size* of a given mammal (e.g. a primate), calculated as the three-quarters power of body size. This is a potentially more meaningful parameter to use than actual body size when any analyses relating to energy utilization are concerned, though it must be remembered that one should ideally correct for actual metabolic requirements (see later).

There is a corollary of the negative allometry of metabolic requirements which is immediately applicable to the question of primate feeding behaviour (Clutton-Brock & Harvey, 1977b; Geist, 1974; Jarman, 1974; Martin, 1979). The higher metabolic requirement per unit body weight of small mammals is generally associated with a need for foods which will liberate energy relatively rapidly. Larger-bodied mammals, on the other hand, can afford to feed on foodstuffs which are relatively low in energy content and/or slow to release their energy content during digestion. Thus, small-bodied primates typically feed on a mixture of small animal prey (mainly arthropods) and plant foods rich in carbohydrates (e.g. gums; fruit), while folivorous habits are generally found with primates of moderate to large body size. The larger primates are, in any case, unable to obtain a very large proportion of their diet in the form of small animal prey, so they are probably constrained to feed on more readily accessible plant foods (fruits and/or leaves). It should be emphasized, though, that larger-bodied primates need not necessarily eat low-energy foods; the point is that they can afford to do so. For example, to take the two largest living primate species, the gorilla (*Gorilla gorilla*) does indeed feed primarily on leafy material, yet the orang-utan is predominantly frugivorous. For the larger-bodied primate species, then, it would appear that there could be two alternative "strategies" — consumption of high-energy foods (fruits), permitting higher energy expenditure in behaviour, or consumption of low-energy foods (leaves), with consequent restriction on the energy available for behaviour. This divergence is confirmed by a plot of the percentage of foliage in the diets of individual primate species against body weight (Clutton-Brock & Harvey, 1977b).

When we turn to the application of allometric considerations to field studies, we find that this kind of approach has been introduced only relatively recently. The earliest papers referred to birds (see Schoener, 1968, 1969) and non-primate mammals (see McNab, 1963). All the papers on allometric aspects of primate behaviour have been published in the last decade (Jorde & Spuhler, 1974; Milton & May, 1976; Clutton-Brock & Harvey, 1977a, b) and to date the question of metabolic scaling has barely been investigated. Before considering specific questions of energy requirements, however, it is worthwhile to review some of the main conclusions which have been reached to date in allometric studies of primate field data:

(1) Feeding group size in primates tends to increase with increasing body size (Clutton-Brock & Harvey, 1977a). However, it should be noted that the three largest-bodied primate species (*Pan troglodytes, Gorilla gorilla, Pongo pygmaeus*) exhibit only small to moderate feeding group sizes. Therefore, there may be a trend towards reduction in feeding group size at the highest body weights among living primates (see Fig. 3).

(2) Home range size shows a tendency to increase in a reasonably regular fashion with body size. This is true whether one divides home range by the number of animals in the feeding group to calculate the theoretical area of home range available per individual (Milton & May, 1976) or whether one plots home range size against the total weight of the population group (Clutton-Brock & Harvey, 1977a, b). A similar regular increase in home range size with body size has been reported for a sample of bird species (Schoener, 1968) and for a sample of non-primate mammals (McNab, 1963).

Clutton-Brock & Harvey (1977a, b) and Milton & May (1976) have also shown that there is an overall tendency for predominantly frugivorous primate species to have larger home ranges at a given population group weight or body size than predominantly folivorous species. (Nevertheless, it must be emphasized that Milton & May indicated ranges of intraspecific variation which considerably reduce the predictive value of this overall separation.)

(3) A similar positive correlation has been suggested between day range length (the average path length covered by individuals of a feeding group during the active period) and feeding group weight (Clutton-Brock & Harvey, 1977a, b). Clutton-Brock & Harvey also indicated that at any given feeding group weight, terrestrial species have longer day ranges than arboreal species and that predominantly

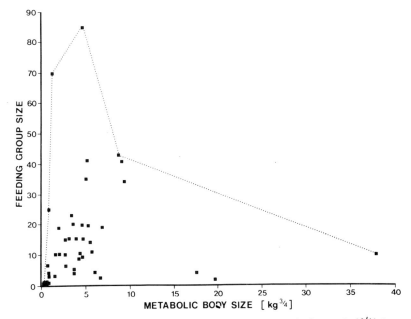

FIG. 3. Plot of feeding group size against metabolic body size (= (body size)$^{3/4}$) for a sample of 44 different primate species. (Data selected from Clutton-Brock & Harvey, 1977b, excluding some special cases such as the gelada baboon and supplemented by additional data on *Saguinus oedipus* and nocturnal prosimians.) The dotted line simply provides a guide to the peak values over the range of body size, emphasizing the fact that the largest feeding group sizes have been reported for species of intermediate body size.

frugivorous primates tend to have longer ranges than predominantly folivorous species.

(4) Population density declines with body weight, but biomass increases with body weight (Clutton-Brock & Harvey, 1977a, b).

(5) With specific reference to feeding behaviour, Clutton-Brock & Harvey (1977b) found that the total time per activity period spent feeding generally increases with body size, but that (regardless of body size) time spent feeding declines with the percentage of foliage in the diet. They also showed that the percentage of the diet represented by the two most commonly eaten plant food species declines with increasing body size, indicating that, as a general rule, large-bodied primates are less selective feeders than small-bodied primates.

All these empirical generalizations are of value (even though all of them are perhaps logically only to be expected) when it comes to a comparative analysis of primate behaviour under natural conditions. Although the comparative approach is not the only one which can be used, it will doubtless prove to be one of the most instructive because the number of factors affecting primate behaviour is so great that studies of single species cannot easily lead to firm conclusions. Clutton-Brock, Harvey & Rudder (1977) and Clutton-Brock & Harvey (1977a) have, for example, demonstrated the value of, and the necessity for, allometric analysis in their discussion of the phenomenon of *sexual dimorphism* in primates. Past discussions of this topic have been confounded by the fact that, other things being equal, sexual dimorphism tends to become exaggerated as the body size of species increases. If this phenomenon is not taken into account, by first examining and allowing for the relationship of sexual dimorphism to body size, it is impossible to discuss the problem clearly. This provides a good illustration of one of the major advantages of the allometric approach: it permits one to identify the scaling effects of body size in a comparison of a number of species of different sizes and to distinguish those effects from others which are of more immediate interest.

There is another way of excluding, or at least minimizing, the effects of body size on any parameter under consideration, which is to conduct comparisons of pairs of primate species of similar body sizes. If such comparisons can be conducted using data collected in areas where the two species are sympatric (i.e. occur together in the same general habitat), one can also reduce variations due to major habitat differences. This approach has been used to advantage in comparing: the ringtail lemur, *Lemur catta*, with the brown lemur, *Lemur fulvus* (Sussman, 1974, 1975); the red colobus, *Colobus badius*, with the black-and-white colobus, *Colobus guereza* (Clutton-Brock, 1974; Struhsaker & Oates, 1975); the purple-faced langur, *Presbytis senex*, with the grey langur, *Presbytis entellus* (Hladik, 1977) and the lar gibbon, *Hylobates lar*, with the siamang, *Symphalangus syndactylus* (Chivers, 1972). All these studies have produced additional confirmation that predominant folivores tend to have smaller home ranges than predominant frugivores, though it is not necessarily the case that the more frugivorous species live in larger social groups. The latter is in fact true for the *Lemur, Colobus* and *Presbytis* species, but not for the hylobatid species. (This could be taken as yet another indication that true monogamous family groups represent a very stable pattern of social organization once developed, as is the case with all living representatives of the family Hylobatidae.) When sufficient paired species comparisons of this kind have been conducted, one can formulate "rules" which can complement or confirm those recognized from broad allometric analyses. There can be no doubt that the use of such rules, generated by comparative examination of quantitative field data, will be considerably more reliable for the interpretation of primate behaviour under natural conditions than the *ad hoc* explanations which may be used in the absence of comparative analysis.

Interpretation of patterns of primate social organization can, however, be given a still firmer basis through explicit reference to metabolic requirements and their relationship to ranging behaviour. Unfortunately, one really requires figures on the active basal metabolic rates of primates under natural conditions in order to establish an allometric relationship between actual metabolic requirements and body size, and the necessary data are simply not available. Nevertheless, some conclusions can be drawn from the metabolic rule established by Kleiber (1932) for basal metabolic rates. Firstly, it can be taken as highly probable that active metabolic requirements involve a *negative* allometric relationship with body size (i.e. $\alpha < 1$). Secondly, it is unlikely that the value of α for the allometric relationship between total metabolic requirements over the day and body size will depart greatly from the value of 0.75. It seems likely that ranging behaviour of any primate species is more likely to be related to metabolic requirements than to body size *per se*. For this reason, it would be useful (for example) to consider data on home range size and on other parameters of primate ranging activity in relation to "metabolic body size" $(= (\text{body size})^{3/4})$ as a first approximation. In any event, because of the likelihood that there is a negative allometric relationship between total daily metabolic requirements and body size, there is a certain danger in using measures which compound uncorrected body size with numbers of animals, as is the case with the "feeding group weight" or the "population group weight" defined and used by Clutton-Brock & Harvey (1977a, b). For example, in their plot of home range against population group weight (Clutton-Brock & Harvey, 1977b), an average group of talapoin (*Miopithecus talapoin*) is indicated as having approximately the same total weight as an average group of howler monkeys (*Alouatta villosa*), i.e. 87.5 kg vs. 92.4 kg. Yet the energetic requirements of the talapoin (as estimated from basal metabolic rate) theoretically exceed those of the howler group by a factor of 43% because of the body size difference. The "metabolic biomass" (= population group size × individual metabolic body size) for an average group of talapoin is 82.6, whereas that for an average group of howler monkeys is only 57.7. Other things being equal, one would accordingly expect a population group of talapoin to require a larger home range than a population group of howlers.

Resorting to first principles, one can clarify the situation by considering an ideal case in which home range area is determined purely by available food resources and in which a number of species of differing average body sizes feed on exactly the same resources, with all other aspects maintained equal. Theoretically, home range size under such conditions should increase linearly with metabolic requirements. A corollary of this is that the biomass for each species should increase with increasing body size, since larger-bodied primates require less metabolic energy per unit of body weight. In practice, these ideal conditions will not be met for a variety of reasons, since large-bodied primates can in general eat a greater proportion of lower-energy foodstuffs such as leaves (Martin, 1979), which are likely to be more abundant, and because there are numerous other factors (e.g. interspecific competition, predation, differential effects of body size on locomotor energy requirements and other aspects of energy utilization) which come into play. However, it is instructive to examine the relationship between home range requirements and body size to see whether metabolic scaling is likely to be the major factor. Since home ranges may overlap to quite different degrees, measures of home range area for individuals or groups may not give an accurate picture of the relationship between available food resources and individual food requirements. An alternative approach is to calculate the "hypothetical individual space" in m^2 by taking the reciprocal of population density and to plot this against body weight (Fig. 4).

The results agree with and expand upon previous analyses, in that there is an overall trend for "hypothetical individual space" to increase with body weight and in that three "grades" can now be recognized. At any given body size, a nocturnal prosimian (feeding on a mixed diet of insects and fruit and/or gums) will typically require a larger individual space than a diurnal primate feeding mainly on fruit, and the smallest individual space is required by a folivore. However, it seems unlikely that the positive correlation between "hypothetical individual space" and body weight can be explained purely on the basis of metabolic scaling, since for frugivores and folivores the 95% confidence limits for the slope on the principal axis exceed the expected value of 0.75. Only for the nocturnal insectivorous prosimians are the confidence limits of the slope just compatible with simple basal metabolic scaling, and this may be attributable to the very small sample size ($N = 6$; see Table III). In other words, there is no firm evidence that individual space requirements in primates are directly scaled to basal metabolic requirements; large-bodied primates require larger areas of "hypothetical individual space" than metabolic requirements alone would indicate. In fact, the same conclusion is reached by plotting the home range area per individual (as defined by Milton & May, 1976) against body size,

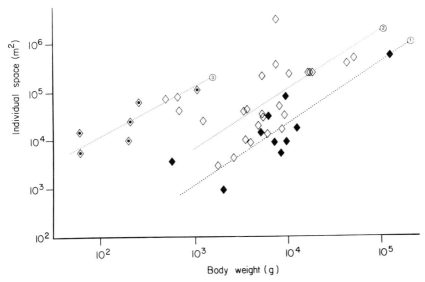

FIG. 4. Plot of "individual space" in m² (the reciprocal of population density) against body weight for 42 primate species (data from Clutton-Brock & Harvey, 1977b, supplemented by data on nocturnal prosimians and *Saguinus oedipus*).

Diamond with dot = nocturnal insectivorous prosimians ($N = 6$).

Open diamond = predominantly frugivorous diurnal primates ($N = 26$).

Black diamond = predominantly folivorous primates ($N = 10$).

The dotted lines indicate reduced major axes for folivores (1), diurnal frugivores (2) and nocturnal prosimians (3). Note that diurnal frugivorous or folivorous lemurs fall within the same ranges as for diurnal simians, well below the nocturnal insectivore line. Note also the considerably greater scatter of the data for diurnal frugivores, reflected in the correlation coefficient (Table III).

with the exception that here the 95% confidence limits for the slope of the principal axis exceed 0.75 for all three groups of primates. Thus, a special explanation must be sought for the fact that large-bodied primates apparently require larger areas per individual than basal metabolic requirements alone would dictate, especially given that lower-energy foods are more commonly utilized by large-bodied species. It seems likely that increased interspecific competition or some similar factor must oblige large-bodied primate species to exploit relatively larger areas per individual than might be theoretically expected on basal metabolic grounds. This should be taken into account in any synthetic theory of the relationship between primate ranging behaviour and food availability.

If food availability is a major determinant of broad patterns of primate social organization, then analysis of actual energetic relationships should provide a more detailed understanding. As yet, there have been relatively few studies which have specifically attempted to

TABLE III

Analysis of ranging requirements of primates with respect to body size

Primate group	Linear regression		Reduced major axis		Major axis[a]			
	Slope	Intercept	Slope	Intercept	Slope	Intercept	r	r^2
1. Individual space								
Nocturnal insectivorous prosimians ($N = 6$)	0.876	+2.373	1.033	+2.010	1.039 (0.616– 1.772)	+1.996	0.848	0.719
Diurnal frugivores ($N = 26$)	0.652	+2.280	1.221	+0.161	1.445 (0.793– 3.116)	−0.673	0.534	0.285
Folivores ($N = 10$)	1.070	+0.023	1.293	−0.842	1.362 (0.905– 2.174)	−1.110	0.827	0.684
2. Home range per individual								
Nocturnal insectivorous prosimians ($N = 6$)	1.342	+1.447	1.760	+0.497	2.050 (1.130– 5.401)	−0.163	0.763	0.582
Diurnal frugivores ($N = 28$)	1.092	+0.793	1.457	−0.553	1.639 (1.202– 2.353)	−1.245	0.749	0.561
Folivores ($N = 10$)	1.268	−0.650	1.460	−1.396	1.542 (1.100– 2.282)	−1.720	0.868	0.754

[a] Figures in brackets indicate 95% confidence limits for the slope of the major axis.
N.B. With these relatively low values for the correlation coefficient (r), there are quite marked differences between the *slopes* determined with the three different statistical techniques.

consider primate energetics (e.g. Charles-Dominique & Hladik, 1971; Coelho, 1974; Coelho, Bramblett, Quick & Bramblett, 1976; Coelho, Bramblett & Quick, 1977), and these have been confined to estimations of energy utilization rather than direct measurements in the field. With the advent of physiological telemetry, it should eventually prove possible to collect direct field data on primate energy utilization (e.g. on costs of locomotion and home range defence). For the present, it seems that the following categories of primate energy utilization are of importance in the adult: routine metabolism, locomotion, feeding, social interaction, reproduction. This sequence probably represents a decreasing order of magnitude of energy utilization; certainly, it is true that resting metabolic requirements will tend to represent the major part of the energy budget of any primate and it is likely that locomotor activity is generally the most important behavioural category with respect to energy requirements. This being the case, some attention must be given to the total path length covered by each primate species in the course of each activity period (i.e. the "day range length" of Clutton-Brock & Harvey, 1977a, b). Clutton-Brock & Harvey, in their analysis of day range length, examined their data with respect to feeding group weight, so their conclusions do not relate to body size as such. In addition, their sample included only two small-bodied nocturnal prosimian species, both of which happen to be unusually sluggish in their locomotor behaviour (viz. *Lepilemur mustelinus* and *Loris tardigradus*). Given recent additional data on small-bodied prosimian species (for *Galago senegalensis moholi* — Bearder & Martin, 1979, and for *Tarsius spectrum* — J. MacKinnon, personal communication), a surprising conclusion emerges when day range length is examined with respect to body size. Since a terrestrial habit is likely to permit increased day range length, as travel on the ground is easier than through the trees, predominantly arboreal primates ($N = 30$) have been analysed separately from semi-terrestrial and terrestrial species ($N = 12$). In both cases, it emerged that there was virtually no correlation between day range length and body size ($r = 0.032$ for arboreal primates; $r = 0.065$ for ground-living primates), and the slope of the principal axis was close to zero in both cases (0.05 and 0.12, respectively). As expected, terrestrial or semi-terrestrial primates tended to travel further overall, with an average day range length of about 2300 m, compared to approximately 950 m for arboreal primates. (*Galago senegalensis moholi* and *Tarsius spectrum* both travel approximately 1.5 km in the course of an average night's activity and Pagés (1978) reports 1.5 km as the upper limit per night for *Microcebus coquereli*.) The fact that day range length does not show a consistent increase with increasing body size in any given habitat (arboreal or terrestrial) is very significant from the point of view of energy consumption. On the one hand, it implies that increasing constraints are placed on locomotor expenditure as body size increases; on the other it means that the larger than expected "individual space" values for large-bodied primates (see above) are not dictated by increased energy requirements for locomotor expenditure. Further, since "individual space" increases with body size while day range (on average) does not, one can conclude that large-bodied primates in general are likely to cover a smaller proportion of their home range in any one activity period than do small-bodied primates.

This latter aspect can be considered in more detail by considering the relationship between day range length and home range perimeter. If the home range area is treated as a perfect circle, the minimum possible perimeter length is indicated by the circumference. The ratio of day range length to this circumference can then be taken as a crude indicator of the extent to which the home range can be covered in one activity period (= the "range traversing index"). Figure 5 shows the distribution of range traversing indices for a sample of 41 primate species for which the necessary parameters have been determined. As expected, the largest values for the index (0.8 and above) are found with certain small-bodied primates (prosimians and some small New World primates), while the three great ape species (chimpanzee, gorilla, orang-utan) all have values below 0.4. Overall, the index is negatively correlated with body size. It is also seen that

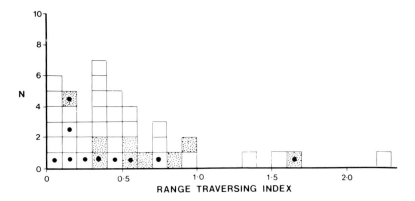

FIG. 5. Frequency histogram of values for the range traversing index (see text) in a sample of 41 primate species (data as for Fig. 4, with one omission). Folivore species (indicated by black dots) typically have low values of the index, with the notable exception of *Lepilemur mustelinus* (extreme right black dot), while species with conspicuous territorial defence (stippled) cover almost the entire range of indices.

folivores (with the exception of *Lepilemur mustelinus*) have low range traversing indices, as might be expected from the relatively low energy value of leaves in their diets. Incidentally, this approach also permits a test of the widespread belief (e.g. see Denham, 1971) that conspicuous territorial behaviour can only be shown by primate species which (in addition to deriving benefit from such behaviour) can effectively patrol their resources (i.e. engage in active defence on range boundaries). If this were true, conspicuous territorial behaviour would only be exhibited by certain primate species with high range traversing indices. However, this is apparently not the case. In fact, *Indri indri* has an index of only 0.15, while *Symphalangus syndactylus* and *Hylobates lar* have intermediate indices of 0.51 and 0.66, respectively. All these species utter characteristic calls which probably have a territorial defence function, so the implication is that territorial defence may be maintained without the need for the territory owner to actively patrol territorial boundaries. This provides a useful example of the way in which quantitative data from a range of primate species may be used to examine specific hypotheses.

REPRODUCTIVE STRATEGIES

A relatively small, but important, part of the energy budget of any primate species is represented by the energetic requirements of reproduction. The partitioning of resources between individual maintenance and production of offspring has been succinctly stated by Gadgil & Bossert (1970: 21):

> Any organism has a limited amount of resources at its disposal, and these have to be partitioned between reproductive and non-reproductive activities. A larger share of resources to reproductive activities, that is, a higher reproductive effort at any age, leads to a better reproductive performance at that age; this may be considered as a profit function. This reproductive effort also leads to a reduction in survival and growth and consequent diminution of the reproductive contribution of the succeeding stages in the life history; this may be considered as a cost function. Natural selection would tend to an adjustment of the reproductive effort at every age such that the overall fitness of the life history would be maximized.

Thus, any explanatory model of primate social organization which does not take into account "reproductive strategies" is bound to be incomplete. However, studies which attempt to incorporate reproductive parameters in the analysis of primate social organization are as yet in their infancy. All that can be done here is to outline some of the approaches which promise to yield useful theoretical insights.

One all-embracing concept which must eventually be integrated with a theory of primate social organization is that of r- and K-selection formulated by MacArthur & Wilson (1967) and neatly reviewed by Pianka (1970). The basic notion can be expressed in terms of a simple curve of population growth and stabilization (Fig. 6). When a population is growing and there is no density-dependent limitation on population size (e.g. no limitation on food resources), it can expand geometrically. The rate of expansion will then be equivalent to the "intrinsic rate of increase" (r), i.e. the per capita rate of net increase in a given environment. On the other hand, once the "carrying capacity" (K) of the environment has been reached, density-dependent limiting factors such as food resources will operate and (apart from random fluctuations) there will be no further consistent expansion of population size. During the growth phase of a species population, selection will favour high reproductive turnover and — since food resources are not limiting — selection pressure for efficiency of resource-utilization will be weak. In other words, a high

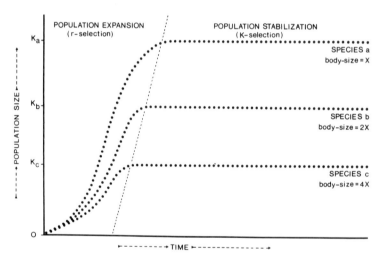

FIG. 6. Simplified illustration of the principle of r- and K-selection. When a population is in the growth stage, geometric increase in population size is permitted. Here, food resources will not be limiting and selection will favour maximum reproductive potential (r-selection) even if resources are utilized somewhat wastefully. When a population is at carrying capacity, selection will favour efficiency of resource utilization (K-selection) and a high level of intraspecific competition will operate. Those species which most typically exist under conditions of population growth (e.g. where catastrophic mortality is common or where colonisation is the rule) will be subject to r-selection. Species which typically occur in stable habitat conditions (e.g. rainforest) will be subject to K-selection.

The three separate graphs illustrate a hypothetical set of species of differing body-sizes (X; $2X$; $4X$) to show that larger-bodied species are likely to attain carrying capacity earlier in a given habitat (assuming that diet and reproductive potential are maintained constant).

intrinsic rate of increase will be favoured and MacArthur & Wilson (1967) use the term "r-selection" to describe evolution under these conditions. By contrast, when a species population has reached the carrying capacity of a given environment, the limiting resource which inhibits further increase in population size (e.g. food availability) will represent a major factor in evolution. Intraspecific competition for resources will be intense and selection will favour efficiency of resource utilization combined with a relatively low reproductive turnover. Here, there will be selection for efficient production of offspring and for adaptations which enhance the survival of those offspring in a highly competitive environment. MacArthur & Wilson (1967) refer to this effect as "K-selection". It had already been pointed out by Dobzhansky (1950) that natural selection in the tropics differed from natural selection in temperate zones, with greater intraspecific competition characterizing the former. Mac-Arthur & Wilson (1967) have generalized this observation in terms of the degree of environmental stability. In very stable habitats (e.g. rainforest, deep sea zones), populations of a given species are likely to exist at carrying capacity most of the time, and K-selection will usually operate. On the other hand, where habitats are unpredictable (e.g. seasonal environments with catastrophic winter mortality or with marked differences between years) or where species are adapted as colonizers of temporary habitats, r-selection will be typical. Pianka (1970) has summarized the differential features of environment, population dynamics and reproductive variables which distinguish r-selection from K-selection, and these have been summarized (with some omissions and modification to apply specifically to mammals) in Table IV. Two features not specifically mentioned by MacArthur & Wilson (1967) or by Pianka (1970) which have been included in the table are body size and brain size. It is now clear that K-selection tends to be associated with large body size and with a relatively large brain size. The association with large body size is to be expected partly because (other things being equal) a large-bodied species will reach the carrying capacity earlier (Fig. 6) and partly because a larger body can provide better buffering against environmental fluctuations. The association with large relative brain size is not quite so obvious, but it is likely that conditions of intense intraspecific competition (K-selection) will favour increased brain size to enhance efficiency of resource location and utilization and to further offspring survival. A recent paper by Eisenberg & Wilson (1978) has shown that in bats the fruit-eating species (restricted to tropical areas) have relatively larger brains than insectivorous/carnivorous species (which occur in temperate zones as well as the tropics). Fruit-eating requires efficient

TABLE IV

Correlates of r- *and* K-*selection (modified from Pianka, 1970)*

	r-selection	*K*-selection
Climate	Variable and/or unpredictable	Fairly constant and/or predictable
Mortality	Often catastrophic, non-directed, density-independent	More directed, density-dependent
Population size	Variable in time, non-equilibrium, usually well below carrying capactiy, unsaturated communities, recolonization frequent	Fairly constant in time, equilibrium at or near carrying capacity, saturated communities, no recolonization required
Intraspecific & interspecific competition	Variable, often lax	Usually keen
Selection favours	1. Rapid development 2. High r_{max} 3. Early breeding 4. Short reproductive intervals 5. Multiple litters	1. Slower development 2. Low r_{max} 3. Late breeding 4. Long reproductive intervals 5. Single litters
Correlates	1. Small body size 2. Short life-span 3. Short reproductive life-span 4. Relatively small brain	1. Large body size 2. Long life-span 3. Long reproductive life-span 4. Relatively large brain
Leads to:	Productivity	Efficiency in use of resources

location of scattered, but locally abundant, food resources, while this is not generally true of the insectivorous/carnivorous species.

As Pianka (1970) and Gadgil & Bossert (1970) have emphasized, *r*-selection and *K*-selection represent opposite ends of a spectrum and the two kinds of selection can only be viewed in relative terms. For example, all vertebrates are typically *K*-selected with respect to all invertebrates (doubtless a reflection of overall body size differences — see Pianka, 1970), while primates as a group are typically *K*-selected with respect to many other mammal groups (Martin, 1975a). Primates exhibit an array of reproductive characters characterized by relatively slow reproductive turnover and elaborate parental care

(Martin, 1975a, b; Rudder, 1979) and it has been argued (Martin, 1975a) that this adaptive complex can be traced to a long period of K-selection in a tropical, arboreal habitat. However, there are primate species which are relatively more K-selected than others, even though the order Primates as a whole exhibits one of the lowest ratios of reproductive effort to overall energy utilization (Rudder, 1979). For example, one can compare the characteristics of the Gabon rainforest bushbaby species *Galago alleni* (Charles-Dominique, 1977b) with those of *Galago senegalensis moholi* living in South African wooded savannah subject to marked seasonal changes (Bearder & Martin, 1979). These two prosimian species are relatively similar in body size, so a direct comparison between them (Table V) is particularly appropriate. It can be seen from the comparison that *G. alleni* best fits the K-selection model, compared to *G. senegalensis moholi*, in terms of greater environmental stability, lower reproductive potential, more intense competition and larger relative brain size (as reflected by the "encephalization quotient" of Jerison, 1973). This example also suggests how r- versus K-selection may influence social organization. Male and female *G. senegalensis moholi* have smaller home ranges than male and female *G. alleni* and the competition between males is less intense, as reflected in the higher adult male: adult female sex ratio, in the lower incidence of injury scars, and in the smaller number of female ranges overlapped by the range of one dominant (Central A) male in *G. senegalensis moholi* (Charles-Dominique, 1977b; Bearder & Martin, 1979). This provides a strong indication of a link between reproductive strategies and social organization, and in this particular case it is possible to make a testable prediction. Northern populations of lesser bushbabies, such as the Sudan form currently recognized as the subspecies *Galago senegalensis senegalensis*, are reported to have a reproductive potential intermediate between that of *G. senegalensis moholi* and that of *G. alleni* (cf. Table V). Butler (1967) reports that female lesser bushbabies in the Sudan can breed twice a year, as is the case with *G. s. moholi*, but typically have a single infant at each birth. Accordingly, one would expect the environmental conditions in the Sudan to be less unpredictable than in South African wooded savannah (though less predictable than in the rainforest of Gabon) and the pattern of social organization should exhibit an intermediate degree of intraspecific competition. If this prediction is borne out, it will confirm the value of considering reproductive strategies in terms of r- and K-selection within an overall theory of primate social organization.

Ultimately, broad patterns of primate social organization should be to some extent explicable in terms of energy budgets considered

TABLE V

r- and K-selection in bushbabies (data from Charles-Dominique, 1977b, and from Bearder & Martin, 1979)

	Galago senegalensis moholi	Galago alleni
Average body weight	200 g	275 g
Climate	Strictly seasonal; rainfall largely restricted to wet season; extreme temperature range. Great fluctuations from year to year	Mild seasonal changes; year-round rainfall; temperature variable within narrow limits. Consistency from year to year
Mortality	Apparently quite heavy; heavy predation by genets observed and high infant mortality in very cold winters	Moderate overall, but intense competition between adult males accompanied by high mortality in this age–sex category
Population size	No long-term data available	No long-term data available
Intraspecific & interspecific competition	Adult males tolerate range overlap with subordinate or sub-adult males. No competing prosimian species; no obvious competitor for main plant food source (gums)	Adult males do not tolerate range overlap with other grown males. Four potentially competing prosimian species; probably compete with numerous species for main plant food source (small fruits)
Reproductive parameters	1. Each female can produce two litters of two infants per year 2. Gestation period = 123 days 3. Breeding strictly seasonal 4. Full motor development of infant at 25 days	1. Each female can produce a single infant a year 2. Gestation period = 140 days 3. Breeding year round 4. Full motor development of infant at 30 days
Correlates	1. Male home range = 10 ha; female home range = 6 ha; 2. Dominant male range overlaps those of 2–5 females 3. Encephalization quotient = 0.86	1. Male home range = 34 ha; female home range = 10 ha; 2. Dominant male range overlaps those of 6–8 females 3. Encephalization quotient = 1.10

in relation to environmental stability. Given selection for a particular reproductive strategy (relative r-selection or relative K-selection) in a habitat with certain characteristics, the typical pattern of social organization should be explicable in terms of patterns of energy availability from food resources and energy utilization for routine metabolic requirements, locomotion, social interaction and reproduction. Under relative r-selection, social behaviour which maximizes reproductive effort without firm constraints on efficiency of limiting resource utilization will be favoured; under relative K-selection, social behaviour will be attuned to a low but consistent reproductive turnover with a premium on adaptations which maximize survival of individual offspring and thus reduce the reproductive effort (in terms of resource consumed) to produce each surviving descendant. Orians (1969) has discussed, with particular reference to birds, how mating strategies may be influenced by resource availability and utilization and has produced cogent reasons for the predominance of polygyny, and the relative rarity of monogamy among mammals generally. It is through such considerations, allied with detailed attention to energetics, that we should eventually be able to come closer to a reliable interpretation of the sporadic occurrence of true monogamous family groups in the primates. One prediction deriving from Orians' discussion is that paternal behaviour should be prominent in monogamous primate species. While it has yet to be shown that definite investment of some kind is present in all monogamous primate species, it is certainly true that the most obvious examples of such behaviour are found among such species (e.g. in Callitrichidae, *Aotus, Symphalangus syndactylus* – see also Kleiman, 1977). Since true monogamous family groups have been found predominantly in rainforest conditions, it would seem that such a social system might be favoured by K-selection combined with a particular energetic equation in which the optimum strategy for the male is to confine his ranging area to that of one female and to co-operate in infant-rearing or infant defence in order to maximize infant survival. The very fact that pair-living males can only produce large numbers of offspring by successful breeding over many years is itself indicative of a link with long-term environmental stability and hence with K-selection. Obviously, a great deal must still be done to provide a proper interpretative basis for primate social organization along these lines, but at least the framework of a synthesis can now be identified.

INTEGRATION OF LABORATORY STUDIES AND FIELD STUDIES

At various points it has become clear that information generated in laboratory studies is vital for the interpretation of broad patterns of primate social organization under natural conditions. Given the observation that food resources are likely to be a key factor in the determination of natural primate behaviour, and the resulting need to translate feeding data into energetic terms, laboratory studies of primate nutrition and metabolic requirements are of prime importance. Integrated field and laboratory studies of primate diets in relation to social organization were pioneered by Hladik & Hladik (Hladik et al., 1971; Hladik & Hladik, 1972) and their results indicate that in future primate field studies it will doubtless become increasingly necessary both to collect quantitative data on feeding behaviour (see Clutton-Brock, 1977b) and to determine the nutrient contents of different foodstuffs. It will also be necessary to conduct direct measurements on primate metabolic requirements, both in the field and in the laboratory. There will in addition be specific issues requiring attention, such as the now very widespread reports of primates including small quantities of soil or similar substances in their diets (geophagy). The exact significance of this latter behaviour has not yet been elucidated, but there are some indications that essential minerals may be obtained in this way (e.g. see Hladik & Gueguen, 1974). Quite sophisticated analyses of primate diets, and of patterns of food availability in the wild, will be required to interpret accurately the relationship between social organization and diet.

The same applies to studies of primate reproduction, another key feature relating to social organization. Certain reproductive parameters (e.g. precise parameters of the oestrous cycle, pregnancy and infant development) can only be determined in the laboratory and must be determined in order to interpret accurately observations made under natural conditions. In future, it will be necessary to determine accurately how patterns of social organization relate to the breeding success of individual males and females and to correlate these in turn with overall reproductive strategies. As yet, there have been few attempted analyses of primate breeding in relation to social organization in the wild (e.g. Rowell, 1967; Dunbar & Dunbar, 1977) and such studies must be considerably expanded. One specific example of the value of integrating laboratory and field studies of primate reproduction is provided by the diagnosis of pregnancy. It is often valuable to determine in the field whether a given primate female is pregnant or not and to predict birth-dates, particularly since births may easily be missed if not expected. Prediction of birth-dates permits advance planning of observations so that births can be followed in detail. Laboratory work on the detection and measurement of chorionic gonadotrophin and oestrogens in the urine of female great apes (chimpanzee, gorilla, orang-utan) has shown that pregnancy can be reliably diagnosed with fresh or deep-frozen samples collected from three weeks post-conception to the time of birth (Martin, Seaton & Lusty, 1976; Martin, Kingsley & Stavy, 1977; Kingsley & Martin, 1979; Seaton & Lusty, 1976; Clift & Martin, 1978). In addition, given a series of a few samples (e.g. six) spread over a part of pregnancy (e.g. two months), it is possible to predict birth-dates with an accuracy of ±3 weeks. In fact, where accurate data on matings are available it is possible to make even more accurate predictions of birth dates, whilst avoiding mistakes due to matings during pregnancy. Since only small urine samples are required (a few drops) and since it is possible to send non-frozen samples by post for fairly accurate diagnosis, it is perfectly feasible to apply this to wild populations. Having perfected the technique with samples from great apes maintained in zoological collections, it was recently possible to diagnose pregnancy (and the approximate stage) with single samples from three chimpanzees in the natural study population at Gombe National Park, Tanzania (Fig. 7). All three chimpanzees have since given birth, confirming the assessment that they were

FIG. 7. A wild-living female chimpanzee ("Winkle"; right of picture) correctly diagnosed as late pregnant on the basis of a single urine sample sent by post. (Photograph taken at Gombe National Park, Tanzania and published with the kind permission of Dr Jane van Lawick-Goodall.)

in late pregnancy. If such studies are extended in the future, they could considerably enhance our understanding of reproductive performance in wild primates and perhaps reduce the length of field studies required to provide accurate data on reproduction.

Also related to the question of reproductive strategies is the need to establish probable family relationships among members of wild primate populations. Since extremely long-term and consistent field studies are needed to establish genealogies by direct observation, some kind of short-cut would be welcome. Modern work on blood groups and on blood protein polymorphisms in primates (e.g. Jolly & Brett, 1973) can theoretically permit the reconstruction of genetic relationships among members of a population, using small blood samples. It is therefore to be expected that primate field studies will in future require collection of blood samples in order to decide questions of relationship (e.g. paternity) relating to reproductive strategies. Here, field and laboratory expertise must be developed in parallel.

Finally, a word should be said about the question of primate parasites. Various suggestions have been made that parasitism may influence patterns of social organization and, although such suggestions have not yet made much headway, field and laboratory studies of parasitic organisms will be required to test them and complete the picture.

DISCUSSION

Throughout this review of primate field studies, the question of genetic control of behaviour has been essentially avoided. This has been done not because the distinction between innate and learned behaviour is unimportant, but because field studies themselves cannot provide direct answers. In any case, many of the links that have been suggested between primate social behaviour and other parameters (e.g. feeding ecology; body size; reproductive strategies) remain valid regardless of whether the behaviour is innate or learned. From the point of view of general models of the relationships between primate social behaviour and such variables, it does not matter whether behaviour is genetically fixed or not. However, if any meaningful discussion of the likely evolutionary history of primate social behaviour is to be conducted, the question of genetic control of behaviour cannot be avoided completely. Observations to date permit two clear statements. On the one hand, intraspecific variability is considerable and it is obvious that any genetic control over

social patterns must be very generalized, permitting flexible adjustment to local habitat conditions. On the other, there are clearly
limits to intraspecific variation and it is possible to think in terms of
broadly "species-specific" patterns of behaviour. For example, all
recent observers agree that gorillas in various study areas essentially
live in "harem" groups, while those who have studied orang-utans
agree on the essentially solitary behaviour of this primate species. It
has also been seen that primates which live in true monogamous
family groups exhibit far less variability than primates with other
types of social organization. Thus, it is fairly obvious that there must
be *some* genetic influence over social behaviour, though considerable
variation is also possible. Consideration of hypothetical ancestor-
dependent relationships is therefore quite valid, provided that it is
kept at a general level. There are many valid reasons for believing
that the ancestral primates were nocturnal and lived in a dispersed
"harem" system and that with the advent of diurnal life in the simian
primates (monkeys and apes) this pattern led to the formation of
gregarious foraging parties with the same basic constitution. Only
under special conditions did special adaptation lead to the emergence
of multi-male groups or true monogamous family groups.

Using a combination of approaches, such as allometric analysis,
paired species comparisons and energetic considerations, certain
limited conclusions have been drawn about the relationships between
primate social organization and feeding ecology. Additional inferences
have been made from the standpoint of reproductive strategies. A
truly synthetic theory will, however, require a multifactorial approach which genuinely integrates all such considerations and also
refers to other aspects not dealt with here (particularly details of
inter-individual interactions). Anti-predator defence requirements
undoubtedly play some part in shaping patterns of primate social
organization, but this topic has been considered in this review
because quantitative data are sadly lacking and most of the discussions in the literature are purely hypothetical. The numbers of
different species competing for particular resources in a given habitat,
and the exact form of competition, are also of importance. Although
some excellent field studies have been conducted specifically to
investigate interspecific competition in relation to primate behaviour
(e.g. Gautier-Hion, 1978; Charles-Dominique, 1977b), a systematic
survey of primate ecological niches and the effects of competitors
will be required before this aspect can be treated in a quantitative
fashion. It is also important to bear in mind special phenomena such
as *polyspecific associations* (temporary or semi-permanent association
of primate groups from different species — Gautier & Gautier-Hion,

1969; Gartlan & Struhsaker, 1972), which can provide unique insights into the factors promoting gregarious behaviour in primates. With polyspecific associations, certain functions of the typical mono-specific primate group (e.g. vigilance with respect to predators; co-operation in foraging) are taken over by a large polyspecific group, while standard intraspecific functions (notably reproduction) are left unaffected. Analysis of such special cases should strengthen overall interpretations of patterns of primate social organization. Given increased integration between field studies and laboratory studies in the future, and given an increasing emphasis on long-term monitoring of primate populations together with problem-oriented, quantitative field studies, the framework of interpretation sketched out in this review should be greatly consolidated in due course.

A synthetic theory of primate social organization under natural conditions is intrinsically valuable in its own right. However, there are also two significant practical extensions. Firstly, natural primate populations are becoming increasingly threatened by destruction of habitat and there are several species or subspecies whose wild popu-lations can now be numbered in the hundreds (e.g. the aye-aye, *Dau-bentonia madagascariensis*; the golden lion tamarin, *Leontopithecus rosalia*; the mountain gorilla, *Gorilla gorilla beringei*). Effective con-servation measures in the future, involving protection and doubtless management of reduced populations in reserve areas, will require detailed knowledge of the natural behaviour and ecology of the pri-mate species concerned. In a quite different direction, any attempt to reconstruct the evolution of human social behaviour and perhaps to gain insights into the modern human condition will be dependent upon a fully synthetic theory of primate social behaviour, ecology and reproduction. Although such an attempt will remain very specu-lative, it is better to base it on a set of empirical rules relating to all non-human primate species, rather than relying on questionable single species comparisons. Only through such a synthetic theory can one begin to approach the common basis for interpretation of pri-mate and human behaviour envisaged by Zuckerman in 1932.

ACKNOWLEDGEMENTS

Thanks are due to the Royal Society and to the Boise Fund for sup-porting the author's field studies of nocturnal prosimians (latterly in association with Dr S. K. Bearder) and for permission to reprint the diagram in Fig. 2. The concepts developed in this paper owe their development to discussions with numerous primate field-workers,

particularly with Dr P. Charles-Dominique, Dr S. K. Bearder and Dr
J. MacKinnon. Thanks also go to Dr Jane van Lawick-Goodall for
permission to use the photograph in Fig. 7, and to Dr T. Clutton-
Brock and Dr P. H. Harvey for permission to discuss in detail the
data so painstakingly compiled in their joint papers. Dr J. MacKinnon kindly permitted reference to unpublished data from his recent
study of *Tarsius spectrum* in Indonesia. Dr M. Hills has provided
valuable advice on statistical aspects of allometry. Finally, particular
gratitude is expressed to the following people for providing detailed
comments on the draft manuscript: Dr T. Clutton-Brock; Dr P. H.
Harvey; Dr. R. Wrangham.

REFERENCES

Altmann, J. (1974). Observational study of behaviour: sampling methods.
    Behaviour **49**: 227–267.
Altmann, S. A. (1967). Preface. In *Social communication among primates*:
    ix–xii. Altmann, S. A. (Ed.). Chicago: Chicago University Press.
Bauchot, R. & Stephan, H. (1969). Encéphalisation et niveau évolutif chez les
    simiens. *Mammalia* **33**: 225–275.
Bearder, S. K. & Martin, R. D. (1979). The social organization of a nocturnal
    primate revealed by radio-tracking. In *A handbook on biotelemetry and
    radio-tracking*: 623–648. Amlaner, C. J. & MacDonald, D. W. (Eds). Ox-
    ford: Pergamon Press.
Benedict, F. G. (1938). Vital energetics: a study in comparative basal meta-
    bolism. *Publs Carneg. Inst.* No. 503: 1–215.
Bramblett, C. A. (1976). *Patterns of primate behavior*. California: Mayfield
    Pub. Co.
Brody, S., Procter, R. C. & Ashworth, U. S. (1934). Basal metabolism, endogen-
    ous nitrogen, creatinine and neutral sulphur excretions as functions of
    body weight. *Mo Res. Bull.* No. 220: 1–40.
Butler, H. (1967). Seasonal breeding of the Senegal galago (*Galago senegalensis
    senegalensis*) in the Nuba mountains, Republic of the Sudan. *Folia prima-
    tol.* **5**: 165–175.
Carpenter, C. R. (1934). A field study of the behavior and social relations of
    howling monkeys (*Alouatta palliata*). *Comp. Psychol. Monogr.* **10**: 1–168.
Carpenter, C. R. (1940). A field study in Siam of the behavior and social relations
    of the gibbon (*Hylobates lar*). *Comp. Psychol. Monogr.* **16**: 1–212.
Carpenter, C. R. (1954). Tentative generalizations of grouping behavior of non-
    human primates. *Hum. Biol.* **26**: 269–276.
Carpenter, C. R. (1962). Field studies of a primate population. In *Roots of
    behavior: Genetics, instinct and socialization in animal behavior*: 286–
    294. Bliss, E. L. (Ed.). New York: Harper.
Charles-Dominique, P. (1975). Nocturnality and diurnality: An ecological inter-
    pretation of these two modes of life by an analysis of the higher vertebrate
    fauna in tropical forest ecosystems. In *Phylogeny of the primates*: 69–88.
    Luckett, W. P. & Szalay, F. S. (Eds). New York: Plenum Press.

Charles-Dominique, P. (1977a). Urine marking and territoriality in *Galago alleni* (Waterhouse 1837 — Lorisoidea, Primates): a field study by radio-telemetry. *Z. Tierpsychol.* 43: 113—138.

Charles-Dominique, P. (Martin, R. D., translator) (1977b). *Ecology and behaviour of nocturnal prosimians.* London: Duckworth.

Charles-Dominique, P. (1978a). Solitary and gregarious prosimians: evolution of social structures in primates. In *Recent advances in primatology.* 3. *Evolution*: 139—149. Chivers, D. J. & Joysey, K. A. (Eds). London: Academic Press.

Charles-Dominique, P. (1978b). Ecologie et vie sociale de *Nandinia binotata* (Carnivores, Viverridés): comparaison avec les prosimiens sympatriques du Gabon. *Terre Vie* 32: 477—528.

Charles-Dominique, P. & Hladik, C. M. (1971). Le *Lepilémur* du Sud de Madagascar: écologie, alimentation et vie sociale. *Terre Vie* 25: 3—66.

Chivers, D. J. (1972). The siamang and the gibbon in the Malay Peninsula. In *Gibbon and Siamang.* 1: 103—135. Rumbaugh, D. M. (Ed.). Basle: Karger.

Chivers, D. J. (1974). The siamang in Malaysia. *Contr. Primatol.* 4: 1—335.

Clift, J. P. & Martin, R. D. (1978). Monitoring of pregnancy and postnatal behaviour in a female lowland gorilla, *Gorilla g. gorilla*, at London Zoo. *Int. Zoo Yb.* 18: 165—173.

Clutton-Brock, T. H. (1974). Primate social organization and ecology. *Nature, Lond.* 250: 539—542.

Clutton-Brock, T. H. (Ed.) (1977a). *Primate ecology: studies of feeding and ranging behaviour in lemurs, monkeys and apes.* London: Academic Press.

Clutton-Brock, T. H. (1977b). Methodology and measurement. In *Primate ecology*: 585—590. Clutton-Brock, T. H. (Ed.). London: Academic Press.

Clutton-Brock, T. H. & Harvey, P. H. (1977a). Primate ecology and social organization. *J. Zool., Lond.* 183: 1—39.

Clutton-Brock, T. H. & Harvey, P. H. (1977b). Species differences in feeding and ranging behaviour in primates. In *Primate ecology*: 557—584. Clutton-Brock, T. H. (Ed.). London: Academic Press.

Clutton-Brock, T. H., Harvey, P. H. & Rudder, B. (1977). Sexual dimorphism, socionomic sex ratio and body weight in primates. *Nature, Lond.* 269: 797—800.

Coelho, A. M. (1974). Socio-bioenergetics and sexual dimorphism in primates. *Primates* 15: 263—269.

Coelho, A. M., Bramblett, C. A. & Quick, L. B. (1977). Social organization and resource availability in primates: A socio-bioenergetic analysis of diet and disease hypotheses. *Am. J. phys. Anthrop.* 46: 253—264.

Coelho, A. M., Bramblett, C. A., Quick, L. B. & Bramblett, S. S. (1976). Resource availability and population density in primates: a socio-bioenergetic analysis of the energy budgets of Guatemalan howler and spider monkeys. *Primates* 17: 63—80.

Crook, J. H. (1965). The adaptive significance of avian social organisations. *Symp. zool. Soc. Lond.* No. 14: 181—218.

Crook, J. H., Ellis, J. E. & Goss-Custard, J. D. (1976). Mammalian social systems: structure and function. *Anim. Behav.* 24: 261—274.

Crook, J. H. & Gartlan, J. S. (1966). On the evolution of primate societies. *Nature, Lond.* 210: 1200—1203.

Dawson, G. A. (1977). Composition and stability of social groups of the tamarin *Saguinus oedipus geoffroyi* in Panama: Ecological and behavioral implications. In *The biology and conservation of the Callitrichidae*: 23—37. Kleiman, D. G. (Ed.). Washington: Smithsonian Institution Press.

Denham, W. W. (1971). Energy relations and some basic properties of primate social organization. *Am. Anthrop.* 73: 77—95.

Dobzhansky, T. (1950). Evolution in the tropics. *Am. Sci.* 38: 209—221.

Dunbar, R. I. M. & Dunbar, E. P. (1977). Dominance and reproductive success among female gelada baboons. *Nature, Lond.* 266: 351—352.

Eisenberg, J. F., Muckenhirn, N. A. & Rudran, R. (1972). The relation between ecology and social structure in primates. *Science, Wash.* 176: 863—874.

Eisenberg, J. F. & Wilson, D. E. (1978). Relative brain size and feeding strategies in the Chiroptera. *Evolution* 32: 740—751.

Ellefson, J. O. (1968). Territorial behavior in the common white-handed gibbon *Hylobates lar* Linn. In *Primates: studies in adaptation and variability:* 180—199. Jay, P. C. (Ed.). New York: Holt.

Freeland, W. J. (1976). Pathogens and evolution of primate sociality. *Biotropica* 8: 11—24.

Freeland, W. J. (1977). Blood-sucking flies and primate polyspecific associations. *Nature, Lond.* 269: 801—802.

Freeland, W. J. & Janzen, D. H. (1974). Strategies in herbivory by mammals: the role of plant secondary compounds. *Am. Nat.* 108: 269—289.

Gadgil, M. & Bossert, W. H. (1970). Life historical consequences of natural selection. *Am. Nat.* 104: 1—24.

Gartlan, J. S. & Struhsaker, T. T. (1972). Polyspecific associations and niche separation of rain-forest anthropoids in Cameroon, West Africa. *J. Zool., Lond.* 168: 221—266.

Gautier, J. -P. & Gautier-Hion, A. (1969). Les associations polyspécifiques chez les Cercopithécidae du Gabon. *Terre Vie* 23: 164—201.

Gautier-Hion, A. (1973). Social and ecological features of talapoin monkey — comparisons with sympatric cercopithecines. In *Comparative ecology and behaviour of primates*: 147—170. Michael, R. P. & Crook, J. H. (Eds). London: Academic Press.

Gautier-Hion, A. (1978). Food niches and coexistence in sympatric primates in Gabon. In *Recent advances in primatology*. 1 *Behaviour*: 269—286. Chivers, D. J. & Herbert, J. (Eds). London: Academic Press.

Geist, V. (1974). On the relationship of social evolution and ecology in ungulates. *Am. Zool.* 14: 205—220.

Glander, K. E. (1975). *Habitat and resource utilization: An ecological view of social organization in mantled howler monkeys.* Ph.D. Thesis: University of Chicago.

Goss-Custard, J. D., Dunbar, R. I. M. & Aldrich-Blake, F. P. G. (1972). Survival, mating and rearing strategies in the evolution of primate social structure. *Folia primatol.* 17: 1—19.

Gould, S. J. (1966). Allometry and size in ontogeny and phylogeny. *Biol. Rev.* 41: 587—640.

Hemmingsen, A. M. (1950). The relation of standard (basal) energy metabolism to total fresh weight of living organisms. *Rep. Steno Mem. Hosp. Nord. Insulinlab.* 4: 7—58.

Hinde, R. A. (1974). *Biological bases of human social behavior.* New York: McGraw-Hill.

Hinde, R. A. (1977). Changing approaches to the social behavior of higher vertebrates. *Spec. Publs Acad. nat. Sci. Philad.* 12: 339—362.

Hladik, C. M. (1977). A comparative study of the feeding strategies of two sympatric species of leaf monkeys: *Presbytis senex* and *Presbytis entellus*. In *Primate ecology*: 324–353. Clutton-Brock, T. H. (Ed.). London: Academic Press.

Hladik, C. M. & Gueguen, L. (1974). Géophagie et nutrition minérale chez les primates sauvages. *C. r. hebd. Séanc. Acad. Sci., Paris (D)* 279: 1393–1396.

Hladik, C. M. & Hladik, C. M. (1972). Disponibilités alimentaires et domaines vitaux des primates à Ceylan. *Terre Vie* 26: 149–215.

Hladik, C. M., Hladik, A., Bousset, J., Valdebouze, P., Viroben, G. & Delort-Laval, J. (1971). Le régime alimentaire des primates de l'île de Barro-Colorado (Panama). Résultats des analyses quantitatives. *Folia primatol.* 16: 85–122.

Horr, D. A. (1975). The Borneo orang-utan: Population structure and dynamics in relationship to ecology and reproductive strategy. In *Primate behavior* 4: 307–323. Rosenblum, L. A. (Ed.). New York: Academic Press.

Jarman, P. J. (1974). The social organization of antelope in relation to their ecology. *Behaviour* 48: 215–267.

Jerison, H. J. (1973). *Evolution of the brain and intelligence*. New York: Academic Press.

Jolly, A. (1972). *The evolution of primate behavior*. New York: Macmillan & Co.

Jolly, C. J. & Brett, F. L. (1973). Genetic markers and baboon biology. *J. med. primatol.* 2: 85–99.

Jorde, L. B. & Spuhler, J. N. (1974). A statistical analysis of selected aspects of primate demography, ecology and social behaviour. *J. anthrop. Res.* 30: 199–224.

Kermack, K. A. & Haldane, J. B. S. (1950). Organic correlation and allometry. *Biometrika* 37: 30–41.

Kingsley, S. R. & Martin, R. D. (1979). A case of placenta praevia in an orang-utan. *Vet. Rec.* 104, 56–57.

Kinzey, W. G., Rosenberger, A. L., Heisler, P. S., Prowse, D. L. & Trilling, J. S. (1977). A preliminary field investigation of the yellow handed titi monkey, *Callicebus torquatus torquatus*, in Northern Peru. *Primates* 18: 159–181.

Kleiber, M. (1932). Body size and metabolism. *Hilgardia* 6: 315–353.

Kleiber, M. (1961). *The fire of life: An introduction to animal energetics*. New York: John Wiley.

Kleiman, D. G. (1977). Monogamy in mammals. *Q. Rev. Biol.* 52: 39–69.

Kummer, H. (1968). Social organization of hamadryas baboons, a field study. *Bibliotheca primatol.* 6: 1–189.

MacArthur, R. H. & Wilson, E. O. (1967). *The theory of island biogeography*. Princeton, New Jersey: Princeton University Press.

MacKinnon, J. (1974). The behaviour and ecology of wild orang-utans (*Pongo pygmaeus*). *Anim. Behav.* 22: 3–74.

Martin, R. D. (1972). Adaptive radiation and behaviour of the Malagasy lemurs. *Phil. Trans. R. Soc. (B.)* 264: 295–352.

Martin, R. D. (1973a). Comparative anatomy and primate systematics. *Symp. zool. Soc. Lond.* No. 33: 301–337.

Martin, R. D. (1973b). A review of the behaviour and ecology of the lesser mouse lemur (*Microcebus murinus*, J. F. Miller 1777). In *Comparative ecology and behaviour of primates*: 1–68. Michael, R. P. & Crook, J. H. (Eds). London: Academic Press.

Martin, R. D. (1975a). Strategies of reproduction. *Nat. Hist., N.Y.* 84: 48—57.

Martin, R. D. (1975b). The bearing of reproductive behavior and ontogeny on strepsirhine phylogeny. In *Phylogeny of the primates*: 265—297. Luckett, W. P. & Szalay, F. S. (Eds). New York: Plenum Press.

Martin, R. D. (1979). Phylogenetic aspects of prosimian behaviour. In *The study of prosimian behaviour*: 45—77. Doyle, G. A. & Martin, R. D. (Eds). New York: Academic Press.

Martin, R. D., Kingsley, S. R. & Stavy, M. (1977). Prospects for coordinated research into breeding of great apes in zoological collections. *Dodo* 14: 45—55.

Martin, R. D., Seaton, B. & Lusty, J. (1976). Application of urinary hormone determinations in the management of gorillas. *A. Rep. Jersey Wildl. Trust* 12: 61—70.

Mason, W. A. (1968). Use of space by *Callicebus* groups. In *Primates: Studies in adaptation and variability*: 200—216. Jay, P. C. (Ed.). New York: Holt.

McNab, B. K. (1963). Bioenergetics and the determination of home range size. *Am. Nat.* 97: 133—140.

Milton, K. & May, M. L. (1976). Body weight and home range area in primates. *Nature, Lond.* 259: 459—462.

Niemitz, C. (1979). Outline of the behavior of *Tarsius bancanus*. In *The study of prosimian behavior*: 631—696. Doyle, G. A. & Martin, R. D. (Eds). New York: Academic Press.

Nissen, H. W. (1931). A field study of the chimpanzee: Observations of chimpanzee behavior and environment in western French Guinea. *Comp. Psychol. Monogr.* 8: 1—122.

Orians, G. H. (1969). On the evolution of mating systems in birds and mammals. *Am. Nat.* 103: 589—603.

Pagés, E. (1978). Home range, behaviour and tactile communication in a nocturnal Malagasy lemur, *Microcebus coquereli*. In *Recent advances in primatology. 3. Evolution*: 171—177. Chivers, D. J. & Joysey, K. A. (Eds). London: Academic Press.

Pianka, E. R. (1970). On *r*- and *K*-selection. *Am. Nat.* 104: 592—597.

Pollock, J. I. (1975a). *The social behaviour and ecology of* Indri indri. Ph.D. Thesis: University of London.

Pollock, J. I. (1975b). Field observations on *Indri indri*: a preliminary report. In *Lemur biology*: 287—311. Tattersall, I. & Sussman, R. W. (Eds). New York: Plenum Press.

Rijksen, H. D. (1978). A field study on Sumatran orang-utans (*Pongo pygmaeus abelli*, Lesson 1827): Ecology, behaviour and conservation. *Meded. Landb. Hoogesch. Wageningen* 78—2: 1—420.

Rowell, T. E. (1967). Female reproductive cycles and the behavior of baboons and rhesus macaques. In *Social communication among primates*: 15—32. Altmann, S. A. (Ed.). Chicago: University of Chicago Press.

Rudder, B. C. C. (1979). *The allometry of primate reproductive parameters.* Ph.D. Thesis: University of London.

Schmidt-Nielsen, K. (1972). *How animals work*. London: Cambridge University Press.

Schoener, T. W. (1968). Sizes of feeding territories among birds. *Ecology* 49: 123—141.

Schoener, T. W. (1969). Models of optimal size for solitary predators. *Am. Nat.* 103: 277—313.

Seaton, B. & Lusty, J. (1976). A new approach to radioimmunoassay methodology for steroid hormones. *Proc. Soc. Endocrinol.* **68**: 36–37 *P.*

Smith, C. C. (1977). Feeding behaviour and social organization in howling monkeys. In *Primate ecology*: 97–126. Clutton-Brock, T. H. (Ed.). London: Academic Press.

Sokal, R. R. & Rohlf, F. J. (1969). *Biometry.* San Francisco: W. H. Freeman & Co.

Stephan, H. (1972). Evolution of primate brains: a comparative anatomical investigation. In *The functional and evolutionary biology of primates:* 155–174. Tuttle, R. (Ed.). Chicago: Aldine-Atherton.

Stephan, H. & Bauchot, R. (1965). Hirn-Körpergewichtsbeziehungen bei den Halbaffen (Prosimii). *Acta Zool., Stockh.* **46**: 1–23.

Struhsaker, T. T. & Oates, J. F. (1975). Comparison of the behavior and ecology of red colobus and black-and-white colobus monkeys in Uganda: a summary. In *Socioecology and psychology of primates*: 103–123. Tuttle, R. H. (Ed.). The Hague: Mouton.

Sussman, R. W. (1974). Ecological distinctions in sympatric species of *Lemur*. In *Prosimian biology*: 75–108. Martin, R. D., Doyle, G. A. & Walker, A. C. (Eds). London: Duckworth.

Sussman, R. W. (1975). A preliminary study of the behavior and ecology of *Lemur fulvus rufus* Audebert 1800. In *Lemur biology*: 237–258. Tattersall, I. & Sussman, R. W. (Eds). New York: Plenum Press.

Sussman, R. W. (Ed.) (1979). *Primate ecology: problem-oriented field studies.* New York: John Wiley & Sons.

Sussman, R. W. & Tattersall, I. (1976). Cycles of activity, group composition and diet of *Lemur mongoz mongoz* Linnaeus 1766 in Madagascar. *Folia primatol.* **26**: 270–283.

Wilson, E. O. (1975). *Sociobiology: the new synthesis.* Cambridge, Mass.: Belknap Press (Harvard University Press).

Winter, J. W. (1975). *The behaviour and social organisation of the brush tail possum* (Trichosurus vulpecula *Kerr*). University of Queensland: Ph.D. Thesis.

Wright, P. C. (1978). Home range, activity pattern, and agonistic encounters of a group of night monkeys (*Aotus trivirgatus*) in Peru. *Folia primatol.* **29**: 43–55.

Zuckerman, S. (1932). *The social life of monkeys and apes.* London: Kegan Paul, Trench & Trubner.

Symp. zool. Soc. Lond. (1981) No. 46, 337–359

Hormones and the Sexual Strategies of Primates

J. HERBERT

Department of Anatomy, University of Cambridge, UK

SYNOPSIS

The hormonal responses and strategies which occur during social interaction in primates or which are related to the structure of primate societies are the subject of this paper. Endocrine events are classified as critical (those which must occur at a given time in the animal's life history to influence behaviour), recurrent (those that can recur at intervals, and which will either reflect or determine behaviour) and contextual (hormonal changes that reflect the social structure or behavioural context in which interactions occur).

The possibility that intra-uterine exposure to testosterone is a critical event in primates (as opposed to rodents) is discussed and the behavioural consequences are considered. Critical behavioural events (e.g., gender assignment in humans) are compared with endocrine ones, and clinical and experimental concepts or models of sexual differentiation are compared. Other critical events (e.g., rejection from the natal group in monkeys) which may occur later during adolescence, and may be associated with endocrine changes, are also discussed.

Recurrent events, such as alterations in sexual interactions accompanying the female's menstrual cycle, are themselves dependent upon different categories of hormonal effects. These include those that change the female's own behaviour or that act principally on the behaviour of the male, but the two categories have different implications for the way that social structure of a primate group is affected, and for the sexual role of either sex, particularly relating to sexual selection or competition.

Similar considerations apply to contextual endocrine events, which may constrain or impair a primate's behavioural responses to its own hormones, or to those in other animals, over a long period of its life. Characteristic hormonal profiles are beginning to be defined in individual monkeys which relate to the monkey's standing in the "dominance" structure of its group, though the form that these profiles take may differ between the sexes in a given species, or between species (depending on the characteristics of the kind of social group the species habitually forms). Finally, consideration is given to hormone-dependent signals passing between primates; to their form, their number, and their relative importance in determining sexual and aggressive interaction.

INTRODUCTION

There are several ways of trying to understand the principles of the social organization of primates. A currently popular one is to try to relate the social structure of a species to its ecology, for instance to the nature and distribution of its food supply (Crook & Gartlan, 1966; Clutton-Brock & Harvey, 1977). Another is to enquire into the significance of the various behavioural strategies adopted within or between primate groups to maximize transmission of particular genes by reproductive processes (inclusive fitness; Hamilton, 1963). Yet a third is to study the group according to the behavioural interactions and the nature of the relationships between individuals (Hinde, 1974; Seyfarth, Cheney & Hinde, 1978) and so define the structure of the group. Each of these approaches addresses a different aspect of primate biology: they are not mutually exclusive or contradictory and the tendency, increasingly, is to combine them. Rather little consideration has been given to the involvement of physiological mechanisms in these behavioural events. Since many behavioural strategies are ultimately directed towards reproductive goals, it is natural to turn to neuroendocrine mechanisms as potentially important in this context. This chapter considers the roles that certain hormones might play in these phenomena, and enquires how endocrine mechanisms might be involved in the behavioural strategies operating in primate societies. A number of physiological mechanisms are available to bring about a behavioural objective; humoral factors may be involved in some, but not others. Are there special reasons why endocrine mechanisms are used in one context, but not another? Are there neuroendocrine strategies to match behavioural ones? Inevitably it will be necessary to refer also to the neural mechanisms on which hormones act, though the compass of this chapter will allow them to be discussed only superficially.

The information available on which this paper has to be based is not only uneven, but it is also unequal. We know far more about the details of the social organization of primates, and the way these differ between species (or between habitats) than we do about the particulars of endocrine function, though the latter may be equally variable between species. Furthermore, endocrine phenomena are normally not studied adequately in the wild, and whereas the study of social organization is legitimately pursued in the laboratory, it is usually directed towards different questions from those occupying workers in the field. So it will be necessary to make a number of simplifications, and to select from both the ethological and endocrine literature to illustrate the points to be made.

ENDOCRINE EVENTS: A CLASSIFICATION

If one examines the endocrine events in a primate's life, they seem to divide into three categories: *critical, recurrent* and *contextual*. Critical events are those that occur only once (or, at most, very rarely) but which can have a decisive influence on a primate's subsequent behaviour. They are usually represented by the secretion (or absence) of a hormone, or class of hormones, at particular moments in the animal's life history. The principal questions, therefore, relate to why the hormone is produced and the nature of its effects.

Recurrent endocrine events are those that occur intermittently or predictably throughout a primate's life or a portion of it, at times determined by the activity of its neuroendocrine apparatus. The effects that such events will have on a primate's behaviour will depend upon its relationships within its group with the other members.

The third category to be discussed here is the contextual endocrine event. Contextual endocrine events are those that determine a primate's basic endocrine status, and it is upon this that the other events, such as recurrent ones, are superimposed. A primate's contextual endocrine condition is therefore determined by both its genetic and developmental history, and, more particularly, by the social context in which it grows up and lives. The latter is the principal focus of the discussion to follow here. An animal's social context will influence the level of secretion of certain hormones, as well as its ability to respond to additional hormones produced, for example, during recurrent endocrine phenomena. This approach owes much to ethological analyses by which it has become evident that social interaction takes place on the basis of established relationships, which may themselves be altered by such interaction, and that the sum of such relationships represents the social structure of the group (e.g. Hinde, 1974).

CRITICAL EVENTS

An event which may be considered critical is the secretion of androgens which occurs during the foetal life of a male (but not a female) in both rhesus monkeys (Resko, 1970) and man (Abramovich & Rowe, 1973). Is this behaviourally important? A direct test of the consequences of this event would be to prevent it in the normal male, for example by prenatal castration (Jost, 1953); however, this has not been possible in primates. So all the evidence rests on giving testosterone to pregnant monkeys and studying their female offspring, assuming that the hormone reaches the foetus in a way that resembles that in the normal male. This remains an assumption.

The results of such treatment are described by Goy and co-workers.

Either testosterone or one of its major intra-cellular metabolites, dihydrotestosterone, has marked effects on juvenile behaviour. In untreated animals, mounting activity and "rough and tumble" play are typically seen more often in the male infant rhesus than the female, but occur more frequently in females exposed to testosterone prenatally (Goy & Resko, 1972). It must be emphasized that this treatment alters the frequency of these types of behaviour, rather than inducing patterns abnormal for female monkeys (Goy, 1978). It is noteworthy that infantile mounting is not decreased by castrating the infant male, in marked contrast to the inhibiting effects of castration on mounting behaviour in adults (Goy, 1979). Similarly, prepubertal boys, despite low levels of testosterone, also show penile erection and masturbate (Green, 1979). So little is known about the function of play that it is difficult to say whether the "rough and tumble" type has any lasting effect on behaviour, though it has been suggested that, in marmosets, this kind of play may be a prerequisite for copulation (Abbott, 1978), or that it may represent some kind of sorting process which determines which males will be ejected from the group subsequently (Goy, 1979). The problem with this latter interpretation is that, in baboons at least, almost all males seem to leave their group (see below).

It is natural to turn to adult sexuality to seek the results of pre-natal hormone treatment. The behaviour of adult primates is sexually dimorphic, though normal adult monkeys can, and do, display those behaviours more characteristic of the opposite sex (Kempf, 1917; Chevalier-Skolnikoff, 1976). We do not know whether androgen-treated infant female rhesus are subsequently unable to exhibit normal feminine responses to males, though such females, after being given testosterone in adulthood, will show more mounting (i.e., male-type) behaviour than ordinary females (Eaton, Goy & Phoenix, 1973). But androgen-exposed human infants are not prevented from behaving sexually as normal females (Money & Ehrhardt, 1972; Ehrhardt & Baker, 1974), a result quite unexpected in view of the "de-feminizing" effects of comparable treatment in rodents (Plapinger & McEwen, 1978). Nevertheless, treated monkeys and humans can, it seems, function sexually as females in the normal way. Similar behavioural results have followed experimental mimicry (in female marmosets) of the surge of testosterone which occurs postnatally in males (Abbott & Hearn, 1979), starting shortly after birth, in humans, rhesus monkeys and marmosets (Forest, Cathiard & Bertrand, 1973; Robinson & Bridson, 1978). It should be recognized, however, that the high plasma titres of testosterone found post-natally may be associated with increased levels of testosterone-binding globulin, which may limit the hormone's biological effect (Forest, Sizonenko, Cathiard & Bertrand, 1974).

Despite these reservations, let us for the moment assume that the secretion of prenatal testosterone is a critical event determining the nature of subsequent sexual responses in primates, and consider the significance of the timing of this endocrine event. As we shall see, there is now ample evidence that both the secretion of hormones, and the effect that this has on the brain, are postnatally modifiable by the physical and behavioural environment of the individual. The intra-uterine foetus is insulated from such factors, even though they might operate through the mother in a highly attenuated fashion. A critical endocrine event, therefore, can occur relatively undisturbed *in utero*, whereas to time it postnatally would expose it to the possibility of the modulations described in more detail below.

This may be a plausible explanation for this particular endocrine strategy, though another is equally arguable. If hormones (or any other factors) are to induce long-lasting effects on brain function, they must operate at a time when the brain is in a condition to make the necessary adjustments (structural or biochemical). Given what we know of the plasticity of the brain, this is most likely to be able to occur during the developmental period. Thus, the endocrine strategy may be constrained by coincident neurobiological events, rather than behavioural ones. But is the perinatal androgen surge a critical event? The evidence so far is unimpressive, though manifestly scanty. Even were it more complete, there are two further problems: are these experiments telling us what we need to know about the development of sexuality; and if early androgen determines "maleness", what does this mean in the subsequent life of the monkey? Studies on humans suggest that there are aspects of sexuality not being tested experimentally in primates. During normal development sexual identity (classifying oneself as male or female), gender role (the preferred behaviour pattern), and erotic preference (for like- or oppositely-sexed partners) can be distinguished and can develop independently (Money & Ehrhardt, 1972). There is no reason why comparable events should not occur in monkeys, though whether perinatal hormones alter some (or all) is not clear in either monkeys or humans, despite claims to the contrary (Dorner *et al.*, 1975). What is clear, however, is that special behavioural techniques, still to be developed, are needed to supply the answers.

Studies on humans, indeed, have suggested that a critical *behavioural* event in this context occurs after birth; that gender identity, at least, depends upon the sex of rearing and may be independent of genetic sex. This means that there is some difference (not yet clearly defined) between the parental response to a supposed male or female child (Money & Ehrhardt, 1972; Ehrhardt & Meyer-Bahlburg, 1979). If this is generally applicable we may imagine, therefore, that sex of rearing, if it is a critical behavioural event, operates on future development (gender identity) without the agency of any endocrine intermediary. Contrary evidence, however, not entirely explicable on the basis of present knowledge, derives from male children born with a deficiency in the conversion of testosterone to dihydrotestosterone. Such children evidently appear and behave as females until puberty, at which time they masculinize both somatically and behaviourally (Imperato-McGinley, Guerrero, Gautier & Peterson, 1974). We need more evidence to clarify these apparent contradictions.

Maternal rejection during infancy may represent another critical behavioural event and has been much studied (see Hinde, 1974). There is no evidence that this depends on endocrine factors in the mother or infant, though it seems highly likely that the profound behavioural response of the infant to rejection (Kaufman & Rosenblum, 1967) may alter its hormone levels in a way now familiar from studies on adults (see below). Whether these endocrine effects, if they occur, have any sort of long-term behavioural consequences remains unknown.

Let us briefly consider another even more difficult point. Suppose that a single endocrine critical event occurring during foetal life determines "maleness". What does this mean? Sexually dimorphic roles vary a good deal in different primate species; for example, in monogamous as compared to polygamous species, or in those primates typically forming a single-male rather than multi-male organization. Limiting the discussion to rhesus monkeys, a male can be differentiated from a female by: showing more aggressive play; reaching puberty later; forming more obvious dominance hierarchies with other males; behaving more aggressively; leaving the natal group to join another; showing less parental behaviour; and tolerating other males less. The list is far from complete. All these behaviours are modifiable by the extraneous factors of individuality or group composition, but the question is whether they are all based upon a single critical perinatal endocrine event, reinforced in later life (perhaps to differing degrees) by gonadal activity. Furthermore, some sex differences in behaviour (e.g., emigration from the natal group) are qualitative in that they seldom occur in female rhesus monkeys. Others (e.g., aggression, dominance, mounting) more resemble the sexually dimorphic traits like play behaviour during the juvenile period, in that they differ quantitatively between the sexes.

PHYSIOLOGICAL MECHANISMS AND COST/BENEFIT ANALYSES

Before considering in more detail the second and third categories of endocrine events (recurrent and contextual), we must discuss the relationship between behavioural strategies of the adult and the physiological methods by which they are brought about. Much interest is currently being taken in the strategies by which animals, particularly primates, seek to maximize their inclusive fitness; that is, the proportion of their genes they or their relatives carry that are passed to succeeding generations. Here we are particularly concerned with reproductive behaviour, the most direct means by which these strategies are expressed. A central contention is that a behavioural strategy has developed as the result of balancing the likely cost (in terms of the possibility of failure or the expenditure of some limited resource) versus the benefit (the chances of increasing genetic contribution to the population pool) of a given course of action (Trivers, 1972; Clutton-Brock & Harvey, 1976).

What concerns us here is the process by which these decisions are brought about, and particularly the part played by neuroendocrine events. Various physiological strategies are possible. Suppose we take, as our example, the emigration from the natal group which occurs characteristically in so many male primates. A primary function, it is supposed, of this process, is to avoid in-breeding between related members of a species, since homozygosity has been shown to be deleterious (see Packer, 1979). Three physiological mechanisms could be devised to bring about behaviour which avoids in-breeding.

(1) An analysis of the situation by using high-level neural processing at various levels of complexity, allowing the animal to estimate directly either the consequences of in-breeding, or of competing within his natal group for females, or of his capacity for survival during transfer. Whilst humans may have developed these abilities in some cases, it is highly unlikely to be found in so complete a form in non-human primates, though the primate brain may allow improved recognition of more immediate (approximate) causative factors by increasing the individual's ability to interpret and assess the behaviour of others (Humphrey, 1976).

(2) The development of a behavioural bias which tends to diminish in-breeding as a secondary consequence; for example, a mechanism that makes female siblings sexually unattractive to males as they reach maturity (Shepher, 1971). This will result in the females of neighbouring troops becoming relatively more attractive and hence drawing males away from their natal group to join others (Packer, 1979), an event made more likely by the adolescent monkey's

developing propensity for sexual interaction as his testes secrete increasing amounts of androgen (Rose, Bernstein, Gordon & Lindsley, 1978).

(3) More immediate humoral mechanisms, which, for example, precipitate behavioural responses from adult males towards adolescents in some species. These responses can be postulated to be dependent upon the development of hormone-sensitive features as the young males grow up and enter puberty, and result in their ejection from the group. Hormones change the appearance and behaviour of adolescent males, and may, in this way, cause them to be no longer tolerated by the resident mature males in the group. This intolerance might also depend upon the hormonal condition of the adult males, i.e., might occur more frequently during the mating season when hormone levels in both adolescent and adult are maximal (Lindburg, 1969; Hall & Mayer, 1967).

These three types of mechanism are not exclusive. A male may weigh up, by using cognitive mechanisms, the likely competition from males of a neighbouring troop in comparison with the relative attractiveness of its females, and be forced to make a choice based upon the behaviour shown towards it by the dominant males of its own group. Furthermore, each of the three categories contains several sub-groups. Whilst assessing directly the cost/benefit ratio of in-breeding or transfer requires very high-order neural processing, at a less sophisticated level the animal can more easily estimate the size of an oppressor's canines, or his own ability to appease potential aggression.

There is no doubt that transfer between groups represents a highly critical behavioural event, and that, in most primate species in which it has been described, it occurs most frequently in males (Boelkins & Wilson, 1972; Mohnot, 1978; Packer, 1979), although there are some species in which females rather than males emigrate (e.g., chimpanzees and gibbons, amongst others) (Harcourt, 1979). Since some baboons transfer more than once (Packer, 1979) it is unlikely that this is dependent upon a single endocrine event such as puberty, in these males at least. Furthermore, the supposition that intolerance toward males is not entirely dependent upon immediate endocrine events, such as puberty, follows from the observation that aggressive behaviour by adult males (for example, langurs) towards other males starts early in the latters' life, before puberty, continues through the juvenile period and is only completed at puberty (Mohnot, 1978). However, it is known that the secretion of testosterone alters both appearance and behaviour (e.g., play) in the adolescent monkey (Dixson & Herbert, 1974; Rosenblum & Bromley, 1978), so that the timing of transfer strongly suggests that this hormone is playing some important causative role. It follows that the secretion of the adolescent's testes may represent another critical event, though one which is somewhat different from that which may occur during perinatal life. Whereas perinatal secretion may permanently alter some behavioural propensity and thus be critical in that sense, puberty may initiate a sequence of events that leads to a male transferring to a second troop in which it may spend the rest of its reproductive life. The first, then, is critical for determining the nature of a behaviour pattern and the second for the context in which it will occur.

RECURRENT EVENTS

A recurrent event is exemplified by the effect that the phase of the female's menstrual cycle has upon sexual interaction (Rowell, 1972). The most extensive analysis, both behavioural and endocrine, has been made on the rhesus monkey (Herbert, 1974, 1977). The cycle itself is regulated by neuroendocrine mechanisms principally dependent upon the secretion of oestrogen and progesterone from the ovary, and the way this alters the action of LHRH on the pituitary (Knobil, 1974). The behavioural findings may apply to some other primates (e.g., gorillas, talapoins and baboons — Nadler, 1976; Scruton & Herbert, 1970; Saayman, 1968) though not necessarily to others (e.g., orang-utans, stump-tailed macaques — Nadler, 1977; Slob, Weigand, Goy & Robinson, 1978). Hormonal changes increase the likelihood of the female's mating, whether this occurs in the context of a monogamous relationship, a one-male system or a multi-male arrangement in which a consortship is formed during the middle part of the female's menstrual cycle. These changes come about, it seems, by two principal mechanisms. The first depends upon hormones acting directly on the female's nervous system to change her behaviour towards the male so that she either solicits him more often (i.e., behaves "proceptively"), or accepts his sexual advances more readily (that is, behaves "receptively") (Beach, 1976). It is important to recognize that both these events will alter not only the male's behaviour but also his own hormone levels. The second mechanism occurs in other, peripheral hormone-dependent structures, such as the vagina or the sexual skin. Hormones cause these tissues to emit stimuli, either visual or olfactory, which increase the female's "attractiveness" for the male (Beach, 1976; Baum, Everitt, Herbert & Keverne, 1977). The female's behaviour, however, may emphasize these non-behavioural stimuli, as when she "presents" to the male (Zuckerman, 1932), thus exposing the perineum from which they originate.

Let us now consider the way that the endocrine system controls these various components and how they relate to the changes in behaviour during the menstrual cycle. A female rhesus monkey's attractiveness is largely regulated by her ovarian hormones, oestrogen and progesterone, acting on her vagina, the first encouraging the formation of odours (pheromones) that stimulate the male, the second antagonizing this effect (Keverne, 1976; Baum, Keverne, et al., 1976; Baum, Everitt et al., 1977). Thus males respond to sexual solicitations by ovariectomized females more readily after the females are given oestrogen, but less readily if progesterone is given as well (Baum, Keverne et al., 1976). Furthermore, the male is more active in trying to mate with oestrogen-treated females (even when they do not solicit him) because of the non-behavioural cues that the females emit. These behaviours, reproduced under more natural conditions, result in the male consorting with the female during the time her oestrogen levels remain high, which, in some species, is correlated with the sexual skin being maximally swollen (Carpenter, 1942; Kaufmann, 1965; Rowell, 1972; Fedigan & Gouzoules, 1978). Since, in baboons and rhesus monkeys, consortships are made and broken during each cycle, they are examples of recurrent endocrine-dependent behavioural events. It is important to recognize that a single hormone, acting on a defined target structure, can initiate a complex sequence of behavioural events. Thus, if progesterone is introduced into the female's vagina, thus limiting its action to the local epithelium, the female's attractiveness declines, and the male mates less eagerly. Females so treated commonly show increased solicitation of the male, presumably attempting to compensate by using behavioural mechanisms for their decreased non-behavioural attractiveness (Baum, Keverne et al., 1976): however, this is not a direct action of progesterone on their brains. In a natural setting, we might imagine the final outcome might be the end of the consortship.

In the rhesus monkey, there is evidence that the female's own behaviour — her proceptivity and receptivity — is regulated by androgens. Removing these hormones by adrenalectomy combined with ovariectomy, or by giving the female antibodies specifically binding testosterone, reduces either or both categories of behaviour (Everitt & Herbert, 1971; Everitt, Herbert & Hamer, 1972; Martensz, Everitt & Newton, 1979). Testosterone (or its precursor, androstenedione) restores or stimulates these behaviours, whereas dihydrotestosterone, a principal metabolite of testosterone in the male, does not (Trimble & Herbert, 1968; Wallen & Goy, 1977). The evidence suggests that these hormones act on the female's brain, particularly the hypothalamus (Everitt & Herbert, 1975). The relative contributions of endocrine-dependent events in male and female to the behaviour between them within a consortship will vary according to circumstances, species and individuals. For example, it has been suggested that, in macaques, changes in the female's attractiveness (a non-behavioural quality) are the most important variable (Eaton, 1973), and thus changes in the male's behaviour exert the most influential control. The participation of androgens in regulating the female's behaviour may explain why some female macaques accept the male throughout the cycle (unlike the females of many non-primate species) since androgen levels, which regulate the female's behaviour, fluctuate less prominently during the cycle than oestrogen levels and are not antagonized by progesterone secreted during the luteal phase. There may be other hormones with similar direct actions on the brain. In particular, LHRH, the hypothalamic releasing hormone, has been said to stimulate receptivity in female rats (Moss & McCann, 1973) and to enhance sexual activity in hypogonadal men (Mortimer et al., 1974); furthermore, heightened prolactin may decrease "libido" in humans (see below).

Thus, two sets of mechanisms, central and peripheral, are separable on the basis of both their site of action and the natures of the hormones primarily involved (Baum, Everitt et al., 1977). A further point is that the behavioural control exerted on the two mechanisms may also differ. If other members of the group are to influence sexual interactions between a pair, they could be operating upon different mechanisms in male and female. Since the female controls interactions with the male by altering her proceptivity and receptivity, it is upon these behaviours that a group's influence might be exerted to modulate the female's role. However, if members of the group are to operate upon the male, then they might need to affect his evaluation of the female's attractiveness. Thus relationships within the group could modify different components of sexual interaction by acting upon different parts of the behavioural repertoire of the participants in a consortship. If group members operate upon the behaviour of the male or the female of a consortship, the result may be the same in so far as that consortship may be affected, but the consequences will be very different for the individuals concerned. Behavioural forces acting upon a male will prevent him from forming another consortship, but this is not true for the female; the likelihood of the female not being fertilized is not necessarily much reduced. This will have a marked effect on the reproductive strategies within a group.

ENDOCRINE STRATEGIES IN THE SOCIAL GROUP

Sexual interaction in primates, as in other species, is regulated by intrasexual selection (e.g., the process by which males compete with each other for access to females), and epigamic selection (e.g., the ability of males to attract the sexual responses of the females) (Huxley, 1938; Crook, 1972). Intra-sexual competition is best developed in males, which, it is proposed, compete for females because the latter's reproductive potential is lower, though females will exert greater care in the choice of males because their investment in the offspring is more (Trivers, 1972). Females, therefore, may also compete with each other to attract those males with preferred qualities (Orians, 1969).

Consider the situation where a number of males are competing for sexual access to the female. The result of intra-sexual competition for this resource, as many others, reflects the dominance structure of the group. If we examine the role of testicular hormones, for example testosterone, in this process, we find that there is general agreement (though not a great deal of clear data) that these hormones may emphasize or develop characteristics of the males (e.g., musculature, canine teeth) which provide some of the equipment used in intra-sexual competition. But there is much less evidence that behavioural correlates, such as "aggressiveness" or the ability to dominate others, are enhanced by testosterone or reduced by castration (Rose, Gordon, Bernstein & Catlin, 1974), or that these treatments influence the outcome of intra-sexual competitiveness (Wilson & Vessey, 1968; Green, Whalen, Butley & Battie, 1972; Mazur, 1976; Dixson & Herbert, 1977; Bielert, 1979). We must be careful to separate the hormonal effects on aggressiveness (i.e., the amount of aggression shown) from those on dominance (the direction and outcome of aggressive episodes), because the former may be specifically accentuated in sexual contexts by the male's gonadal hormones when females are sexually attractive (Zuckerman, 1932; Wilson & Boelkins, 1970; Keverne, Meller & Martinez-Arias, 1978; Teas, Taylor & Richie, 1978). This is because the males are drawn into conflict with each other for a limited resource and become involved in the costs and benefits of competing for it, rather than there being any direct change in their "aggressiveness" *per se* induced by their own testicular hormones.

CONTEXTUAL EVENTS

The dominance hierarchy is one way of describing a set of relationships upon which recurrent events, such as consortships, take place.

Recent evidence shows that the dominance structure not only influences the outcome of sexual competition but also alters the hormone levels of the individuals concerned, providing us with an example of the third category, the contextual endocrine event. These important changes in hormone levels have only recently received their due share of attention. Being dominant or subordinate has effects on the endocrine state of both sexes. In the male talapoin, for example, testosterone levels are higher and prolactin is lower in the dominant animal than in the subordinate; exposing the male to sexually attractive females raises cortisol in both (Keverne *et al.*, 1978; Keverne, 1979). Subordinates not only have elevated prolactin levels but their testosterone levels do not rise when they are in the company of attractive females; neither do they mate (Eberhart, Keverne & Meller, 1980). Subordinate female talapoins also show higher cortisol and prolactin levels than dominant ones, the hyperprolactinaemia being sufficient in some cases to prevent LH being discharged in response to oestrogen, thus rendering these females relatively or absolutely infertile (Bowman, Dilley & Keverne, 1978). A comparable phenomenon may exist in humans (Besser, 1979). Reduced fertility in subordinate monkeys has been described both in the wild and in captivity (Dunbar & Dunbar, 1977; Abbott & Hearn, 1979).

Do such hormonal changes play any part in the process of sexual competition? Subordinate males may have reduced plasma testosterone in order to diminish the aggression-provoking qualities they possess for the dominant male. In such a case the hormonal change may be part of the subordinate's adaptation to minimize the cost (e.g. danger) of staying in a social group. A rather different explanation is also plausible, though equally untested: that subordinates' testosterone levels are reduced by more dominant males to enhance the latters' own relative sexual attractiveness for the females. This would help tilt the balance in favour of the victor of intra-sexual competition, and would represent one advantage of being dominant. Furthermore, hormonal changes in the male may also add to the female's ability to discriminate dominant from subordinate males, assuming the former to be preferable as mates.

In situations where there is to be competition between females for males, there are at least two possible strategies by which less successful females can be prevented from contributing to the gene pool. The first is to prevent their mating activity by behavioural means; this is described, for example, in female marmosets which remain within the family group but which are inhibited from mating by more dominant females (Abbott & Hearn, 1978). The second is for subordinate females to be able to mate but to be less fertile, and an example given for talapoins is described above. The two strategies may have similar reproductive results, but vastly different behavioural ones. In the first, the females will be effectively withdrawn from all sexual competition. In the second, males may compete for access to subordinate females, but with the prospect of reduced reproductive benefit if they succeed. To avoid winning a female with reduced fertility, males need to compete more vigorously for more dominant females, and one study on rhesus monkeys shows, indeed, that high-ranking females tend to be preferred to lower-ranking ones (Perachio, Alexander & Marr, 1973) and there is evidence that this may be true in talapoins. However, the fact that female talapoins commonly dominate males complicates the issue (Meller & Keverne, in press). In situations where their sexual behaviour is suppressed, females may need to leave the group in order to mate, and this mechanism, there-fore, might be particularly apposite for monogamous species, such as marmosets.

The social function of prolactin in the context described here is worth emphasizing. We have seen that elevated levels can cause infertility in monkeys and in humans. In human males, hyperpro-lactinaemia has been associated with impotence and decreased sexual interest (Besser et al., 1978). If this is true of monkeys (and there is no experimental evidence yet), then this hormone may provide a direct humoral means by which a subordinate male adapts to the presence of attractive but inaccessible females, a situation which might otherwise provoke a maladaptive internal conflict. Alterna-tively, raised prolactin in the subordinate may represent a mechanism which functions to the advantage of the dominant male, because it reduces the likelihood of the subordinate mating surreptitiously, by rendering him relatively uninterested in copulation. At this stage, we have no evidence favouring one interpretation over the other, and cannot therefore determine whether socially induced hyperprolactin-aemia occurs for the benefit of the subordinate or the dominant male. Though similar functions can also be ascribed to testosterone (see above), the two explanations can be reconciled, since it is now known that prolactin lowers the plasma levels of testosterone in monkeys (Herbert, unpublished) and may interfere with its action.

EPIGAMIC COMPETITION AND SIGNALS

It has often been pointed out (e.g., Crook, 1972) that male primates are amongst the most decorated of the mammals. This coloration usually appears at puberty and can, in some cases at least, be ascribed to the hormonal events occurring at this time (Zuckerman & Parkes, 1939). Castration or reduced testicular activity may diminish genital coloration in some species (e.g., rhesus monkeys) but not in others (e.g., talapoin monkeys) (Dixson & Herbert, 1974). Whilst we know that the sexual behaviour of male primates falls after castration or when testicular hormones are low (Dixson & Herbert, 1977; Phoenix, 1978; Mendoza, Lowe, Resko & Levine, 1978), there is less evidence for their epigamic effectiveness being reduced, though female talapoins sexually solicit dominant, castrated males more frequently when these males have been treated with testosterone (Dixson & Herbert, 1977). However, it is not yet clear whether this is a consequence of the changes in the male's sexual behaviour, or of alterations in endocrine-dependent non-behavioural signals from him. The amount of behavioural control exerted by the females over sexual interactions seems to vary between species. If a female exercises a high degree of behavioural control over her interactions with the male, then hormonal changes in her might be expected to have a more pronounced effect on determining this interaction than in females of species whose behaviour is characteristically controlled principally by the male. This has been found in the apes, at least in captivity, in which the animals were only allowed to interact for restricted periods; sexual behaviour between gorillas, in which the female largely controls mating activity under these conditions, exhibits a prominent mid-cycle peak whereas in the orang-utan, in which the male exerts most control, there is little fluctuation (Nadler, 1976, 1977). However, it should be pointed out that if interactions were mainly controlled by non-behavioural hormonally induced changes in the female ape's attractiveness (as may be the case in the rhesus monkey — see above), then the opposite result might have been predicted.

It is now relevant to consider more closely the signals by which hormonal changes are translated into behavioural events, or by which behaviours come to alter endocrine secretion. A well-known complication is that a signal used in one behavioural context, say a sexual one (e.g., the "presentation" posture), may be modified to serve another purpose (in this example, the reduction of aggression) (Kempf, 1917; Zuckerman, 1932). A second is that genital displays in, for example, squirrel monkeys are apparently used to establish or maintain dominance ranks rather than to transmit sexual information (Ploog, Blitz & Ploog, 1963). The way that hormones influence either the frequency or the appearance of such displays will thus depend upon the context in which they are given. Furthermore, the behavioural significance of the relationship between a hormone and its dependent signal may vary in like manner.

It hardly needs pointing out that signals can be transmitted through a number of sensory modalities, and that while olfactory or visual pathways are the ones most commonly employed in the behaviours considered here, tactile stimuli from the genitalia during copulation itself can have powerful effects on subsequent interactions (Herbert, 1973). Considerable work has been directed towards assigning relative importance to these various signals and to the channels through which they operate, but two other aspects will be considered here: whether they are all behaviourally equivalent and why there are a number of signals transmitting hormone-dependent behaviourally important information. If we compare olfactory and visual signals passing from the female rhesus to the male, we see that the important difference between them is not only that they occupy different modalities but that the transmitting animal (in this case the female) can more easily exert behavioural control over visual than over olfactory signals. Proceptive signals usually involve some movement whereby the female attracts the male, characteristically by directing his attention towards her genitalia. The appearance of the female's genitalia may also be altered by swelling or coloration which depend upon her hormonal state. But the point is that the direction in which these somatic signals are given, and the degree to which her behaviour emphasizes them, can be made relatively specific towards a given male. Furthermore, the neural processing of these signals within the male is likely to involve relatively higher (cortical) regions of the visual brain, subject also to extensive neural modulation before being transmitted to those parts of the brain (the limbic areas) primarily concerned with sexual behaviour. The directionality of olfactory signals, on the other hand, cannot be controlled by the female to the same extent. Furthermore, neuro-anatomical evidence suggests that olfactory signals may gain more direct access to the limbic part of the male's brain (Keverne, 1976; Herbert, 1977). These considerations therefore dictate that a female exercising maximal choice during the selection of a mate will concentrate on emitting visual rather than olfactory signals. Alternatively, she will have to control the male's access to her by inhibiting his attempts to mount (e.g., by refusing him); however, this may carry risks of aggressive responses from him, as well as limiting her role in exercising choice of partner. There is also the point that the neural mechanisms by which a male modulates his response to visual signals are different from those influencing his reactions to olfactory ones. In the first case, initial processing is likely to occur in those parts of the brain (the cortex) which are particularly well developed in primates, and where the relatively sophisticated analysis of social signals and responses may occur (Humphrey,

1976). In the second, because the signal has more direct access to the limbic areas of the brain, cortical control, if it is to be exerted, must occur as a secondary procedure.

Competition between females for the sexual attentions of the males also suggests why there are so many signals, occupying so many sensory channels between the participants. A long-standing puzzle is why some female primates possess inflatable sexual skins, whereas others do not; some have brightly coloured genitalia, others do not (see Crook, 1972; Clutton-Brock & Harvey, 1976). As a result of epigamic selection, however, animals compete to influence the behaviour of others. Thus, the more information a female can signal to a male, the greater her chances of success. More significantly, the more she can occupy his sensory input channels the more likely she is to limit his choice of behavioural response, based as it is largely on information being received by his sensory receptors, and thus direct his behaviour towards her. Perhaps we ought not to consider the distribution and function of sexual skin swellings in isolation but in conjunction with the occurrence of other signals, including behavioural postures or olfactory cues, and ask why different species have employed different physiological strategies to fulfil a common purpose. As Clutton-Brock & Harvey (1976) suggest, it may be that a species making less use of one channel (for example, olfaction) has compensated by increased activity in another, though, as we have seen, the various channels are not necessarily equivalent so far as either female or male is concerned.

This discussion of mating activity in primates shows that both recurrent and contextual events are prevalent in the adult, and that both relate hormones to behaviour. Recurrent events are driven by endocrine changes, but the effects they have are modulated by the behavioural relationships within the group, which in turn can induce socially important hormonal patterns. Both, in turn, operate upon the individual characteristics of each monkey which can be determined or influenced by the critical events of development and early post-natal life.

Throughout this chapter there are examples where several neuro-endocrine strategies are available to a monkey to achieve a common goal; to maximize its reproductive success *vis-à-vis* its competitors. The puzzle is why some are chosen and others not, how this varies in different species or in the same species under different circumstances. One aspect which is considered here only briefly is the way that these strategies are related to the cerebral equipment of the animals from which they originate or towards which they are directed. This is likely to become a major area of interest in the future. We need to

know more about the way different species choose particular neuroendocrine mechanisms and whether these vary according to circum-
stance or objective.

ACKNOWLEDGEMENTS

I am most grateful to David Chivers, Tim Clutton-Brock, Alan
Dixson, Rachel Meller and Michael Simpson for their comments on
this chapter. The work from this laboratory described here is supported by an MRC Programme Grant. I wish to record my indebted-
ness to Solly Zuckerman for his help and encouragement during the
earlier days of my work.

REFERENCES

Abbott, D. H. (1978). Hormones and behaviour during puberty in the marmoset.
In *Recent advances in primatology* 1. *Behaviour*: 497—499. Chivers, D. J.
& Herbert, J. (Eds). London and New York: Academic Press.
Abbott, D. H. & Hearn, J. P. (1978). Physical, hormonal and behavioural aspects
of sexual development in the marmoset monkey, *Callithrix jacchus*. *J.
Reprod. Fert.* 53: 155—166.
Abbott, D. H. & Hearn, J. P. (1979). The effect of neonatal exposure to testo-
sterone on the development of behaviour in female marmoset monkeys.
In *Sex, hormones and behaviour*: 299—327. Porter, R. & Whelan, J.
(Eds). (*Ciba Fdn Symp*. No. 62) Amsterdam: Elsevier.
Abramovich, D. R. & Rowe, P. (1973). Foetal plasma testosterone levels at mid-
pregnancy and at term: relationship to foetal sex. *J. Endocr.* 56: 621—
622.
Baum, M. J., Everitt, B. J., Herbert, J. & Keverne, E. B. (1977). Hormonal basis
of proceptivity and receptivity in female primates. *Arch. sex. Behav.* 6:
173—191.
Baum, M. J., Keverne, E. B., Everitt, B. J., Herbert, J. & de Vrees, P. (1976).
Reduction of sexual interaction in rhesus monkeys by a vaginal action of
progesterone. *Nature, Lond.* 263: 606—608.
Beach, F. A. (1976). Sexual attractivity, proceptivity and receptivity in female
mammals. *Horm. Behav.* 7: 105—138.
Besser, G. M. (1979). In Discussion of Keverne, E. B. (1979) (which see).
Besser, G. M., Yeo, T., Delitala, G., Jones, A., Stubbs, W. A., Wass, J. A. H. &
Thorner, M. O. (1978). Clinical neuroendocrine relationships in normal
and disordered prolactin secretion. In *Central regulation of the endocrine
system*: 457—472. Fuxe, K., Hökfelt, T. & Luft, R. (Eds). New York:
Plenum.
Bielert, F. (1979). Androgen treatment of young male rhesus monkeys. In
Recent advances in primatology 1. *Behaviour*: 485—488. Chivers, D. J. &
Herbert, J. (Eds). London and New York: Academic Press.
Boelkins, C. M. & Wilson, A. P. (1972). Intergroup social dynamics of the Cayo
Santiago rhesus (*Macaca mulatta*) with special reference to change in group
membership by males. *Primates* 13: 125—140.

Bowman, L. A., Dilley, S. & Keverne, E. B. (1978). Suppression of oestrogen-induced LH surges by social subordination in talapoin monkeys. *Nature, Lond.* 275: 56—58.

Carpenter, C. R. (1942). Sexual behavior of free-ranging rhesus monkeys (*Macaca mulatta*). *J. comp. Psychol.* 33: 113—162.

Chevalier-Skolnikoff, S. (1976). Homosexual behaviour in a laboratory group of stumptail monkeys (*Macaca arctoides*): form, contexts, and possible social functions. *Arch. sex. Behav.* 5: 511—527.

Clutton-Brock, T. H. & Harvey, P. H. (1976). Evolutionary rules and primate societies. In *Growing points in ethology*: 195—237. Bateson, P. P. G. & Hinde, R. A. (Eds). Cambridge: University Press.

Clutton-Brock, T. H. & Harvey, P. H. (1977). Primate ecology and social organisation. *J. Zool., Lond.* 183: 1—40.

Crook, J. H. (1972). Sexual selection, dimorphism and social organisation in the primate. In *Sexual selection and the descent of man 1871—1971*: 231—281. Campbell, B. (Ed.). Chicago: Aldine.

Crook, J. H. & Gartlan, J. S. (1966). On the evolution of primate societies. *Nature, Lond.* 210: 1200—1203.

Dixson, A. F. & Herbert, J. (1974). The effects of testosterone on the sexual skin and genitalia of the male talapoin monkey. *J. Reprod. Fert.* 38: 217—219.

Dixson, A. F. & Herbert, J. (1977). Gonadal hormones and sexual behaviour in groups of adult talapoin monkeys (*Miopithecus talapoin*). *Horm. Behav.* 8: 141—154.

Dorner, G. (1979). Hormones and sexual differentiation of the brain. In *Sex, hormones and behaviour*: 81—112. Porter, R. & Whelan, J. (Eds). (*Ciba Fdn Symp.* No. 62) Amsterdam: Elsevier.

Dorner, G., Rohde, W., Stahl, F., Krell, L. & Masius, W. G. (1975). A neuro-endocrine predisposition for homosexuality in man. *Arch. sex. Behav.* 4: 1—8.

Dunbar, R. I. M. & Dunbar, E. P. (1977). Dominance and reproductive success among female gelada baboons. *Nature, Lond.* 266: 351—352.

Eaton, G. G. (1973). Social and endocrine determinants of simian and pro-simian female sexual behaviour. In *Primate reproductive behaviour*: 104—123. Montagna, W. (Ed.). Basle: Karger.

Eaton, G. G., Goy, R. W. & Phoenix, C. H. (1973). Effects of testosterone treatment in adulthood on sexual behaviour of female pseudo hermaphrodite rhesus monkeys. *Nature, New Biol.* 242: 119—120.

Eberhart, J. A., Keverne, E. B. & Meller, R. E. (1980). Social influences on plasma testosterone in male talapoin monkeys. *Horm. Behav.* 14: 247—266.

Ehrhardt, A. A. & Baker, S. W. (1974). Fetal androgens, human central nervous system differentiation and behaviour sex differences. In *Sex differences in behaviour*: 33—51. Friedman, R. C., Richart, R. M. & Vande Wiele, R. L. (Eds). New York: Wiley.

Ehrhardt, A. A. & Meyer-Bahlburg, H. F. (1979). Psychosexual development: an examination of the role of prenatal hormones. In *Sex, hormones and behaviour*: 41—57. Porter, R. & Whelan, J. (*Ciba Fdn Symp.* No. 62) Amsterdam: Elsevier.

Everitt, B. J. & Herbert, J. (1971). The effects of dexamethasone and androgens on sexual receptivity of female rhesus monkeys. *J. Endocr.* 51: 575—588.

Everitt, B. J. & Herbert, J. (1975). The effects of implanting testosterone pro-pionate into the central nervous system on the sexual behavior of adrenalectomised female rhesus monkeys. *Brain Res.* 86: 109–120.

Everitt, B. J., Herbert, J. & Hamer, J. D. (1972). Sexual receptivity of bilaterally adrenalectomised female rhesus monkeys. *Physiol. Behav.* 8: 409–415.

Fedigan, L. M. & Gouzoules, H. (1978). The consort relationship in a troop of Japanese macaques. In *Recent advances in primatology* 1. *Behaviour*: 493 –495. Chivers, D. J. & Herbert, J. (Eds). New York & London: Academic Press.

Forest, M. G., Cathiard, A. M. & Bertrand, J. A. (1973). Evidence of testicular activity in early infancy. *J. clin. Endocr. Metab.* 37: 148–151.

Forest, M. G., Sizonenko, P. C., Cathiard, A. M. & Bertrand, J. A. (1974). Hypophysogonadal function in humans during their first year of life. 1. Evidence for testicular activity in early infancy. *J. clin. Invest.* 53: 819–828.

Goy, R. W. (1978). Development of play and mounting behaviour in female rhesus virilised prenatally with esters of testosterone or dihydrotestosterone. In *Recent advances in primatology*. 1. *Behaviour*: 449–462. Chivers, D. J. & Herbert, J. (Eds). New York & London: Academic Press.

Goy, R. W. (1979). In Discussion of paper by R. Green: In *Sex, hormones and behaviour*: 75. (*Ciba Fdn Symp.* No. 62) Porter, R. & Whelan, J. (Eds). Amsterdam: Elsevier.

Goy, R. W. & Resko, J. A. (1972). Gonadal hormones and behavior of normal and pseudohermaphroditic female primates. *Rec. Progr. Horm. Res.* 28: 707–733.

Green, R. (1979). Sex-dimorphic behaviour development in the human: prenatal hormone administration and postnatal socialisation. In *Sex, hormones and behaviour*: 53–80. (*Ciba Fdn Symp.* No. 62) Porter, R. & Whelan, J. (Eds). Amsterdam: Elsevier.

Green, R., Whalen, R., Butley, B. & Battie, C. (1972). Dominance hierarchy in squirrel monkeys. Role of gonads and androgen on genital display and feeding order. *Folia primatol.* 18: 185–195.

Hall, K. R. L. & Mayer, B. (1967). Social interactions in a group of captive patas monkeys (*Erythrocebus patas*). *Folia primatol.* 5: 213–236.

Hamilton, W. D. (1963). The evolution of altruistic behavior. *Am. Nat.* 97: 354–356.

Harcourt, A. H. (1979). Social relationships between adult male and female mountain gorillas in the wild. *Anim. Behav.* 27: 325–342.

Herbert, J. (1973). The role of the dorsal nerves of the penis in the sexual behavior of the male rhesus monkey. *Physiol. Behav.* 10: 293–300.

Herbert, J. (1974). Some functions of hormones and the hypothalamus in the sexual activity of primates. *Progr. Brain Res.* 41: 331–347.

Herbert, J (1977). Hormones and behaviour. *Proc. R. Soc. Lond.* (B.) 199: 425–443.

Hinde, R. A. (1974). *Biological bases of human social behavior.* New York: McGraw-Hill.

Humphrey, N. K. (1976). The social function of intellect. In *Growing points in ethology*: 303–317. Bateson, P. P. G. & Hinde, R. A. (Eds). Cambridge: University Press.

Huxley, J. S. (1938). Darwin's theory of natural selection and the data subsumed in it, in the light of recent research. *Am. Nat.* 72: 416–433.

Imperato-McGinley, M., Guerrero, L., Gautier, T. & Peterson, R. E. (1974). Steroid 5-reductase deficiency in man as an inherited form of male pseudohermaphroditism. *Science, Wash.* 186: 1213–1215.

Jost, A. (1953). Problems of fetal endocrinology: the gonadal and hypophyseal hormones. *Rec. Progr. Horm. Res.* 8: 379–418.

Kaufman, I. C. & Rosenblum, L. A. (1967). The reaction to separation in infant monkeys: anaclitic depression and conservation withdrawal. *Psychosom. Med.* 29: 648–675.

Kaufmann, J. H. (1965). A three year study of mating behavior in a free ranging band of rhesus monkeys. *Ecology* 46: 500–512.

Kempf, E. J. (1917). The social and sexual behavior of infrahuman primates with some comparable facts in human behavior. *Psychoanal. Rev.* 4: 127–154.

Keverne, E. B. (1976). Sexual receptivity and attractiveness in the female rhesus monkey. *Adv. Study Behav.* 7: 155–200.

Keverne, E. B. (1979). Sexual and aggressive behaviour in social groups of talapoin monkeys. In *Sex, hormones and behaviour*: 271–297. (*Ciba Fdn Symp.* No. 62) Porter, R. & Whelan, J. (Eds). Amsterdam: Elsevier.

Keverne, E. B., Meller, R. E. & Martinez-Arias, A. (1978). Dominance, aggression and sexual behaviour in social groups of talapoin monkeys. In *Recent advances in primatology* 1. *Behaviour*: 533–547. Chivers, D. J. & Herbert, J. (Eds). London: Academic Press.

Knobil, E. (1974). On the control of gonadotrophin secretion in the rhesus monkey. *Rec. Progr. Horm. Res.* 30: 1–46.

Lindburg, D. G. (1969). Rhesus monkeys: mating season mobility of adult males. *Science, N.Y.* 166: 1176–1178.

Martensz, N. D., Everitt, B. J. & Newton, M. V. (1979). Passive immunisation against testosterone and sexual behaviour of female rhesus monkeys. *Acta Endocr., Kbh.* (suppl.) 225: 247.

Mazur, A. (1976). Effects of testosterone on status in primate groups. *Folia primatol.* 26: 214–226.

Meller, R. E. & Keverne, E. B. (In press). Female dominance and sexual interactions in captive talapoin monkeys. *Anim. Behav.*

Mendoza, S. P., Lowe, E. L., Resko, J. A. & Levine, S. (1978). Seasonal variations in gonadal hormones and social behavior in squirrel monkeys. *Physiol. Behav.* 20: 515–522.

Mohnot, S. M. (1978). Peripheralisation of weaned male juveniles in *Presbytis entellus*. In *Recent advances in primatology* 1. *Behaviour*: 87–91. Chivers, D. J. & Herbert, J. (Eds). New York: Academic Press.

Money, J. & Ehrhardt, A. A. (1972). *Man and woman, boy and girl.* Baltimore: Johns Hopkins University Press.

Mortimer, L. H., McNeilly, A. S., Fisher, R. A., Murray, M. A. F. & Besser, G. M. (1974). Gonadotrophic-releasing hormone therapy in hypogonadal men with hypothalamic and pituitary dysfunction. *Br. Med. J. 1974* (4): 617–621.

Moss, R. L. & McCann, S. M. (1973). Induction of mating behavior in rats by luteinising hormone-releasing factor. *Science, Wash.* 181: 177–179.

Nadler, R. D. (1976). Sexual behavior of captive lowland gorillas. *Arch. sex. Behav.* 5: 487–502.

Nadler, R. D. (1977). Sexual behavior of captive orang-utans. *Arch. sex. Behav.* 6: 457–476.

Orians, G. H. (1969). On the evolution of mating systems in birds and mammals. *Am. Nat.* 103: 589–603.

Packer, C. (1979). Intertroup transfer and inbreeding avoidance in *Papio anubis*. *Anim. Behav.* 27: 1–36.

Perachio, A. A., Alexander, M. & Marr, L. D. (1973). Hormonal and social factors affecting evoked sexual behavior in rhesus monkeys. *Am. J. phys. Anthrop.* 38: 227–232.

Phoenix, C. H. (1978). Steroids and sexual behaviour in castrated male rhesus monkeys. *Horm. Behav.* 10: 1–9.

Plapinger, L. & McEwen, B. S. (1978). Gonadal steroid–brain interactions in sexual differentiation. In *Biological determinants of sexual behaviour*: 153–218. Hutchison, J. S. (Ed.). London: Wiley.

Ploog, D. W., Blitz, J. & Ploog, F. (1963). Studies on social and sexual behavior of the squirrel monkey (*Saimiri sciureus*). *Folia primatol.* 1: 29–66.

Resko, J. A. (1970). Androgen secretion by the fetal and neonatal rhesus monkey. *Endocrinology* 87: 680–687.

Robinson, J. A. & Bridson, W. E. (1978). Neonatal hormone patterns in the macaque. I. Steroids. *Biol. Reprod.* 19: 775–778.

Rose, R. M., Bernstein, I. S., Gordon, T. P. & Lindsley, J. G. (1978). Changes in testosterone and behavior during adolescence in the male rhesus monkey. *Psychosom. Med.* 80: 60–70.

Rose, R. M., Gordon, T. P., Bernstein, I. S. & Catlin, S. F. (1974). Androgens and aggression: a review and recent findings in primates. In *Primate aggression, territoriality and xenophobia*: 275–304. Holloway, R. L. (Ed.). New York: Academic Press.

Rosenblum, L. A. & Bromley, L. J. (1978). The effects of gonadal hormones on peer interactions. In *Recent advances in primatology* 1. *Behaviour*: 161–164. Chivers, D. J. & Herbert, J. (Eds). New York: Academic Press.

Rowell, T. E. (1972). Female reproductive cycles and social behaviour in primates. *Adv. Study Behav.* 4: 69–105.

Saayman, G. S. (1968). Oestrogen, behaviour and the permeability of a troop of chacma baboons. *Nature, Lond.* 220: 1399–1400.

Scruton, D. M. & Herbert, J. (1970). The menstrual cycle and its effect upon behaviour in the talapoin monkey (*Miopithecus talapoin*). *J. Zool., Lond.* 142: 419–436.

Seyfarth, R. M., Cheney, D. L. & Hinde, R. A. (1978). Some principles relating social interactions and social structure among primates. In *Recent advances in primatology* 1. *Behaviour*: 39–51. Chivers, D. J. & Herbert, J. (Eds). London: Academic Press.

Shepher, J. (1971). Mate selection among second generation Kibbutz adolescents and adults: incest avoidance and negative imprinting. *Arch. sex. Behav.* 1: 293–307.

Slob, A. K., Weigand, S. T., Goy, R. W. & Robinson, J. A. (1978). Heterosexual interactions in laboratory housed stumptailed macaques (*Macaca arctoides*): during the menstrual cycle and after ovariectomy. *Horm. Behav.* 10: 193–211.

Teas, J., Taylor, H. G. & Richie, T. L. (1978). Seasonal influences on aggression in *Macaca mulatta*. In *Recent advances in primatology* 1. *Behaviour*: 573–576. Chivers, D. J. & Herbert, J. (Eds). London: Academic Press.

Trimble, M. R. & Herbert, J. (1968). The effects of testosterone or oestradiol upon the sexual and associated behavior of the adult female monkey. *J. Endocr.* 42: 171–185.

Trivers, R. L. (1972). Parental investment and sexual selection. In *Sexual selection and the descent of man 1871–1971*: 136–179. Campbell, B. (Ed.). Chicago: Aldine.

Wallen, K. & Goy, R. W. (1977). Effects of estradiol benzoate, estrone, and propionate of testosterone or dihydrotestosterone on sexual and related behaviors of ovariectomised rhesus monkeys. *Horm. Behav.* 9: 228–248.

Wilson, A. P. & Boelkins, C. (1970). Evidence for seasonal variation in aggressive behaviour in *Macaca mulatta. Anim. Behav.* 18: 719–725.

Wilson, A. P. & Vessey, S. H. (1968). Behavior of free-ranging castrated rhesus monkeys. *Folia primatol.* 9: 1–14.

Zuckerman, S. (1932). *The social life of monkeys and apes.* London: Kegan Paul.

Zuckerman, S. & Parkes, A. S. (1939). Observations on secondary sexual characteristics in monkeys. *J. Endocr.* 1: 430–439.

Symp. zool. Soc. Lond. (1981) No. 46, 361–388

Primate Specialization in Brain and Intelligence

R. E. PASSINGHAM

Department of Experimental Psychology, University of Oxford, UK

SYNOPSIS

The primates are compared with other animals in their brains, skill and intelligence. The prosimians are not found to differ from other mammals in the size of their brains or the extent of the neocortex. In simian primates, on the other hand, the brain is larger than in any land mammal when the effects of body size are controlled for; and the neocortex forms a greater proportion of the whole brain than in other mammals matched for brain size. It is also shown that the pyramidal motor system is larger in primates, which have hands, than in carnivores and rodents with paws, or ungulates with hooves. It is argued that it has not yet been convincingly shown that any non-primate mammal, including sea mammals, can excel the monkeys and apes on discrimination learning tasks. The variety of ways in which monkeys and apes use tools is also taken to reflect their pre-eminence in intelligence.

INTRODUCTION

I want to examine the credentials of the primates. In calling them "Primates" Linnaeus (1758) implied that they formed the highest order of animals. But, setting aside a vested interest in promoting our relatives, do we have any valid reason for rating them as pre-eminent amongst animals in their brains and abilities?

The construction of scales is not in fashion, now that the notion of evolutionary scales has been thoroughly discredited (Hodos & Campbell, 1969; Martin, 1973). If the time of origin of modern groups is to be the guide, there is no reason to assign the monkey to a different level than the rat or cat. If degree of relationship to man is to be the criterion, the bushbaby must be raised above the dolphin. Few would be happy with either of these conclusions.

How, then, are we to proceed, if evolution is not to provide the yardstick? We must find some new measure which is independent of the branchings of the evolutionary tree. To put it in the current jargon we must look at "grades", not "clades" (Gould, 1976). We might, for example, rate different fish on the efficiency with which they swim, or compare fish with sea mammals in this respect. Or we might, as here, consider the elaboration of the nervous system and levels of intelligence. Just as we grade people in intelligence, so too may we compare the intellectual levels of different animals, given that we have to hand a valid and reliable measure of intelligence.

There is, of course, a literature on what distinguishes primates from other animals. If our interest were in classification we would need to concentrate on those few features which are unique to primates (Martin, 1968). But, if our primary interest is not in taxonomy, we may legitimately take into account characteristics which are shared with other mammals, but thought to be particularly well developed in primates. The consensus on what these features are is well represented by the list drawn up by Le Gros Clark (1971), and extended by Napier & Napier (1967). For our purposes the following two statements are of especial note. Le Gros Clark (1971: 42) claims that in primates there is a "progressive expansion and elaboration of the brain, affecting primarily the cerebral cortex and its dependencies". Secondly, he states that in primates there is "an enhancement of the free mobility of the digits, especially the thumb and big toe (which are used for grasping purposes)". These features are important, because a large brain might be thought to be indicative of intelligence, and a flexible hand of manual skill.

In discussing the brain, the hand and intelligence I hope to touch on three of the many interests which Professor Zuckerman has pursued in his scientific career. The comparison of the brains of different animals requires an understanding of allometric relations and of how shapes can be measured. The dexterity with which the hand is used is determined not only by the anatomy of the forelimb, but also by the neurological mechanisms that control the movements of the hand. Finally, to assess the intelligence of primates we must study their behaviour in the wild and in captivity. Professor Zuckerman has contributed to all three fields: the measurement of shape (Zuckerman et al., 1973); the physiology of the motor areas of the neocortex (Zuckerman & Fulton, 1941); and the scientific investigation of primate behaviour (Zuckerman, 1932, 1933).

BRAIN

Brain Size

Of the organs of the body the brain is the only one that is claimed to be unusually large in primates (Stahl, 1965). The claim is not that the primate brain is unparalleled in its absolute size; for the brain of the elephant is four times larger than the human brain and 14 times as large as the brain of the chimpanzee (Crile & Quiring, 1940). Due allowance has to be made for the allometric relations between brain and body size. Jerison (1973) has done this, and he finds that, when the effects of differences in body size have been removed, the vertebrates fall into two main groups, the fishes and reptiles with relatively small brains, and the birds and mammals with larger ones. Overall the brain is as well developed in birds as in mammals.

For the purposes of comparing different mammals Jerison (1973) calculates an "encephalization quotient", or EQ for short, which relates the brain size of each species to the size expected for an average mammal of the same body weight; the average mammal is defined as having a quotient of 1. Figure 1 gives the values of EQ that Jerison (1973) provides for various mammalian orders. If we compare their mean EQs we can distinguish three groups: the insecti-

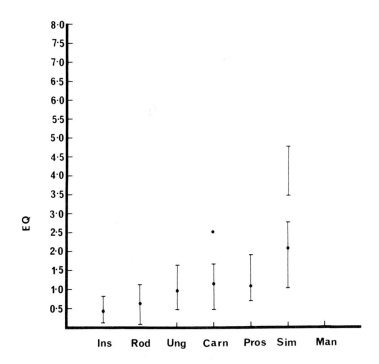

FIG. 1. The encephalization quotients (EQ) for insectivores (Ins), rodents (Rod), ungulates (Ung), carnivores (Carn), prosimians (Pros), simians (Sim), and man. The mean EQ and the range are given for each group. The range given for carnivores excludes the one high value shown as a dot. The range for simians is divided, the upper range representing the values for two cebus monkeys (*Cebus albifrons, Cebus apella*). Data from Jerison (1973).

vores and rodents with low EQs; the ungulates, carnivores and pro-simians with higher values; and the simian primates, the monkeys and apes, with the highest EQ of all. Jerison gives an EQ of 1.3 for an elephant, well within the range for carnivores and ungulates. As can be seen there is an overlap between the groups, but the overlap be-tween the simians and the carnivores and ungulates is not extensive. If we discount one suspiciously low value given by Jerison (1973) for a langur (*Presbytis cristatus*) there are only three out of 25 ungulates which have an EQ within the range for 48 simians, and only two out of 15 carnivores. Statistical comparisons between the groups show the monkeys and apes to have significantly higher EQs (Jerison, 1973).

This relatively clear-cut result is encouraging, because it has been achieved in spite of the errors inherent in the estimation of encephal-ization indices based on brain and body weight. A more reliable reference point than body size can be obtained by taking the size of the spinal cord or the medulla in the hindbrain (Passingham, 1978). In primates, for example, indices based on body weight rank some small monkeys such as the cebus monkey above apes (Fig. 1); the largest apes, the orang-utan (*Pongo pygmaeus*) and the gorilla (*Gorilla gorilla*), are also placed well below many of the monkeys (Stephan, 1972). But when the size of the medulla is used instead of body weight the results are less anomalous; now there is a clear separation between the great apes and the monkeys (Passingham, 1975). The division between the simians and other land mammals might be even more definite if such measures were used.

The validity of encephalization indices which use body weight is particularly dubious in the case of sea mammals. Stephan (1972) gives indices for seals and toothed whales, and some of the values exceed those for the monkeys and apes. He states that some of the seals and whales may have been juveniles, and if so the comparison is biased because their brains may have been nearer to adult size than their bodies. But there is another important objection. The encephal-ization index is supposed to measure the amount of brain tissue in excess of that needed for locomotion and for regulating bodily func-tions; but it cannot be assumed that the demands on the brain in this respect are the same in sea and land mammals. Even if we consider only land mammals, bodies of the same size may have very different composition in different species, varying greatly in the proportions of skin, muscle, bones and fat (Grand, 1977). The differences must be even greater if we compare land and sea mammals; so that we can-not assume that identical body weights for a land and sea mammal can be treated as equivalent for purposes of comparisons between

their brains. It would be better to relate the size of their brains to
their spinal cords.

Brain Proportions

Brains vary not only in size but also in shape. The shape of the brain
reflects, amongst other things, the relative proportions of the different parts. For example, a brain with an extensive neocortex will not
look like a brain in which the neocortex forms a smaller proportion
of the total brain tissue. This consideration leads us to ask whether
simian brains are simply larger versions of the brains of other mammals or whether they also differ in basic structural organization.

Let us take the example of the development of the neocortex
relative to the whole brain. Jerison suggests that in the larger brained
mammals the size of the brain can probably be taken as giving an
accurate estimate of the relative amount of neocortex. He supports
this by arguing that on a log–log plot of neocortical volume against
body weight a single regression line can reasonably be fitted to the
data for the mammals, other than the rodents and the one marsupial
for which there is information. These same data are plotted in Fig. 2.
As in Jerison's (1973) analysis a separate regression line has been
computed for the rodents, but two further regression lines have been
fitted rather than one, a line for the simians and another for the prosimians and carnivores. The lines for these three groups differ in
intercept; and a line of similar slope fitted through the value for the
single marsupial would clearly differ in intercept again.

This analysis indicates that a brain of the same size would have
proportionately more neocortex in a monkey or ape than in a prosimian or carnivore, and the same would be true in a comparison
between a prosimian or carnivore and a rodent. The difference in
the proportions of neocortex to total brain in simians and prosimians
is confirmed by the more extensive data collected on primates by
Stephan, Bauchot & Andy (1970). Irrespective of brain size, prosimians have proportionately less neocortex than monkeys and apes,
though much more than insectivores (Passingham, 1975). The
magnitude of the difference between simian primates and other
advanced mammals can be illustrated with the example of the cebus
monkey (*Cebus*). We can estimate from Fig. 2 that the neocortex
would form 39.6% of the brain in a carnivore with a brain of the
same size as the cebus monkey. This is appreciably less than the value
of 53% which is found for the monkey.

We can check this general finding by using the data collected by
Mangold-Wirz (1966). She measured the size of different parts of the

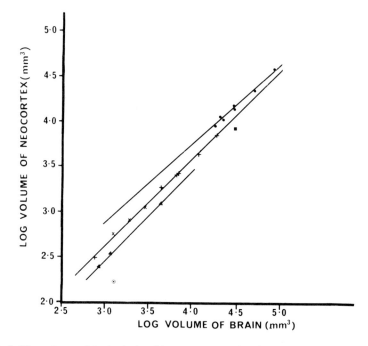

FIG. 2. The volume of the brain (mm³) and neocortex (mm³) in different mammals. ⊙ = 1 marsupial, ▲ = 3 rodents, ■ = 1 ungulate, + = 6 carnivores, X = 3 prosimians, ● = 7 simian primates not including man. Taking N as neocortical volume (mm³) and B as brain volume (mm³) the formulae for the least squares regression lines are: for simians, $\log N = 0.22 + 0.88 \, X \log B$; for prosimians and carnivores, $\log N = -0.35 + 0.99 \, X \log B$; for rodents, $\log N = -0.61 + 1.02 \, X \log B$. Data from Harman (1947).

brain in a large sample of mammals. The parts were separated by gross dissection, and she included the white matter in her measure of the weight of the cerebral hemispheres dorsal to the rhinal fissure. She gives the actual weights of the brains in the different mammals, but only presents her data for parts, such as the cerebral hemispheres, in the form of indices. However, the weight of the hemispheres can be reconstructed from two indices, the Total Index (TI) for the whole brain and the Neopallial Index (NI) for the cerebral hemispheres; the method is explained in the footnote to Fig. 3. In this figure the log of the weight of the cerebral hemispheres is plotted against log brain weight, and regression lines have been fitted to the points for the animals of each mammalian order. It can be seen that the cerebral hemispheres are larger in the monkeys and apes than in carnivores or ungulates with brains of a similar size, and that they are least well developed in rodents and insectivores. To take one example, the crab-eating monkey (*Macaca fascicularis*) had a brain of 63 g,

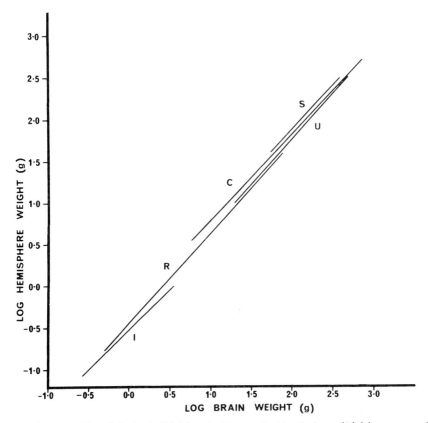

FIG. 3. The weight of the brain (B) (g) and of the cerebral hemispheres (H) (g) as measured dorsal to the rhinal fissure in different mammals. I = 4 insectivores, R = 18 rodents, U = 25 ungulates, C = 20 carnivores, S = 6 simian primates not including man. The formulae for the least squares regression lines are: for insectivores, $\log H = -0.52 + 0.98\,X \log B$; for rodents, $\log H = -0.44 + 1.07\,X \log B$; for ungulates, $\log H = -0.38 + 1.08\,X \log B$; for carnivores, $\log H = -0.25 + 1.02\,X \log B$; for simians, $\log H = -0.14 + 0.99\,X \log B$.

The data have been reconstructed from indices provided by Mangold-Wirz (1966). She calculated a Total Index (TI) as T/B and a Neopallial Index (NI) as N/B, where $T =$ total brain weight, $N =$ the weight of the neopallium, and $B =$ basal stem complex. For any animal the value for the basal stem complex was the size of the brainstem that would be predicted for an insectivore matched for body size (see Jerison, 1973). Since NI/TI $= (N/B)/(T/B)$, it simplifies to N/T; and since Mangold-Wirz (1966) gives the values for T the values for N can be derived.

roughly the same as the jackal (*Canis aureus*) with a brain of 64 g; and this even though the jackal was three and a half times as heavy. But the cerebral hemispheres formed 77.8% of the monkey's brain, and only 60.2% of the jackal's brain. In the elephant with a brain 79 times larger the cerebral hemispheres made up only 69.3% of the total brain.

It has become evident that in describing the brains of mammals we must take two factors into account. Firstly, we must specify the size of the brain when the effects of body size have been removed. But then we must state which group of mammals we are discussing, since brains of identical size vary in structural organization between the different orders. To express it another way, the various mammalian groups have different versions of the brain, quite apart from any variation in size. The simian primates have brains which are not only larger but also specialized in the degree of neocortical development. In both respects they are pre-eminent.

SKILL

It has been remarked that "life's aim is an act, not a thought". In primates the brain's executor is the hand. A scale of digital dexterity has been devised by Heffner & Masterton (1975), and on this measure primates excel over all other mammals, including the racoons. All primates can pick up an object in one hand, unlike rodents such as the squirrels which must use both paws. But there is a marked difference in the variety of movements that the various primates can perform with their hands. The prosimians and most of the New World monkeys use a standard grip irrespective of the size of the object they are handling (Bishop, 1964). All the fingers close round the object together, so that it is held in the palm of the hand. There are only a few New World monkeys that are capable of more discrete movements, such as the spider monkey (*Ateles*) and woolly monkeys (*Lagothrix*) which can pick up small objects in a scissor grip between two fingers. Old World monkeys and apes are much more proficient at moving their fingers independently; all of them can grasp small objects by opposing thumb and forefinger so as to grip the objects at their finger tips (Napier, 1961). They can also move a single finger while holding the others still. This can be shown by requiring them to push one finger through a small hole so as to push a grape off a ledge (Welles, 1975). In simian primates the brain has a remarkable instrument at its command.

We might reasonably expect to find in primates a specialized neural apparatus for controlling the hand, and our expectations are fulfilled. There are two aspects of the motor system that it is profitable to consider. The motor area of the neocortex has its most direct influence on the movements of the body by way of large fibres which it sends to lower regions through the pyramidal tract. We may consider primates and other animals both in the number of pyramidal fibres and in the sites at which they terminate in the brain stem and spinal cord.

Heffner & Masterton (1975) deny that the number of pyramidal fibres is clearly related to manual dexterity in mammals. But their analysis does not take into account the body size of the animals, and thus they fail to correct for the tendency for large animals to require a large pyramidal system irrespective of digital dexterity. When body size is controlled for, a different picture emerges, as can be seen from Fig. 4, in which the number of pyramidal fibres is plotted against body weight. Regression lines have been fitted to three groups of mammals, and the intercepts are clearly very different. The lowest line represents the ungulates, the middle line the rodents and carnivores, and the upper line the monkeys, apes, and man. It will be immediately apparent that these results fit a simple picture; the ungulates have hooves, rodents and carnivores have paws, and primates have hands. The number of fibres is directly related to the mobility of the digits, which is very restricted in a hoof, less so in a paw, and best developed in a hand. When matched for body size the chimpanzee has 3.6 times as many pyramidal fibres as would be expected of an ungulate of the same size, and 1.8 times as many as a carnivore.

This elegant picture has been painted by omitting one disfiguring feature. We have a count of the pyramidal fibres in a sea mammal, the seal (*Callorhinus ursinus*). This has 1.6 times as many fibres as would be predicted for a carnivore of the same weight, and nearly as

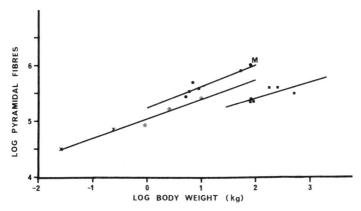

FIG. 4. The number of fibres in the pyramidal tract and body weight (kg) in different mammals. ■ = 6 ungulates, X = 2 rodents, ☉ = 3 carnivores, ● = 6 simian primates, M = man. The formulae for the least squares regression lines are: for ungulates, log fibres = 4.85 + 0.29 X log body weight; for rodents and carnivores, log fibres = 5.06 + 0.35 X log body weight; for simian primates excluding man, log fibres = 5.25 + 0.39 log body weight. Data from Heffner & Masterton (1975).

many as would be expected for a monkey or ape. Yet the seal has minimal digital dexterity (Heffner & Masterton, 1975). There are two ways out of the dilemma: either swimming makes greater demands on the motor system than walking or climbing; or, as argued above, body size may provide an invalid basis for comparison of the brains of sea and land mammals. The anomalies over the size of the pyramidal system and of the brain are both solved at the same time if the second of these two proposals is correct. When the seal is matched for body weight with a land mammal, the size of its pyramidal system is a poor guide to its digital dexterity; perhaps then the size of the brain, when assessed in the same way, gives an inflated estimate of the animal's intelligence.

The site of termination of the pyramidal fibres is known in a variety of mammals (Petras, 1969; Philips, 1971). The movements of the limbs are directly controlled by the motor neurones of the spinal cord. In mammals with paws, such as cats, the pyramidal fibres influence the motor neurones only indirectly via other intervening nerve cells. In the racoon, which has a hand, a few fibres appear to terminate on the motor neurones themselves. Direct pyramidal connections are found in the spider monkey and woolly monkey which have some independence of finger movements, but not in the squirrel monkey which does not (Harting & Noback, 1970). In Old World monkeys and apes, which have opposable thumbs and agile fingers, there are many direct connections from the pyramidal tract on to the motor neurones, more so in the chimpanzee than in the rhesus monkey (Kuypers, 1964). Heffner & Masterton (1975) rate different mammals according to the level of the spinal cord at which the pyramidal fibres terminate, and find a close relation between digital dexterity and the degree of penetration of these fibres down the spinal cord.

Those fibres in the pyramidal tract which terminate on the motor neurones come from the parts of the motor area of the neocortex that control the movements of the hand and foot (Kuypers & Brinkman, 1970). If the pyramidal tract is cut a rhesus monkey (*Macaca mulatta*) recovers in general remarkably well, and is able to walk and climb with little trouble. The most striking symptom is an inability to pick up food from a slot by inserting the forefinger and thumb into the groove in a precision grip (Lawrence & Kuypers, 1968). If the motor area is removed in a rhesus monkey the animal is capable only of a stereotyped grasping movement with the whole hand; in other words it is reduced to the level of manual competence characteristic of the squirrel monkey (Passingham, Perry & Wilkinson, 1978). That this impairment reflects an inability to move one finger independently of the others can be demonstrated by requiring the monkeys to insert one finger in a small hole to push a grape off a ledge. If the motor and somatosensory areas of the neocortex are removed in one hemisphere rhesus monkeys can extend the forefinger of the hand controlled by the intact hemisphere, but they are totally unable to do so when required to push the grape with one finger of the affected hand (Passingham & Perry, in preparation).

The effects of damage to the pyramidal system suggest the following speculation. Just as the sensory areas of the neocortex are primarily concerned with the analysis of fine detail, so the motor area appears to be mainly concerned with the production of detailed movements. Monkeys with damage to sensory areas are poor at making fine discriminations; whereas monkeys with damage to the motor area are unable to command discrete movements, moving their fingers indiscriminately. The large neocortex of simian primates provides them both with very detailed information about the world, and also with the capacity to act upon it in a great variety of ways. A concern with detail is a prerequisite for an intelligent understanding and control of the world.

INTELLIGENCE

If we restrict the comparison to land mammals four facts about primate anatomy can be briefly stated. (1) Simian primates have a brain which is larger than that of other mammals, when due allowance has been made for body size. (2) Their neocortex is better developed than in mammals with brains of the same absolute size. (3) The pyramidal system, which originates from the neocortex, is larger in simian primates than in other mammals; and the fibres influence the motor neurones more directly in those monkeys and apes that have the ability to move their fingers independently. (4) No special claims can be made for the prosimians. The prosimians have sometimes been called "the lower primates", and in terms of the level of their brain organization this assessment appears to be justified.

What, then, is the functional significance of the specialization of the monkeys and apes in brain and hand? The most interesting characteristic of their brains is the selection for a high proportion of neocortex. In rhesus monkeys damage to the association areas of the neocortex disrupts the animal's ability to learn various tasks (Butter, 1968). We might therefore expect to find that in animals there was a relation between the amount of neocortex and their efficiency at solving problems.

Learning Set

One measure that has been widely used is the animal's ability to form a "learning set". It is given a series of visual discrimination problems; on each of these a pair of objects is presented and the animal is allowed a few trials to learn under which of the two objects food is to be found. The pairs of objects vary from problem to problem, but each problem obeys the same rule: the food remains under the same object throughout the problem. If the animal can detect the rule it can solve later problems much more quickly than the initial ones, so benefiting from its previous experience. The animal is said to have formed a "learning set" if its performance improves over the series of problems. That this measure may have some validity as a measure of intelligence is indicated by results obtained on children given a similar series of problems. The performance of the children was found to be related to their intelligence as measured on an independent test of IQ (Harter, 1965). If we compare the scores of different primates we find a correlation between the degree of improvement across the series and the amount of neocortex as related to the size of the medulla (Passingham, 1975).

But do simian primates perform better on this measure than other mammals? The relevant literature has been reviewed by Hodos (1970) and Warren (1973, 1974). Hodos argues that there is overlap between the performance of monkeys and other mammals; and Warren throws doubt on the validity of the comparisons by stressing the many sources of bias and error involved in making them. But one fact is clear: it is not reasonable to compare the results of studies carried out on different species unless the training conditions used were essentially the same. Yet it turns out that different workers have used a variety of methods in training animals on a series of visual discrimination problems (Kintz, Foster, Hart & O'Malley, 1969). Some present each problem to the animal for six trials only (e.g. Miles, 1957); some give more trials (e.g. Zeigler, 1961), or start with more trials and then gradually reduce them until only six trials are given on each problem (e.g. Kamil & Hunter, 1970); and some start by training the animals to a criterion level on each problem (e.g. Doty, Jones & Doty, 1967). Yet we know that the efficiency with which monkeys improve on such problems is determined more by the number of trials than the number of problems given (Miles, 1965); and monkeys also learn more quickly if they are trained to a criterion rather than given a fixed number of trials on each problem (DeVine, 1970). But Hodos (1970) presents the data by number of problems given rather than number of trials. This procedure is invalid because it compares the performance of animals which differ greatly in the length of experience they have had on each of the problems.

There are only a few studies which have used a standard procedure, presenting the animals with six trials on each problem. The results are plotted in Fig. 5 in terms of performance on trial two (T2) as a function of the number of problems given. The three species of monkey out-perform the four species of non-primate mammal; on the last 400 problems there is no overlap in performance between the marmosets (*Callithrix jacchus*) and the cats (Meyers, McQuiston & Miles, 1962). Of the monkeys the rhesus monkey is clearly the best, achieving a level of over 85% correct on T2 after only 392 problems, that is 2352 trials. The three rodents perform less well than the cat, the only carnivore for which there are comparable data.

All the other studies vary greatly in the procedures used, and comparisons can only be drawn with confidence between different species tested in the same study, and thus under the same conditions. If we consider the primates first, we know the squirrel monkey to be inferior to the spider monkey (*Ateles geoffroyi*) and cebus monkey (*Cebus albifrons*) (Shell & Riopelle, 1958). Of the Old World monkeys the rhesus monkey and mangabey (*Cercocebus torquatus*) perform equally well (Behar, 1962), although surprisingly in one study langurs (*Presbytis entellus*) learnt more quickly than rhesus monkeys (Manocha, 1967). The data on great apes are not satisfactory. The apes tested have usually been young, and much less mature than the

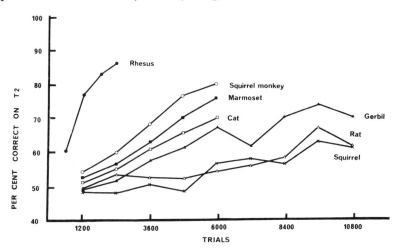

FIG. 5. Percentage correct on trial 2 (T2) as a function of the number of problems and trials in different mammals. Data on the rhesus monkey and marmoset from Miles & Meyer (1956), the squirrel monkey from Miles (1957), the cat from Meyers *et al.* (1962), the gerbil from Blass & Rollin (1969) and the rat and squirrel from Rollin in Warren (1965).

monkeys with which they are compared; yet we know that infant monkeys are very much poorer at learning on discrimination tasks than are mature animals (Harlow, Harlow, Rueping & Mason, 1960). In one study two infant gorillas were tested, and it was tentatively suggested that their performance was better than that of immature rhesus monkeys trained in another laboratory (Fischer, 1962). In another study great apes were directly compared with macaques, and the monkeys did as well as the apes (Rumbaugh & McCormack, 1967). But all but two of the great apes were immature, and the animals were tested with an apparatus which the authors admitted to be inefficient for training the young apes. It is also difficult to motivate great apes to take an interest in such tasks. That chimpanzees can do better than rhesus monkeys is indicated by a study carried out by Hayes, Thompson & Hayes (1953). They directly compare the data for one of their chimpanzees with data on rhesus monkeys tested by Braun (in Harlow, 1951). The final level reached by the monkeys was attained by the chimpanzee in roughly one sixth of the trials.

The data given in Fig. 5 for non-primate mammals can be amplified in two respects. The performance of the rat and squirrel is if anything better than that of the tree shrew (*Tupaia glis*) (Leonard, Schneider & Gross, 1966). Though cats are the only carnivores yet tested with six trials per problem, one study compared cats with other carnivores (Doty *et al.*, 1967). Minks (*Mustela vision*) and ferrets (*Mustela furo*) were found to learn more quickly than skunks (*Mephitis mephitis*) or cats. Unfortunately the results cannot be compared with those for the primates shown in Fig. 5, because for the first 200 problems the animals were trained to a criterion of 20 out of 25 correct responses and only then given 400 problems for six trials each. Data are not provided on the number of trials the animals were given, but they must have greatly exceeded 7400, that is very many more than 5000 on the first 200 problems and 2400 on the last 400. Yet Hodos (1970) compares their performance with that of monkeys trained on the same number of problems, although the number of trials represented by those problems was vastly greater for the carnivores. This procedure is quite invalid. One other carnivore has been tested, the racoon (*Procyon*) (Johnson & Michels, 1958). The racoons reached a higher level than did the marmosets trained by Miles & Meyer (1956) when they had completed the same number of trials. But the racoons were trained to a criterion on each problem, and this is known to be a much more efficient training procedure than the conditions of six trials per problem used with the marmosets. No useful comparison can therefore be drawn between the data for racoons and marmosets.

There are a few studies on birds. Of these the one on chickens by Plotnik & Tallamico (1966) has been suggested to be unreliable (Kamil & Hunter, 1970). The results of a study by Zeigler (1961) on pigeons cannot be assessed in relation to the data in Fig. 5 because six problems were given to a criterion, and a further 120 problems were given for 24 trials rather than six trials; furthermore scores on T2 alone are not given. Again, therefore, Hodos (1970) distorts the picture when he treats the pigeon as the equal of the spider monkey, even though the pigeon had experienced four times as many trials on each problem.

The other two birds for which we have information are the mynah bird and the blue jay. Kamil & Hunter (1970) trained mynah birds (*Gracula religiosa*), giving eight problems to a criterion, then 20 problems for 25 trials, 42 problems for 15 trials, 230 problems for 10 trials and 700 problems for six trials, making 1000 problems in all. On the last 100 problems the birds performed at 72% correct on T2, but this was after 7630 trials, not counting the trials on the first eight problems. Hunter & Kamil (1971) used a similar procedure for the blue jay (*Cyanocitta cristata*), giving 25 trials for the first 25 problems, 15 trials for the next 35 problems, 10 trials for 90 problems, and finally six trials for each of 550 problems, making up 700 problems in all. The birds performed at 73% correct on T2 over the last 100 problems, experiencing 5350 trials in all. That birds should improve on discrimination problems should not come as a surprise when we remember that the brains of birds and mammals are equally well developed in size (Jerison, 1973). Sadly we cannot directly compare the performance of the birds and mammals, because the method used for training the birds is unlike any so far used for mammals. Even though the method is probably very efficient, we can say that the mynah bird and blue jay are no match for a monkey such as the rhesus monkey, which can attain a level of well over 85% correct on T2 after only 2352 trials given in problems of six trials each (Miles & Meyer, 1956).

We set out to discover whether the specialized brains of monkeys and apes equipped them with a superior capacity to solve problems. We used the speed of acquisition of a learning set as a measure of that capacity, on the grounds that it is still the only measure on which there are enough comparative data. And we found the data to be more orderly than had been suggested by Hodos (1970). No non-primate mammal has yet been convincingly shown to perform more effectively than monkeys or apes. It would be absurd to deny that future studies might show that there are carnivores which can rival the marmoset or squirrel monkey; but at the moment the relevant comparative studies have not been done.

We have reached this conclusion in spite of all the errors and distortions that are involved in any attempt to measure animal intelligence by studying learning sets (Warren, 1973). But there is one possible source of bias that should concern us. As Warren (1973) points out, the ability to solve visual discriminations may be determined not only by intellectual capacity but also by the development of the visual system. Monkeys and apes have better visual acuity and colour perception than carnivores, and nocturnal rodents have very poor visual acuity and lack any form of colour perception (Walls, 1967; DeValois & Jacobs, 1971). The conclusion that we have formulated must remain therefore tentative until proper comparative data are available for the performance of animals given a series of discriminations between sounds or smells.

But there is, as yet, no convincing evidence that monkeys and apes owe their superiority to their eyesight alone. Rhesus monkeys are better than cats at learning single auditory discriminations (Wegener, 1976). The poor performance of rats on visual discriminations cannot be attributed solely to their poor vision, since squirrels and tree shrews do no better, even though they are diurnal and have very much better vision (Walls, 1967; DeValois & Jacobs, 1971) (Fig. 5). It is true that rats can, under some conditions, show a marked improvement in performance over a series of smell discriminations. Slotnik & Katz (1974) report two experiments: in one the rats were trained for 100 trials on each of 16 problems; in the other the rats were tested to a criterion on each of 16 problems, and were also punished for making errors. There are several points to be made. First, under these very favourable conditions rats might well improve considerably over a series of visual problems. Secondly, in neither experiment are proper data given on per cent correct on T2. And finally we should not suppose that, let us say, marmosets would do worse under the same conditions. It would be absurd to demote the monkeys and apes on evidence as tenuous as this.

Tools

We have taken as a rough measure of general intelligence the readiness with which an animal grasps a rule that applies to a series of problems. We turn now to practical abilities, the use to which intelligence is put in changing the world. Primates start with a great advantage, as the hand is a more versatile instrument than the paw or mouth. Indeed we might speculate that the hand gives greater scope for intelligent actions, and thus promotes selection for the intelligence with which to use it.

In the previous section we examined the ability to make a variety of sensory discriminations; in this section we enquire into the ability to produce a variety of different acts. We shall restrict our interest to the versatile use of objects as tools. Tool use is reported to occur in the wild in 16 out of the 31 genera of the monkeys and apes (Beck, 1975). But no prosimian uses tools, and there is only one non-primate mammal that does so. This is the sea otter (*Enhydra lutrus*) which cracks open abelones against stones which it holds on its chest as an anvil (van Lawick-Goodall, 1970). There are reports of the use of tools in the wild by some species of birds (van Lawick-Goodall, 1970; Alcock, 1972); however, the number of cases of tool use in simian primates greatly outnumbers the cases for any other animal group.

But it is not, perhaps, the number of occasions on which tools are used that should most impress us, but the variety of purposes to which they are put. There are two main uses that birds make of tools. The first is for breaking eggs: the Egyptian vulture (*Neophron percnopterus*) and the black-breasted buzzard (*Hamirostra melanosterna*) drop stones on large eggs to crack them (van Lawick-Goodall, 1970). The second use is in probing for insects: the woodpecker finch (*Cactospiza pallida*) uses cactus spines to dislodge insects, and this is also done occasionally by the mangrove finch (*Cactospiza heliobates*), and very infrequently by the ground finch (*Geospiza*) and warbler finch (*Certhidea olivacea*) (Alcock, 1972). Brown headed nuthatches (*Sitta pusilla*) also use bark to dislodge pieces of bark from trees so as to uncover insects (van Lawick-Goodall, 1970). The bowerbird (*Ptilonorhynchus violaceus*) puts pieces of bark to a different use, holding a piece in its beak when painting the bower and thus preventing the charcoal mixture from dripping out (van Lawick-Goodall, 1970).

Monkeys and apes use tools in much more varied ways. This can be appreciated from Table I, which sets out the purposes for which tools are used and the means by which these ends are achieved. It lists 14 different actions performed by monkeys and apes, of which 12 have been seen in the wild and the remaining two only in captivity. The most impressive feature is the variety of the actions which are carried out with tools by chimpanzees, baboons, macaques and cebus monkeys. The chimpanzee has a repertoire of 11 actions in the wild, quite outclassing a bird with just one relatively stereotyped performance. Furthermore monkeys and apes are capable of using the same object for several different purposes. Stones can be used as missiles, for cracking hard fruit or for wiping blood; and sticks serve in clubbing, digging and probing into the nests of insects. We might

TABLE I

Tool use in primates: actions performed by non-human primates in using tools.

Action	Tool	Aim	Chimpanzee	Baboon	Macaque	Cebus	Other apes	Other monkeys
Drop	Branches, etc.	Hit or scare intruder	W		W	W	2W	6W
Throw	Stones, etc.	Hit or scare intruder	W	W	C	W	2C	2W, 2C
Club	Sticks	Hit or scare intruder	W			C	1C	
Pound	Stones	Open fruit, nuts	W	W	W	W		
Dig	Sticks	Open nests, dig up roots	WC	W				2W
Lever	Sticks	Open container (of food)	W	C			1C	
Insert	Twigs, grasses	Probe for insects, honey	W				1C	
Sponge up	Leaves, rope	Take up water to drink						
Wipe	Stones, leaves	Clean self	W		W		2C	
Wash	In water	Clean food	W	W	W		1W	
Reach out	Sticks	Touch object to investigate	W					
Rake in	Sticks	Reach and draw in	W	C	C	C	2C	1C
Prop up	Branches	Reach by climb "ladder"	C				1C	
Stack	Boxes	Reach by mount boxes	C					

W = observed in the wild; C = observed only in captivity. Under "other apes and monkeys" the numbers give the number of species in which the action has been observed. Based on Beck (1975) and Warren (1976).

summarize by saying that monkeys and apes can put objects to intelligent use.

The variety of ways in which monkeys and apes use objects can be directly measured by supplying captive animals with an object such as a piece of rope with a knot in it, and then observing what the animals do with it. This has been done by Parker (1974) who devised an index of diversity of behaviour by recording the parts of the body used and the actions performed. A great variety of actions could be carried out with the rope: pulling, mouthing, waving, twisting, rubbing and so on. The mean number of body parts used by the great apes was 33, nearly twice as many as the gibbon and over seven times as many as the monkeys. The great apes also performed a very great number of different actions; the mean was 38, about twice as many as the gibbons, and nearly five times as many as the monkeys. A measure of the combinations of body part and action was also calculated, and on this too the great apes far excelled the other primates. Furthermore their 30 most frequent combinations accounted for only 64—73% of their combined actions.

A test of this sort has never been given to mammals other than primates. Four lemurs (*Lemur catta, Lemur macaco*) were included in the study, and they performed as well as some of the monkeys. But the test would also be appropriate for mammals without hands. Glickman & Sroges (1966) did test the reactions of a variety of zoo mammals to novel objects, but they did not take note of what the animals did with them. This is unfortunate because, as they comment: "it seems probable that one of the main contributions of a complex brain will be to increase the variety of things that an animal does with an object, or to extract novelty from situations that would not be apparent to a less complicated organism." (Glickman & Sroges, 1966: 182).

Sea Mammals

In our discussion so far of the abilities of different mammals we have deliberately omitted all mention of the dolphin and of other sea mammals. This neglect might seem unwarranted, since dolphins are widely regarded as very clever animals. Indeed, as Wilson (1975) argues, claims are sometimes made on their behalf which are so lacking in sobriety as to be of very dubious scientific worth. We are obliged to attempt to assess the intelligence of dolphins, and to compare them in this respect with apes and monkeys.

How have dolphins come to acquire a reputation that would be envied by a chimpanzee? They are easily tamed, but so are dogs;

they are ready performers in shows of public entertainment, but so are many circus animals; they help their fellows, but so do wild dogs (*Lycaon pictus*) or baboons (Wilson, 1975). We are hardly to be convinced by indirect evidence such as this.

We need proper standardized tests which the dolphin can be given, so that the scores can be compared with those obtained by land mammals. No doubt appropriate tests are difficult to devise, because land and sea mammals are built in such different ways. Yet we find, to our amazement, that not one single test has been given to dolphins which could afford a proper comparison with primates and other mammals. The situation is the same as with the elephant. Rensch (1957) reports that the elephant can learn and remember 20 visual discriminations; but the conditions of testing were quite unlike those used with primates. Under other conditions rhesus monkeys have been taught 72 visual discriminations, which they were able to remember for up to 120 days (Strong, 1959); and under yet other conditions chimpanzees can learn by rote the correct response to each of 24 rows of four symbols (Farrer, 1967). The scientific study of the comparative intelligence of animals cannot be pursued in such a disorganized way.

The approach to the dolphin has been as idiosyncratic as with the elephant. Kellogg & Rice (1966) taught a series of visual discriminations under conditions under which the dolphin could not make errors, and then gave a further series in which one or other of the old stimuli was retained for each new pair. The authors admit that there is no comparable work on other animals! Herman, Beach, Pepper & Stallins (1969) gave a dolphin a series of visual discrimination problems for 12 trials each at first, and later six or eight trials for each problem. The dolphin only scored 57.2% on T2 across all the problems, but the authors provide a reasonable explanation of why it did not do better.

In a later experiment Herman & Arbett (1973) had greater success in teaching a dolphin, but in this case the dolphin was given a series of auditory discriminations. It is tempting to suppose that the different outcomes reflect the greater facility of the dolphin in handling auditory rather than visual cues. But it is not obvious that this interpretation is correct, since the two studies differed in many other respects. In the auditory study the dolphin, which had received previous training on auditory discriminations, was first given many problems and trials with the negative stimulus deliberately reduced in salience to give the animal an additional cue. It then learnt many further problems with 50, 20 or 10 trials on each problem; and on each of these the dolphin was guided as to the correct stimulus on T1 by means of a reduction in the salience of the negative cue. On all these trials the dolphin was also allowed to correct its errors when it made them. A final set of 111 problems was given for 12 trials each, and these were given without allowing the animal to correct its errors or assisting it on T1 by manipulating the salience of the negative cue. On T2 the dolphin reached a level of 86.5% correct. It is not possible to reconstruct from the report the total number of trials it took for the dolphin to achieve this, but it must have been many thousands. It would be rash to predict that a monkey could not do as well or better, if it were trained so extensively, and under such very favourable conditions devised especially to promote rapid learning. To belabour the point: it is not whether an animal can score 90% correct on T2 that is the issue. Under careful tutelage many a mammal might reach this level. What we take to be the mark of intelligence is the speed with which the rule is grasped without help from outsiders.

But it is not on tasks such as this that the dolphin has built its reputation, but rather on the complexity of its communication. So impressed have some authors been that they have even pondered the question of whether the dolphin uses a language, akin to human whistled languages (Busnel, 1966). But what are the facts? The bottlenose dolphin has a repertoire of 16 or so whistles, roughly the same number of calls as the chimpanzee and gorilla. It has some ability to copy sounds, matching the number of syllables spoken by producing the same number of bursts (Lilly, Miller & Truby, 1968). But many song birds learn parts of their song by copying the songs of adults (Thorpe, 1961; Nottebohm, 1972), and parrots and mynah birds can copy human speech relatively exactly (Sebeok, 1965; Nottebohm, 1976). There is even a suggestion that macaques (*Macaca fuscata*) may have some ability to imitate calls (Green, 1975).

It is difficult to lay down satisfactory criteria which we can use in deciding if an animal possesses language. But that the animal must be able to name is not controversial. Yet there is only one experiment which purports to demonstrate the ability of dolphins to name by sound. Bastian (1967) taught a male and a female dolphin to press the paddle on the right when shown a continuous light, and the paddle on the left when shown an intermittent one. Conditions were then altered so that the two animals were in separate tanks. The female could still see the cues, but the male had to press one of two paddles before the female, even though it had no guidance from visual cues. It was found that the male could still make the correct choice but only if it was able to hear the female. It is easy to assume, as did Bastian (1967), that the crucial noises that the male must hear are the whistles and pulse trains emitted by the female. But it was quite as likely that the noises that the male relied on were those made by the female when swimming. She approached the right and left paddles by quite different routes, and the male could have made his own choice on the basis of which of the two paddles it knew the female to be approaching. This explanation might be incorrect, but until it has been shown to be so we are entitled to remain unconvinced that dolphins use sounds to name.

The position reached is much like that deplored by Thorndike (1911) when he commented on the standard of evidence that was available at that time on the abilities of animals. As he commented "most of the books do not give us a psychology, but rather a *eulogy*" (Thorndike, 1911: 22). What would he have said today of some of the writings on dolphins (Lilly, 1961, 1967; Stenhuit, 1969)? He complained that the reputation of animals was founded on the equivalent of travellers' tales. "Dogs get lost hundreds of times and no one ever notices it or sends an account of it to a scientific magazine. But let one find its way from Brooklyn to Yonkers and the fact becomes a circulating anecdote" (Thorndike, 1911: 24). The rank which we have assigned to monkeys and apes is based on their performance as judged from comparative studies using experimental tests. They cannot be demoted by force of enthusiastic anecdote or ill-designed experiment.

Yet, we must admit that to carry out good research on dolphins and other sea mammals is a difficult task indeed. If they are studied at sea there are problems enough even in identifying the animals and keeping track of their movements (Wursig, 1979). It took van Lawick-Goodall (1968) and Fossey (Fossey & Harcourt, 1977) very many years to accustom chimpanzees and gorillas to their presence, and thus make significant discoveries about their social and intellectual life. It is now 48 years after the publication of Professor Zuckerman's (1932) book on *The social life of monkeys and apes*; and by now our knowledge of their brains, skill and abilities is extensive. But we still have to wait for a pioneering account of this quality on dolphins. Until we possess such a document I must admit that I would back a chimpanzee against a dolphin any day. The credentials of monkeys and apes are sound. They deserve their Linnaean name as primates.

ACKNOWLEDGEMENTS

I am grateful to Professor L. Weiskrantz for helpful comments on the manuscript. The work was supported by MRC grant 971/1/397/B.

REFERENCES

Alcock, J. (1972). The evolution of the use of tools by feeding animals. *Evolution* 26: 464–473.

Bastian, J. (1967). The transmission of arbitrary environmental information between bottlenose dolphins. In *Les systemes sonars animaux, biologic et bionique.* 803–873. Busnel, R. G. (Ed.). Jouy-en-Josas: Laboratoire de Physiologie Acoustique.

Beck, B. B. (1975). Primate tool behavior. In *Socioecology and psychology of primates*: 413–447. Tuttle, R. H. (Ed.). Hague: Mouton.

Behar, I. (1962). Evaluation of cues in learning set formation in mangabeys. *Psych. Rep.* 11: 479–485.

Bishop, A. (1964). Use of hand in lower primates. In *Evolutionary and genetic biology of primates*: 133–225. Buettner-Janusch, J. (Ed.). New York: Academic Press.

Blass, E. M. & Rollin, A. R. (1969). Formation of object discrimination learning sets by mongolian gerbils (*Meriones unguiculatus*). *J. comp. Physiol. Psychol.* 69: 519–521.

Busnel, R. G. (1966). Information in the human whistled language and sea mammal whistling. In *Whales, dolphins and porpoises*: 544–568. Norris, K. S. (Ed.). Berkeley: University of California Press.

Butter, C. (1968). *Neuropsychology.* California: Wadsworth.

Crile, G. & Quiring, D. P. (1940). A record of the body weight and certain organ and gland weights of 3690 animals. *Ohio J. Sci.* 40: 219–259.

DeValois, R. L. & Jacobs, G. H. (1971). Vision. In *Behavior of nonhuman primates*: 107–157. Schrier, A. M. & Stollnitz, F. (Eds). New York: Academic Press.

DeVine, J. V. (1970). Stimulus attributes and training procedures in learning-set formation of rhesus and cebus monkeys. *J. comp. Physiol. Psychol.* 73: 62–67.

Doty, B. A., Jones, C. & Doty, R. A. (1967). Learning set formation by minks, ferrets, skunks and cats. *Science, N.Y.* 155: 1579–1580.

Farrer, D. N. (1967). Picture memory in the chimpanzee. *Perc. Motor Skills* 25: 305–315.

Fischer, G. F. (1962). The formation of learning sets in young gorillas. *J. comp. Physiol. Psychol.* 55: 924–925.

Fossey, D. & Harcourt, A. H. (1977). Feeding ecology of free-ranging mountain gorilla (*Gorilla gorilla beringei*). In *Primate ecology*: 415–447. Clutton-Brock, T. (Ed.). London: Academic Press.

Glickman, S. E. & Sroges, R. W. (1966). Curiosity in zoo animals. *Behaviour* 26: 151–188.

Gould, S. J. (1976). Grades and clades revisited. In *Evolution, brain and behavior*: 115–122. Masterton, R. B., Hodos, W. & Jerison, H. (Eds). New Jersey: Erlbaum.

Grand, T. I. (1977). Body weight: its relation to tissue composition, segment distribution and motor function. 1. Interspecies comparisons. *Am. J. Phys. Anthrop.* 47: 211–240.

Green, S. (1975). Dialects in Japanese monkeys: vocal learning and cultural transmission of locale-specific vocal behavior. *Z. Tierpsychol.* 38: 304–314.

Harlow, H. F. (1951). Primate learning. In *Comparative psychology* 3rd edn: 183–238. Stone, C. P. (Ed.). New York: Prentice-Hall.

Harlow, H. F., Harlow, M. K., Rueping, R. R. & Mason, W. A. (1960). Performance of infant rhesus monkeys on discrimination learning, delayed response and discrimination learning set. *J. comp. Physiol. Psychol.* 53: 113–121.

Harman, P. J. (1947). Quantitative analysis of the brain-isocortex relationship in Mammalia. *Anat. Rec.* 97: 342.

Harter, S. (1965). Discrimination learning set in children as a function of intelligence and mental age. *J. exp. Child Psychol.* 2: 31–43.

Harting, J. K. & Noback, C. R. (1970). Corticospinal projections from the pre- and postcentral gyri in the squirrel monkey (*Saimiri sciureus*). *Brain Res.* 24: 322–328.

Hayes, K. J., Thompson, R. & Hayes, C. (1953). Discrimination learning set in chimpanzees. *J. comp. Physiol. Psychol.* 46: 99–104.

Heffner, R. & Masterton, B. (1975). Variation in form of the pyramidal tract and its relationship to digital dexterity. *Brain Behav. Evol.* 12: 161–200.

Herman, L. M. & Arbett, W. R. (1973). Stimulus control and auditory discrimination learning sets in the bottlenose dolphin. *J. exp. Analysis Behav.* 19: 379–394.

Herman, L. M., Beach, F. A., Pepper, R. C. & Stallins, R. B. (1969). Learning set formation in the bottlenose dolphin. *Psychon. Sci.* 13: 98–99.

Hodos, W. (1970). Evolutionary interpretation of neural and behavioral studies of living vertebrates. In *Neurosciences* 2: 26–38. Schmidt, F. O. (Ed.). New York: Rockefeller University Press.

Hodos, W. & Campbell, C. R. G. (1969). Scala naturae: why there is no theory in comparative psychology. *Psychol. Rev.* 76: 337–350.

Hunter, M. W. & Kamil, A. C. (1971). Object discrimination learning set and hypothesis behavior in the northern blue jay (*Cyanocitta cristata*). *Psychon. Sci.* 22: 271–273.

Jerison, H. J. (1973). *Evolution of the brain and intelligence.* New York: Academic Press.

Johnson, J. I. & Michels, K. M. (1958). Learning sets and object–size effects in visual discrimination learning by racoons. *J. comp. Physiol. Psychol.* 51: 376–378.

Kamil, A. C. & Hunter, M. W. (1970). Performance on object discrimination learning set by the greater hill mynah (*Gracula religiosa*). *J. comp. Physiol. Psychol.* 73: 68–73.

Kellogg, W. N. & Rice, C. E. (1966). Visual discrimination and problem solving in a bottlenose dolphin. In *Whales, dolphins and porpoises*: 731–753. Norris, K. S. (Ed.). Berkeley: California University Press.

Kintz, B. L., Foster, M. S., Hart, J. O. & O'Malley, J. J. (1969). A comparison of learning sets in humans, primates and sub-primates. *J. genet. Psychol.* 80: 189–204.

Kuypers, H. G. J. M. (1964). The descending pathways to the spinal cord: their origin and function. *Progr. Brain Res.* 11: 178–202.

Kuypers, H. G. J. M. & Brinkman, J. (1970). Precentral projections to different parts of the spinal intermediate zone in the squirrel monkey. *Brain Res.* 24: 29–48.

Lawrence, D. G. & Kuypers, H. G. J. M. (1968). The functional organization of the motor system in the monkey. *Brain* 91: 1–14.

Le Gros Clark, W. E. (1971). *The antecedents of man.* 3rd edn. Edinburgh: University Press.

Leonard, C., Schneider, G. E. & Gross, C. G. (1966). Performance on learning set and delayed response tasks by tree shrews (*Tupaia glis*). *J. comp. Physiol. Psychol.* 62: 501–504.

Lilly, J. C. (1961). *Man and dolphin.* New York: Doubleday.

Lilly, J. C. (1967). *The mind of the dolphin: a nonhuman intelligence.* New York: Doubleday.

Lilly, J. C., Miller, A. M. & Truby, H. M. (1968). Reprogramming of the sonic output of the dolphin: sonic burst count matching. *J. acoust. Soc. Am.* 43: 1412–1424.

Linnaeus, C. (1758). *Systema natura per regna tria naturae.* T.1. Editio Decima. Holmiae: Laurentii Salvii.

Manocha, S. N. (1967). Discrimination learning in langurs and rhesus monkeys. *Perc. Motor Skills* 24: 805–806.

Mangold-Wirz, K. (1966). Cerebralisation und Ontogenesmodus bei Eutherien. *Acta Anat.* 63: 449–508.

Martin, R. D. (1968). Towards a new definition of primates. *Man* 3: 377–401.

Martin, R. D. (1973). Comparative anatomy and primate systematics. *Symp. zool. Soc. Lond.* No. 33: 301–337.

Meyers, W. J., McQuiston, M. D. & Miles, R. C. (1962). Delayed response and learning set performance of cats. *J. comp. Physiol. Psychol.* 55: 515–517.

Miles, R. C. (1957). Learning set formation in the squirrel monkey. *J. comp. Physiol. Psychol.* 50: 356–357.

Miles, R. C. (1965). Discrimination learning sets. In *Behavior of nonhuman primates*: 51–95. Schrier, A. M., Harlow, H. F. & Stollnitz, F. (Eds). New York: Academic Press.

Miles, R. C. & Meyer, D. R. (1956). Learning set in marmosets. *J. comp. Physiol. Psychol.* 49: 219–222.

Napier, J. R. (1961). Prehensibility and opposability in the hands of primates. *Symp. zool. Soc. Lond.* No. 5: 115–132.

Napier, J. R. & Napier, P. H. (1967). *A handbook of living primates.* London: Academic Press.

Nottebohm, F. (1972). The origin of vocal learning. *Am. Nat.* 106: 116–140.

Nottebohm, F. (1976). Vocal tract and brain: a search for evolutionary bottlenecks. *Ann. N.Y. Acad. Sci.* 280: 643–649.

Parker, C. E. (1974). Behavioral diversity in ten species of nonhuman primates. *J. comp. Physiol. Psychol.* 87: 930–937.

Passingham, R. E. (1975). The brain and intelligence. *Brain Behav. Evol.* 11: 1–15.

Passingham, R. E. (1978). Brain size and intelligence in primates. In *Recent advances in primatology* 3: 85–86. Chivers, D. & Joysey, K. A. (Eds). London: Academic Press.

Passingham, R. E., Perry, V. H. & Wilkinson, F. (1978). Failure to develop a precision grip in monkeys with unilateral cortical lesions made in infancy. *Brain Res.* 145: 410–414.

Petras, J. M. (1969). Some efferent connections of the motor and somatosensory cortex of simian primates and felid, canid and procyonid carnivores. *Ann. N.Y. Acad. Sci.* 167: 469–505.

Philips, C. G. (1971). Evolution of the corticospinal tract in primates with special reference to the hand. *Int. congr. Primatol.* 3 (2): 2–23.

Plotnik, R. J. & Tallamico, R. B. (1966). Object quality learning set formation in the young chicken. *Psychon. Sci.* 5: 195–196.

Rensch, B. (1957). The intelligence of elephants. *Scient. Am.* 196: 44–49.

Rumbaugh, D. M. & McCormack, C. (1967). The learning skills of primates: a comparative study of apes and monkeys. In *Progress in primatology*: 289–306. Starck, D., Schneider, R. & Kuhn, H. J. (Eds). Stuttgart: Fischer.

Sebeok, T. A. (1965). Animal communication. *Science, Wash.* 147: 1006–1014.

Shell, W. F. & Riopelle, A. J. (1958). Progressive discrimination learning in platyrhine monkeys. *J. comp. Physiol. Psychol.* 51: 467–470.

Slotnik, B. M. & Katz, H. M. (1974). Olfactory learning set formation in rats. *Science, Wash.* 185: 796–798.

Stahl, W. R. G. (1965). Organ weights in primates and other mammals. *Science, N.Y.* 150: 1039–1041.

Stenhuit, R. (1969). *The dolphin: cousin to man.* London: Dent.

Stephan, H. (1972). Evolution of primate brains: a comparative anatomical investigation. In *The functional and evolutionary biology of primates*: 155–174. Tuttle, R. (Ed.). Chicago: Aldine-Atherton.

Stephan, H., Bauchot, R. & Andy, O. J. (1970). Data on size of the brain and of various parts in insectivores and primates. In *The primate brain*: 289–297. Noback, C. R. & Montagna, V. (Eds). New York: Appleton-Century-Crofts.

Strong, P. N. (1959). Memory for object discriminations in the rhesus monkey. *J. comp. Physiol. Psychol.* 52: 333–335.

Thorndike, E. L. (1911). *Animal intelligence.* New York: Macmillan.

Thorpe, W. H. (1961). *Bird song.* London: Cambridge University Press.

van Lawick-Goodall, J. (1968). The behaviour of free-living chimpanzees in the Gombe Stream Reserve. *Anim. Behav. Monogr.* 1: 161–311.

van Lawick-Goodall, J. (1970). Tool-using in primates and other vertebrates. In *Advances in the study of behavior* 3: 195–249. Lehrman, D. S., Hinde, R. A. & Shaw, E. (Eds). New York: Academic Press.

Walls, G. L. (1967). *The vertebrate eye and its adaptive radiation.* New York: Hafner.

Warren, J. M. (1965). Primate learning in comparative perspective. In *Behavior of nonhuman primates* 1: 249–281. Schrier, A. M., Harlow, H. F. & Stollnitz, F. (Eds). New York: Academic Press.

Warren, J. M. (1973). Learning in vertebrates. In *Comparative psychology: a modern survey*: 471–509. Dewsbury, D. A. & Rethlingshafer, D. A. (Eds). New York: McGraw-Hill.

Warren, J. M. (1974). Possibly unique characteristics of learning by primates. *J. Hum. Evol.* 3: 445–454.

Warren, J. M. (1976). Tool use in mammals. In *Evolution of brain and behavior in vertebrates*: 407–424. Masterton, R. B., Bitterman, M. E., Campbell, C. B. G. & Hotton, N. (Eds). New Jersey: Erlbaum.

Wegener, J. G. (1976). Some variables in auditory pattern discrimination learning. *Neuropsychologia* 14: 149–159.

Welles, J. F. (1975). The anthropoid hand: a comparative study of prehension. *Int. Congr. Primatol.* 5: 30–33.

Wilson, E. O. (1975). *Sociobiology: the new synthesis.* Cambridge, Mass.: Harvard University Press.

Wursig, B. (1979). Dolphins. *Scient. Am.*: 108–119.

Zeigler, H. P. (1961). Learning-set formation in pigeons. *J. comp. Physiol. Psy-*

chol. 54: 252—254.

Zuckerman, S. (1932). *The social life of monkeys and apes.* London: Kegan Paul.

Zuckerman, S. (1933). *Functional affinities of man, monkeys and apes.* London: Kegan Paul.

Zuckerman, S., Ashton, E. H., Flinn, R. M., Oxnard, C. E. & Spence, T. F. (1973). Some locomotor features of the pelvic girdle in primates. *Symp. zool. Soc. Lond.* No. 33: 71—165.

Zuckerman, S. & Fulton, J. F. (1941). The motor cortex in *Galago* and *Perodicticus. J. Anat.* 75: 447—456.

Symp. zool. Soc. Lond. (1981) No. 46, 389—395

Closing Address

R. J. HARRISON

Anatomy School, University of Cambridge, UK

I am honoured to have been asked to give the final contribution to this Symposium. My own original contributions to primatology comprise a mere two papers, one of which years ago incurred the wrath of Lord Zuckerman when he was the Sands Cox Professor of Anatomy in the University of Birmingham. My interest has always been in marine mammals, and while it has been gratifying to me that Drs Widdowson, Joysey and Passingham have referred to cetaceans when discussing growth, taxonomy and intelligence, this cannot be the reason for my presence. What I am here to do, however, is to remind you that this Symposium was organized in honour of Lord Zuckerman's 75th birthday. It was divided into four sections in which the topics discussed covered some but not all of those on which Lord Zuckerman has worked. I shall indicate to you later just how relevant some of these topics are to his past activities, but first I would like to make a few comments of my own on what has taken place during these two days.

I have been impressed by the excellence of the anatomy that has been presented, some quite new, by Moore, Ashton, Oxnard, Lewis, Holmes, and others, on both fossil and present-day forms, even if the objects of study have not always been primates. Indeed it was Professor J. Z. Young who commented at a previous Symposium here on *The primates* (1963) on the importance of accurate morphology but pointed out that the real problem was in getting to grips with the difficulty of defining a particular function and formulations of greater or lesser functions within a major one. There is indeed philosophical difficulty in breaking down any function into its component parts. Trouble arose in the 1963 discussions over the term brachiation, and what was, if it could exist, semi-brachiation. On this occasion it has been walking and whether a footprint is acceptable evidence for plantigrade locomotion that nearly got us into difficulties. Perhaps the Symposium on Vertebrate Locomotion to be held here jointly with the Anatomical Society in March 1980 will attempt to clarify matters of when a function is merely an attribute or when it is a scientific reality capable of being quantified.

To my mind similar uncertainties attend both the definition and quantification of what has become known as "a behaviour". As a student said to me in a viva: "the animal is now a black box, administer so much of a hormone A and out comes behaviour No. 5". Should we continue to study primate behaviour in the wild? Certainly, not everything is known even about any one species but there are disadvantages to such studies, as Lord Zuckerman has pointed out on many occasions. It needs time, and much of it, money and travel facilities just to get there and exist, more time to verify results; and meanwhile back home science advances while our observer remains ignorant. My research students are going in increasing numbers to Malaysia, to participate in a joint venture with Malaysian Universities in a study of primates. It is certainly good fun for those taking part but I wait with some reservations until I learn of beneficial scientific results, and not just of holiday benefits for young naturalists. Better perhaps to study primate behaviour in a centre nearer home? Yet as Dr Herbert is continually and correctly telling me, facilities for such studies are lamentably unsatisfactory and insufficient. Certainly, as I shall point out later, our Monkey Hill provided Lord Zuckerman with the opportunity to be the inaugurator of all such studies, and as has been impressed on me by many, including my professorial colleague Oliver Zangwill: "Solly Zuckerman did it all then and nothing much has happened since".

All of which brings me to my next point, which concerns the usefulness of Primate Centres. Lord Zuckerman has frequently discussed the possible value, and the propriety, of establishing such centres for the study of biological problems using selected primate species. I wonder how different our Symposium would have been had one or more Primate Centres existed in this country. A distinguished American scientist once told me not to worry about the value of research coming out of Primate Centres because their primary purpose was in training recruits for medical schools, for space research and to staff various government agencies. Primate Centres can obviously be used for conservation of primates, for breeding and improving stocks, but many zoos have been doing this for years and quite successfully. My impression is that, except in specific investigations where primates are essential, not a little research making use of them could be done on other forms, and does, I ask, this not also apply to some, if not many, aspects of behavioural studies?

Next there is this vexed question of whether there is such a thing as animal "intelligence". I am frequently asked whether cetaceans are intelligent, almost in a tone of voice that suggests if not, why not? Once again there would seem to be difficulties with definitions and quantification. I remind you that Lord Zuckerman discussed these difficulties in his *Functional affinities of man, monkeys and apes* (1933). There he wrote: "any simple comparison of the 'intelligence' of different sub-human Primates is more or less impossible". Tool-using, problem solving, communication, memory are all manifestations of active nervous systems but when can they said to be used intelligently: only in the human sense when there is appreciation of the probability of what will follow any action. Would it not be better to go back and look again at the brain, its structure and organization, and develop new methods of investigation? Lord Zuckerman thought much the same in 1933. Has our fundamental knowledge really increased all that much since then?

I remind you again that this Symposium has been held in honour of Lord Zuckerman's 75th birthday and to celebrate his contributions to science and especially to primatology. I have evidence that he has contributed 685 items as books, chapters, papers, articles, forewords, discussions and abstracts over the period 1926 to the end of 1978. Of these numerous articles nearly 300 are original contributions to biological science and 145 are on various aspects of primatology. Altogether he has published in at least 151 different journals, books, encyclopedias and newspapers; of these 46 are established, scientific periodicals. His papers have appeared most frequently in *Nature* (26 articles and letters and 43 reviews), in *Journal of Endocrinology* (41) and then, I am gratified to report, in *Journal of Anatomy* (34). Over the years he has collaborated with over 60 different individuals and, with one of the organizers of this Symposium, Eric Ashton, in at least 23 contributions. We have almost all his publications in Cambridge, in the original journals or as reprints. Of the others some have been easy to trace but one has eluded me until today when a copy was sent post haste from the British Library (Cambridge had only a limited run of this learned publication called *Lilliput*). This interesting article considers whether monkeys are men in miniature and is illustrated with etchings from *Monkeyana* by Thomas Landseer (1827). Dr Herbert would probably be shocked to see the clothes selected to be worn by an elegant ape or a dandy of a monkey (does not fashion have a firm grip on attitudes on many subjects?). Dr Passingham might also have noted some of Lord Zuckerman's remarks on intelligence in man differing from that in monkeys in an awareness of the consequences of his actions, and how difficult it is to assess any such awareness in non-human primates, or to show that monkeys "comprehend the significant characters of different social relations".

The first of Lord Zuckerman's publications was in 1925 in the

South African Journal of Science (22: 525) and was concerned with the investigation of the floor of a rock shelter on a farm in District Middelburg, Cape. Fortunately for science, the weather was hot, the findings only animal bones and the enthusiasm of a would-be palae-ontologist waned. He returned to the study of the anatomy of ba-boons, encouraged by a lecturer who was to launch him on his career as a primatologist. This work grew to such an extent that when he came to London shortly afterwards he was able to present his find-ings to the Zoological Society on this very day, the first of June, in 1926, and at about the hour that I am speaking now, exactly 53 years later. The paper he read, with Sir John Bland-Sutton in the chair, was called "Growth changes in the skull of the baboon" (*Pro-ceedings of the Zoological Society of London* 1926: 843–873) and it gives us a clear indication of his future interests. Even though he was at that time a clinical student at University College Hospital, and of only two months' standing, he laid down the guide-lines for prima-tology for the next five decades and in some respects probably for many a future decade. When reading this paper he took to task Th. H. Huxley for what were considered inaccurate or unsupported state-ments. On returning to his seat, a Fellow who had arrived during the reading of his paper turned to him and said to his consternation, "My name is Huxley, and I would like to have a few words with you about what you have just said". This was, of course, Julian Huxley with whom Lord Zuckerman later had a close friendship.

In 1928 Zuckerman published his second paper on primates, with a note on the umbilical cord of a human foetus coiled round the neck intervening (*British Medical Journal*, 1927). His next primate contribution was on age changes in the chimpanzee and was an altogether more polished and scholarly work than that on the ba-boon skull. Again it indicated future interests in that it also dealt with primates, growth, measurement, age changes, sex differences and reproduction. All these are topics which the organizers of this Symposium had in mind when inviting the speakers who have fasci-nated us over the past two days. This paper also had short notes on the Taungs skull, remarking that the measurements of this skull of a young *Australopithecus* differed from those of a skull of a modern chimpanzee only in the variance of one dimension, and in that one respect hardly significantly. In my view, what a pity that this opinion of a percipient medical student was not taken as the last word — it would have saved us all much apparently pointless perturbation.

His interest in primates led to Zuckerman's appointment as Pro-sector to this Society. While carrying out routine post-mortems on baboons he observed a young corpus luteum in the ovary from a female believed to be in mid-cycle. This observation initiated an important series of papers on menstrual cycles in primates published in the *Proceedings of the Zoological Society of London* from 1930 onwards in which A. S. Parkes was soon to collaborate. The work on reproductive organs and cycles naturally aroused Zuckerman's curiosity about primate behaviour and relationships between the sexes. Not only had he personal experience of observing monkeys in the wild in South Africa but there on his primatological doorstep was Monkey Hill in our own Zoological Gardens with its established baboon colony. He gathered together all the existing information and together with his own observations produced his classic book *The social life of monkeys and apes* in 1932. It was dedicated to R. C. B. who was the lecturer from Liverpool University, Mrs Ruth Bisbee, who when on leave to teach in South Africa had stimulated him to work on primates. She had made him re-write first drafts of papers, a salutary task he has subsequently set others when writing up their own work. In this book, which I am delighted to hear is to be re-issued with an updating commentary, Zuckerman propounded concepts of primate social life which included those of territoriality, aggression, dominance and sex ratio, and the importance of diet. Several of today's speakers have drawn attention yet again to the relevance still of these concepts in primate behaviour, and they appertain just as much to other animals' social interactions. Dr Martin's remarks about density of population, range and freedom of travel and availability of resources could apply to cetological studies. Indeed, a distinguished physician on seeing Dr Martin's headings on the blackboard remarked to me subsequently: "Someone has been talking about Zuckerman's experiments on the effects of bombing". In a way this accidental remark may have more significance than intended, in that Man is busy "bombing" species on to the endangered list.

Zuckerman continued to work on primate reproduction after he moved to Oxford University and began to investigate the effects of hormones on various organs. I show a slide, lent to me by Dr Rowlands, of the participants in what must have been the most remarkable of early meetings of endocrinologists, the first International Conference on Sex Hormones, Paris, 1937. It is remarkable in that there are gathered together almost all those who made the first contributions on this topic. Present are Edgar Allen, P. Ancel, Max Aron, S. Aschheim, Pol Bouin, R. Collin, R. Courrier, Ruth Deanesly, L. Desclin, Charles Dodds, Ch. Hamburger, Carl Hartman, F. L. Hisaw, Marc Klein, Alan Parkes, Idwal Rowlands, Hans Selye, A. E. Severinghaus, and of course S. Zuckerman. His papers at that meeting were on the interaction of ovarian hormones in the menstrual cycle and on uterine bleeding after neural lesions, an indication of an interest in nervous control of reproduction although he expressed hesitation (in 1938) at ascribing changes in the menstrual cycle to some hypo- thalamic factor. A member of our audience here today told me when working in the same department at Oxford before the Second World War what a revelation it was to realize how Zuckerman's work, together with Parkes, indicated how hormones could stimulate hitherto unresponsive organs, such as the uterus masculinus (*Journal of Anatomy* 1936, 70: 323).

It is not possible here to comment on Lord Zuckerman's very con- siderable contributions to the Allied effort during the Second World War. He has given us a fascinating account of his involvement in many activities in his autobiographical volume *From apes to warlords*. We look forward to the second volume describing his experiences from 1946. However, we suspect it was his use during wartime of biometrical analysis and statistics of probability that led him to apply these techniques to primate anatomy in both living and fossil forms. In this he collaborated most successfully with Eric Ashton, who has just given us a brief historical account of how this work has progressed since the first joint paper was presented at the Inter- national Anatomical Congress, 1950, Oxford, and then in many published subsequently in *Philosophical Transactions of the Royal Society of London* (B) and *Proceedings of the Royal Society of Lon- don* (B). I show you a few slides of gatherings at that Oxford meeting which include W. E. Le Gros Clark, and G. W. Corner whose work complemented that of Lord Zuckerman, and at a Birmingham Ana- tomical Society meeting in 1957 where several of his collaborators were present. One is "Sandy" (A.P.D.) Thomson, whose death sadly occurred recently. He and Zuckerman continued the investigation of nervous control of reproduction by a series of intriguing experiments on both vascular and nervous connexions. The post-war period also saw the development of outstanding work on the ovary with the help of a team of investigators which included Anita Mandl and S. H. Green. It is a pity perhaps that matters ovarian have not been in- cluded in this Symposium, but then we have just had them all "wrapped up" in the second edition (1977) of *The Ovary*, edited by Lord Zuckerman and Barbara Weir, to which a good few in this room have contributed by review and by personal research.

Finally, one has to mention the essential part Lord Zuckerman has played as scientific adviser to Government and as member or chair- man of committees of national importance. I am not qualified to comment on these activities but I did once encounter one of his "problems" in the form of the tanker *Torrey Canyon* when pursuing my own research. Later, at the time of its going aground on the Seven Stones reef, a cartoon by "Jak" appeared in an evening paper. It carried the legend: "Mind you, Solly Zuckerman's got a lot of other good ideas". In honour of your 75th birthday, to mark the occasion of this Symposium, and for all your good ideas, the participants would like you, Lord Zuckerman, to accept this silver medallion as an expression of our appreciation for all your many, many contributions to primatology.

Symp. zool. Soc. Lond. (1981) No. 46, 397—406

Reply by Lord Zuckerman

It is anything but easy for me to add a pendant to Professor Harrison's concluding remarks, and particularly to have to do so after so laudatory a review of my scientific career. I have been deeply honoured by the whole occasion, and immensely grateful not only to those by whom it was organized but to the many old colleagues who have participated. The presence here of some of my oldest scientific friends has moved me greatly.

There is Sir Alan Parkes, my senior in years, whom I met very soon after I came to England in 1926, at a time when he was helping lay the foundation stones of the modern science of reproductive physiology. His presence reminds me that this year, 1979, marks the centenary of the birth of the late Professor F. H. A. ("Tibby") Marshall of Christ's College, Cambridge, the College to which Alan Parkes went at the end of the First World War. One of Marshall's mentors had been Walter Heape, as Marshall in turn was Parkes's, and I do not exaggerate when I say that the transformation of reproductive physiology into a respectable field of enquiry was mainly the achievement of these three men. Marshall's *Physiology of reproduction*, of which the third edition was edited by Alan Parkes, is a work to which all of us who have worked in the field owe more than we could ever repay.

J. Z. Young is here, another of my oldest friends. We first met in the rooms of Sir Arthur Tansley, then Professor of Botany in Oxford. I remember one of Tansley's dinner parties, at which we both were, together with a young undergraduate called Peter Medawar who surprised me by the way he stood up to the great Schrödinger, another of Tansley's guests, in a discussion of certain aspects of the philosophy of science. In retrospect I should not have been surprised.

Professor Cave is here. He and I must be among the oldest surviving members of the Anatomical Society.

I am particularly gratified by the presence of my old colleagues from Birmingham, pupils in whose subsequent success I take pride. And sitting here in front of me is Frank Yates, the doyen of statistical science in this country. He taught me and my colleagues when statistics are appropriate in the analysis of biological data and when not. My debt to him — and here I am speaking for all my old pupils — is enormous.

I have found this Symposium not just a perspective. For me it is also a retrospective picture of primate biology. My interests in the subject have been catholic. Each of the four sessions has represented for me a powerful and continuing interest. The advantage of being 75 is that during these past few days I have been able to listen to what has happened in fields which were in their infancy when I started out on a research career.

During an interval between sessions, I was asked whether, when in 1926 I gave my first paper to the Zoological Society — a paper to which reference has been made by more than one speaker — I should have been able to predict that more than 50 years later there would be an occasion such as this. My answer was clearly "no". I should have been tempting fate at the time if I had tried to predict even five years ahead. I had no idea where my steps were going to lead me, any more than I could have supposed that Alan Parkes, who was then working on the changes X-irradiation induced in the mouse ovary, would be working at the close of his scientific life on the reproductive physiology of the turtle.

Basic research is unpredictable both in its results and in its wider consequences. As I once wrote, a researcher is often like one of those ancient explorers who had set out to find the North-West Passage, and who ended by discovering the New World. If I had known for certain what was going to emerge from any piece of work on which I was engaged, I should never have embarked on it. One starts from what is already known, and the foundations on which one builds are always added to by the new things which one discovers oneself or learns from others. To give one example — like everyone of my generation I was brought up to believe that oogenesis is a continuous process. I accepted the proposition as an act of faith. Only after years of apparently fruitless work based implicitly on the classical assumption did I realize that the only way to make sense of the results that were emerging from my experiments was to reject the proposition completely. My disbelief was justified. Today it is agreed that oogenesis in the mammal is not a continuous process, but that it ceases well before puberty, and in most species shortly after birth. That is the way it often is with science. We inherit the wisdom, good or bad, of the past.

In a recent review I read that science before the Second World War was what the writer called very "amateurish". It is possible that those who will be attending meetings in this hall 20, 30 years hence, might also regard the science of today, the papers to which we have been listening, as amateurish. But they would be wrong. It does not matter in what age we live — the quality of new knowledge has always to be judged within its own time frame. Establishing a piece of knowledge is never easy. The intellectual effort involved can only be related to the state of knowledge from which one begins. Obviously it is a waste of effort to rediscover what is already established or to embellish with triviality what is already known. Past belief can only be upturned by new and verifiable discovery.

I am immensely appreciative of all the laudatory and flattering remarks that have been made about the value of my contributions to the fields that have been discussed at this meeting. Professor Young, Professor Barrington, Dr Herbert and now Professor Harrison, have spoken more than generously. Professor Harrison has reminded me of two or three of my contributions which I had completely forgotten. I particularly appreciated Professor Barrington's reminder that I myself had launched the series of Symposia which have been published these past 20 years by the Zoological Society. I am sure that this Symposium will fit well into the series.

I have learnt a great deal these past two days, not only from those whom I had the privilege to launch on their research careers but also from the other research workers who have contributed. My only regret is that two of my old colleagues, Professor Krohn and Dr Anita Mandl, have not been able to be with us, the one because of illness, the other because she is abroad.

Obviously I shall not be expected to summarize the papers which have been presented. But I might be allowed to refer in a sentence or two to some of the ideas that were stimulated in me by what I heard. Dr Widdowson regaled us with a fascinating series of observations about rates of physical growth. I found myself asking how soon it would be before the stages of physical development which she described would be correlated with stages in the development of behaviour. The techniques which Tanner and Healy described in their paper on "Size and shape in relation to growth and form" seemed to me to be just as applicable to growth changes in internal organs, such as the brain, as to the surface characters with which they were dealing. Professor Moore, an old pupil, resurrected my interest in the spheno-ethmoidal angle. It gave me great pleasure to hear about the considerable advances that he has made in our understanding of the growth of the face in primates, advances that are not only of fundamental scientific interest but which clearly also have considerable clinical applications. Professor Ashton reviewed the story of the australopithecines. I was pleased that he began by telling the meeting that the focus of the studies of the Birmingham school was always on the anatomical characteristics of the fossils, and that we were not concerned to participate in any — what I shall call — public competitive auction about the importance to our ideas of human phylogeny of this or that fossil. We all need to remember that the primate fossils about which so much is written nowadays could be laid out on this long desk. I once made a rough calculation of the number of Hominoidea which must have walked this Earth since they first appeared in the Pliocene. The answer had to be expressed in units of millions. It has always seemed extraordinary to me, therefore, that while only a minute sample of the Hominoidea which ever existed are available for study — an infinitesimal fraction of one per cent — we still find that as each new fossil is unearthed, its finder publishes in *Nature* or the popular press yet another statement about the unique significance of his prize in the line of human descent. It stands to reason that the remains that have been found — separated as they are in time by hundreds of thousands of years and in space by hundreds of miles — do not necessarily constitute a lineage. How could they? Before we start speculating about our origins, let us at least be clear about the anatomical facts. Ashton has made that point abundantly clear.

He referred to an occasion when a systematic statistical error crept into calculations we had made of the variance of certain dental dimensions, and he also told us that when the error was corrected, the conclusions that had been drawn from the first analysis still stood. He implied that he was responsible for the error. Let me say here and now — that was not the case. If anyone was responsible, I was, but let me also confess, it was not I who devised the particular equation which we used. I shall not reveal the name of the statistician who did, and whose error was discovered by Dr Yates and Dr Healy.

Professor Oxnard's paper revealed for us a truly significant extension of work which he started in my laboratory. His demonstration of the relationship of the internal architecture of the vertebral bodies to lines of force struck me as having not only considerable practical applications in clinical medicine, but also to be a very useful tool in discussions of posture in the great apes — and possibly also in the small number of fossil primates whose vertebrae have been found.

Professor Lewis's paper on the "Functional morphology of the joints of the evolving foot" was equally revealing. It showed once again that when an objective anatomical view is brought to bear on the problem, it turns out that claims that the fossil australopithecines walked like man are very insecurely, even fallaciously, based. His paper also added weight to the conclusion which other studies have indicated, namely that these creatures possessed a unique combination of anatomical characters, as unique as those one finds in man or in any one of the living great apes.

The session dealing with "Reproductive biology and neuroendocrinology" opened with a reminder by Professor Holmes of the way fashion changes in the study of the pituitary gland. As he told us, the neural lobe was first the focus of attention. People then became overwhelmingly interested in the anterior lobe — the pars distalis. He has now demonstrated all too clearly that the pars intermedia literally remains *terra incognita*. Indeed, I doubt if we know any more about its functions today than we did when Professor Hogben was working on the melanophore hormone which has its source in this part of the gland. As I listened to Professor Holmes, I was reminded that it was in Parkes's laboratory, shortly after I arrived at University College, that I first met Lancelot Hogben, then on a visit from South Africa. Hogben was at the time opening up the field of comparative endocrinology, and was particularly interested, as is Professor Holmes now, in the functions of that still neglected lobe of the pituitary gland.

Then came Professor Finlayson's account of neuroendocrine functions in the insect, a field of study which was a focus of the work of the late Sir Francis Knowles when he was a member of my staff in Birmingham, and with whom Professor Finlayson was associated. He told us that the concept of neurosecretion cannot be confined to the endocrine organs — a proposition which, if I remember correctly, was enunciated by Professor Young many years ago, in the days when neuroendocrinology first became a fashion. But the apparent simplicity of the neurosecretory model in the insect contrasts sharply with the complex relationship between collections of neurones in the hypothalamus and the pituitary — as was made only too evident by Dr Cross in his paper. Dr Cross reminded us that the real sexual differences that exist in the visceral functions of the brain are under the influence of sex hormones — a point which excited the attention of Dr Beach in the United States many years ago when he studied the micturition reflex in dogs. He also showed us how difficult it is to be certain about the specificity of the same population of hypothalamic neurones, not only in terms of identification, but also in terms of what one might call their chemical differentiation. I noted that he never once used the word "neurosecretion" during his talk.

The memory I shall carry away with me of Alan Parkes's paper on reproduction in the green sea turtle is of these large creatures swimming across the Atlantic from the shores of Brazil to Ascension Island — a return journey of 2000 miles — finding the beach where they had laid the year before, digging a hole two feet deep, and then laying 2000 eggs, and all this without, so far as anyone can tell, ever feeding during the course of the whole process. One would dearly love to know about the energy balance in these animals. Parkes also suggested that pheromones play a significant part not only in the mating activities of these creatures, but possibly also in the way they navigate and then locate the particular beach where they lay their eggs. What he told us may have been "good old natural history", as he described his paper at the end, but he also provided a powerful reminder that many major problems for research can only be derived from a first-hand knowledge of the life of the creatures one is studying.

The final session dealt with "Primate sociology and behaviour", and I was struck by several points in Dr Martin's opening paper. He told us that behaviour in monkeys is more variable than is their morphology; that behaviour in monkeys or prosimians or apes which the field-worker might find in the forest is not necessarily antisocial, and that the feeding and breeding relationships in groups of monkeys that are encountered in the wild are not necessarily the same. He also made the point that "pair-living primates" are rare and that the few that are known, such as the gibbon, are creatures of the forests. When it came to Dr Herbert's turn, I was not surprised to hear that the endocrine status of different individuals in a group of monkeys or apes determines the social reactions of the entire group. But what was surprising and new was to learn that the social status of a monkey also determines the intensity of its response to a hormonal stimulus. It was also interesting for me to see the curves he threw on the screen showing the trend of the variations in the concentrations of steroid in the blood serum that occurs during the course of a menstrual cycle. I found it gratifying that measurements of today accord fully with the hypothesis which I put forward as long ago as 1929, in this Society's then meeting room.

Perhaps I should also refer to the talapoin monkey about which Dr Herbert spoke. The sex-skin phenomenon of this creature was discovered, not in Birmingham, as he believed, but here, again in the Zoological Society. I remember too that a mature female of the species happened to die, in I think 1929, at the critical moment in the cycle to provide one of the best photomicrographs ever of the endometrium at the moment of break-down.

Needless to say, there was much more that I learnt these past two days than just the points to which I have referred in my rapid survey. So let me turn to the perspectives and prospectives in primate biology as I see them. And allow me to preface my remarks with some general observations.

Over the 50 years in which I have been either directly or indirectly engaged in research, there has not only been a great increase in the number of people trained in one or other branch of science, but an even greater growth in the volume of resources which goes to scientific investigation. It costs far, far more in real terms to keep a scientist busy and happy these days than it did when I started work. Regretfully, I have to confess that I do not believe that there is any close correlation between these measures of growth on the one hand and the emergence of major seminal ideas that transform the body of knowledge and so the direction of enquiry on the other. This century, and mainly its first 50 years, that is to say before the post-war upturn in scientific activity, have witnessed the birth of genetics; of the formulation of the general and special theories of relativity; of the definition of hormonal action; of a basic understanding of immunological processes; of the emergence of nuclear physics; of molecular biology — to mention a few of the big developments of science. Today's generation of younger scientists takes these things for granted. So they do the very powerful new tools of scientific enquiry that have been developed in the years since the end of the Second World War — for example, to name only some of the more outstanding, the computer, the laser, the electron microscope, the mass spectrometer, techniques for measuring nuclear resonance and radioimmunological assays. As I see it, while new techniques make scientific enquiry possible where it was not possible before — ideas still come before techniques. To use a new technique in the investigation of an idea which derives from some spurious concept does not make the idea any sounder than it was at the start. This thought came to me as I listened to Dr Joysey's fascinating paper on "Molecular evolution and vertebrate phylogeny in perspective". As a basis he had to use the phylogenetic trees which have been drawn on the basis of speculation about the evolutionary relationship of one group of primates to another. Over the years the differences between one suggested tree and another have been so trivial that I always find myself wondering what the excitement is all about. He took geological time into account. He then demonstrated that the possibility of the occurrence of parallel evolution came up whichever way he chose to arrange the results of his chemical analysis of the structure of the myoglobins of 40 different species of mammal. By changing the assumptions ever so slightly, the most unlikely conclusions resulted. His was a very revealing — and to me, witty — paper and, as he also told us, a costly study in terms of computer time. It is worth noting that with the techniques and computers now available, he was able to type 40 myoglobins in far less time than it took Perutz to reveal the basic structure of the one that was first unravelled.

Dr Joysey's was the most recent in a series of lessons about primate phylogeny to which I have listened in this Society. The conclusion I have drawn from all is that speculation about man's origins is less a branch of science than a game. The broad outline of the story has changed not at all over the years, and since the possibility of parallel evolution can never be ruled out, there never will be a way of stating unequivocally that man or any one of the apes — extant or extinct — had not evolved more or less independently of his fellow creatures over the millions of years they have existed. It is useful to remember that a generation cycle in a monkey or ape lasts something between five and ten years. Let us assume ten. A century would account for ten generations; a thousand years, a hundred; a million years, a hundred thousand; — so there would have been something like half a million generations or more in the period that has elapsed since the days of the East African and Ethiopian fossil australopithecines of which we read so much. Who knows what changes, what parallel changes, took place over such a span of generation-time in isolated breeding groups of apes — or ape-like men, call them what you like. It took only 100 generations to bring about significant morphological change in the African green monkey that now lives on the West Indian island of St Kitts. Ordinary hard-headed breeders of livestock would, I fear, raise their eyebrows in surprise were they to listen to some of the naïve speculations about human ancestry that emanate from assumedly learned scientific circles.

In introducing today's opening session, Professor Barrington referred to what he called my critical probing at a meeting on comparative endocrinology that had taken place at Cold Spring Harbour in the early 1950s. Those here today who have worked with me will need no reminding that it has always been my habit to embark with a sceptical attitude on any enquiry in which I have been concerned, whether as teacher or participant. Fashion, as Professor Holmes has reminded us in his historical account of research into the pituitary gland, always seems to play a big part in biological enquiry. Tutors or research directors turn the attention of Ph.D. students to fields where progress is being made or where it seems to be made. Unfashionable fields or those where problems are real but intractable, are neglected, sometimes for decades, until interest is restimulated because of an advance in knowledge in some collateral field, or because of the development of some new technique of enquiry. Whichever way interest is restimulated, the questions then chosen for research and to which new techniques of enquiry seem relevant must be formulated scientifically. Computers, radioimmunoassay methods, the possibility of registering the action-potentials of a single neurone

— which Barry Cross has demonstrated to us today — modern statistical methods; these are all critical to progress. But to be used with profit to the growth of the body of knowledge they all need to be applied in a relevant way to real questions.

I say this because I am not sure that I would regard some of the issues with which Dr Passingham dealt under the title "Primate specialization in brain and intelligence" as constituting real scientific questions. Tests to establish primate "intelligence" should surely be specific to the lives of the creatures themselves. Dr Passingham indicated that, after reaching the conclusion that monkeys and apes are the cleverest of mammals, the question arises about their position relative to the dolphin or porpoise. But what have monkeys to do with porpoises or dolphins? He referred to the chimpanzee Lana, the creature celebrated in the popular press for her ability to communicate in American sign language. But her abilities in this respect have already been matched in work with pigeons. Experiments on language in the great apes have a long way to go before they prove anything. All animals communicate — usually by means undiscernible to us, within the social worlds in which they exist. What, for example, do we yet know for sure about the migratory or homing mechanisms in the house-martin?

In my view, students embarking on studies of the behaviour of monkeys or apes must begin from the basis of a general knowledge, not only of the biology of these creatures but also of vertebrates in general. I think Professor Harrison was right to question the wisdom of sending young students — perhaps even those who have never before studied a single monkey — into the wilds to make observations on the ways the creatures organize their social lives. He put a question — "would it not be preferable to divert the resources expended in this way to establish large primate breeding colonies in this country where the creatures could be studied under proper supervision?" I seem to recall that Dr Martin said that there were something like 83 expeditions of one sort of another now planned and known to the Primate Society of Great Britain. I confess that I share Professor Harrison's view. It would be better to set up primate colonies where first-class people could work in a properly endowed academic environment, as opposed to sending untrained people to study in isolation in an environment devoid of scientific criticism. Only too often it is a case of people finding answers without knowing what the questions are. As G. K. Chesterton's Father Brown said: "It isn't that they can't see the solution. It is that they can't see the problem." Public money is scarce and should not be spent on redisovering the wheel. Before the uninitiated get to the wilds, let them study with the Martins of this world, and learn how to gather relevant quantifiable information — and also when the use of statistical analysis adds to knowledge, and when not. As Healy put it yesterday, the statistician has a great deal to learn from the biologist, and he hoped that the biologist had something to learn from the statistician. That he certainly has.

So, in conclusion, what shall I say? I think that I would like to start my scientific life all over again. I wish it was possible to be treated with the juvenile hormone, even the juvenile hormone of the cockroach described to us by Professor Finlayson — in spite of his suggestion that I was already under the influence of a sufficient quantity. I would like the chance of retreading the fields that I have already covered, to pick up errors, to identify lost questions, to start re-learning from my old pupils — to learn from those who have become more up-to-date than I am in the subjects which have been discussed at this Symposium. But they would have to accept me as a pupil just as I am. Nothing, I fear, will suppress my critical probing. Nothing, I fear, will change my ingrown attitude of critical disbelief — until I am satisfied both by question and answer.

Finally, once again let me thank those who organized this Symposium in my honour; the Chairmen of the Sessions; the contributors, and those who came to this Symposium. I am proud to have been the cause of this meeting, and I am glad to have learnt that some of the things which I have uncovered have been of use to others in their work.

June 1979   LORD ZUCKERMAN

Author Index

Numbers in italics refer to pages in the References at the end of each article

Subject Index

Numbers in italics refer to figures